The Obsolescence of the Human

posthumanities

CARY WOLFE, SERIES EDITOR

75 *The Obsolescence of the Human*
Günther Anders

74 *Language Machines: Cultural AI and the End of Remainder Humanism*
Leif Weatherby

73 *Hermes II: Interference*
Michel Serres

72 *The Abyss Stares Back: Encounters with Deep-Sea Life*
Stacy Alaimo

71 *Prosthetic Immortalities: Biology, Transhumanism, and the Search for Indefinite Life*
Adam R. Rosenthal

70 *The Memory of the World: Deep Time, Animality, and Eschatology*
Ted Toadvine

69 *Hermes I: Communication*
Michel Serres

68 *Nietzsche's Posthumanism*
Edgar Landgraf

67 *Subsurface*
Karen Pinkus

66 *Making Sense in Common: A Reading of Whitehead in Times of Collapse*
Isabelle Stengers

65 *Our Grateful Dead: Stories of Those Left Behind*
Vinciane Despret

64 *Prosthesis*
David Wills

63 *Molecular Capture: The Animation of Biology*
Adam Nocek

(continued on page 346)

The Obsolescence of the Human, Volume 1

On the Soul in the Age of the Second Industrial Revolution

Günther Anders

Translated by Christopher John Müller
Edited by Christopher John Müller and Christian Dries

posthumanities 75

University of Minnesota Press
Minneapolis
London

Originally published in German as Günther Anders, *Die Antiquiertheit des Menschen. Über die Seele im Zeitalter der zweiten industriellen Revolution*,

Published by the University of Minnesota Press
111 Third Avenue South, Suite 290
Minneapolis, MN 55401-2520
http://www.upress.umn.edu

ISBN 978-1-5179-1264-2 (hc)
ISBN 978-1-5179-1265-9 (pb)

A Cataloging-in-Publication record for this book is available from the Library of Congress.

Printed on acid-free paper

Contents

Translator's Preface

With this volume, an English translation of Günther Anders's arguably most influential work is for the first time available in full. First published under the title *Die Antiquiertheit des Menschen* in 1956, *The Obsolescence of the Human* opens radical and uncannily prescient perspectives on the ongoing "'technification' of our being,"[1] our "apocalypse blindness," and life among increasingly capable machines. I would encourage readers (especially those unfamiliar with Anders's work) to skip ahead to the main text, as Anders's voice speaks with an immediacy that requires no supplementary commentary. Yet it is precisely this immediacy—Anders's strong "presence" in the text—that also creates significant challenges when translating his work. The writing style by which it is achieved—to cite Herbert Marcuse's apt assessment—is "*difficil*," that is, "tricky."[2] Marcuse makes this comment in a 1955 letter, which discusses the would-be translation of some of Anders's essays into English; yet, although these appeared soon after, no English edition of *Die Antiquiertheit des Menschen* ever followed.[3] The reason for this was most likely not just the style in which the book was written. Anders nevertheless echoes Marcuse's sentiment in a 1979 interview, in which he notes how those émigré authors who had "cultivated their own style of writing"—he names Bertolt Brecht, Thomas Mann, Alfred Döblin, and tentatively adds himself to the list—had "much greater difficulties than some others" to write in English or oversee translations of their own work.[4] And as if to prove a point, Anders employs the eminently untranslatable word *Eigensprachler* to refer to this group of writers who had "personalized language" to such an extent that they were unable or unwilling to leave the medium of their native tongue.

This opening account might give the impression that Anders indulged in some inaccessible "private language." Nothing, however, could be further from the truth. The use of diary entries, the inclusion of poems, the persistent linguistic wit and wordplay, the constant commentary on a given word choice—to name just a few traits of Anders's personalized approach to German (and to philosophical argumentation)—have the rare trait of clarifying rather than obfuscating. More precisely, they clarify by distorting our sight. And this, as I explain in a moment, is even true of Anders's at times highly convoluted syntax. In all its different aspects, the voice in which *The Obsolescence of the Human* addresses its readers is thus a key component of what Anders himself refers to as his method of "philosophical exaggeration," of creating "overstated portrayals pointing in the direction of the truth."[5] Like the motifs of the occasionally absurd images that Anders's writings create, the style in which these images are linguistically presented is itself designed to reconfigure the way the reader affectively and rationally relates to the subject matter depicted. Put otherwise, *The Obsolescence* is not only activist in its content but also in its form, the latter attempting to activate readers' imaginations as well as their critical and intellectual resistance—to both the perspectives the text opens and the world these perspectives portray.

The principal aim of this translation is to approximate the overall effect of Anders's exciting and eccentric use of German. A good example of the difficulties that arise when trying to recreate its effects in English can be found at the very beginning of the introduction, where Anders invokes our "*postliterary illiteracy*" while leading us into the text. The use of this phrase forms part of a protracted anecdote in which Anders recounts how he, the author of *The Obsolescence,* was allegedly heckled and interrupted while giving a talk about the very material we are about to read. The illiteracy in question, the anecdote implies, is not the product of a lack of education, interest in literature, or ability to read, but of an entirely mundane addition to the world: the supposedly unmediated access provided by photorealistic images in TV, newspapers, and magazines. These images, Anders notes, create a sense of frictionless participation, inclusion, and connection—a feeling that is false precisely because it lacks friction. "In showing the world," the talk supposedly concluded with its heckle-inducing logic, "images conceal it."

By entangling us readers in a small controversy and providing an account of the hostility with which his ideas were apparently met, Anders's opening makes use of a narrative approach that is frequently deployed in *The Obsolescence.* Yet, rather than simply being mobilized to pitch the "explanatory capacities" of the old technology of text against the deceptive power of increasingly pervasive image technologies, the style in which this anecdote is written quite literally resists any simple reading. In the original, the entire anecdote forms one single sentence, which in the most recent German edition (2018) runs on for twenty-two lines. To further add to the complexity, this sentence doesn't address us directly in the first person (as the text otherwise frequently does) but in indirect reported speech. A literal translation would thus open something like this: "At the conference just mentioned, the following happened to the writer of these lines: he was giving an account—in the context of what he calls '*postliterary illiteracy*'—of today's flood of images: of the fact, therefore, that . . ." (and at this point, the sentence continues for another nineteen lines). The phrase *postliterary illiteracy*—which the talk had supposedly mobilized to mark the advent of a new regime of power in which broadcast technologies can be used to "exclude" us from the world by means of a false sense of immediacy, inclusion, empowerment, and insight—is thus embedded in a syntactical structure that makes this "illiteracy" an entirely tangible condition, one that is "immediately" felt and textually recreated. It is this kind of linguistic friction that Anders's style of writing continually modulates, for it serves to dramatize and defamiliarize the normalized, invisible, user-friendly world of devices, gadgets, and machines—as well as the repressed impacts and effects they have on us and the world. And rather than excluding himself from this condition, Anders, through his textual persona, seems intent on documenting, discovering, and staging it on our behalf to confront us with a technological world that is familiar yet portrayed as something we are struggling to perceive, read, and comprehend.

It was this aspect of Anders's work—its peculiar estrangement effect—that posed the greatest challenge while translating. In the Anglo-American context, as Lawrence Venuti has influentially noted, translation has traditionally been guided by the ideal of "invisibility." "The translator," as Venuti puts it, "works to make their work 'invisible,' producing the illusory effect of transparency that simultaneously masks its status as an illusion." That is to say, the translated text is meant to feel "natural, i.e., not translated."[6] Yet, in contexts where friction and idiosyncratic uses of language are instrumental to a given style, as is the case with Anders, translation almost seems to amount to the opposite. It entails developing strategies to somehow transpose the often "unnatural" feel of the source text. As a rule, I chose accuracy over elegance. Wherever practically feasible, I have tried to retain the idiosyncrasies of Anders's style, such as the eccentric use of textual emphases or the use of the em dash after the full stop (which is mimicked here at the end of this sentence).[7]— I have also tried to retain the frequent reliance on passive verb forms, double negations (e.g., "it is not impossible"), and the like, as these seem designed to function as activating devices, requiring us to work at the positive image ourselves while reading.

The same approach, however, could not be adopted for the syntax. As indicated above, Anders's texts have a tendency to be written in long sentences comprising interlaced subclauses that gradually build up to a concluding point, resolving the linguistic tension and clarifying the meaning. This approach is impractical and highly counterintuitive in English, and it certainly does not have a similarly cathartic effect. As Anders's textual distortions are not barriers to meaning and comprehension but their very vehicle, I sought to recreate the vivid reading experience of the German original. I usually divided long sentences into shorter ones, at times rearranging the resulting sequence of sentences and opening with the concluding point to invoke the image that is then given substance. I hope this has helped create a version of the text with a "comfortably different" feel (to mobilize Benjamin Nickl's term here).[8] The same applies to the rendition of the textual persona through which Anders addresses us as readers.

As hinted at in the comment about Anders's use of indirect reported speech, *The Obsolescence* frequently positions us as viewers, either watching the drama of Anders's thought unfold or witnessing it "directly" through the lens of diary entries and anecdotes. And this at times really is a drama—a kind of "written" broadcast or film in which we are carried from scene to scene, and from one of Anders's encounters with his contemporaries or certain given technologies to the next.[9] This not only contributes to the popularizing and accessible feel of his text, but at times also seems to channel, appropriate, and redirect the affective world of machine interaction, broadcast technologies, and media formats the text seeks to confront. And the textual persona addressing us has multiple registers. It speaks in the voice of a moralist and apocalyptic doomsayer, a phenomenologist highly versed in the language of philosophy, a poet and fable writer, a humorous "caricartoonist"[10] depicting scenes of everyday life, a chronicler of émigré experience who is all too accustomed to being relegated to the position of an outsider who views the world through the lens of embarrassment and shame . . . And yet, in German, despite this multiplicity and through all these voices, the textual persona retains, or rather *gains*, a clear and distinct identity that can leave the impression that one is in a conversation with its author. To maintain coherence between these multiple registers across an entire book, a number of strategies were adopted.

The German text frequently plays with the multiplicity of meaning of given words and cleverly switches between them. To activate this multiplicity, single German words were at times translated into more than one word to make all these meanings available, and this has usually been recorded in a translator note. Anders's reliance on the word *Mensch* (human being) presents the translator with a major challenge. In the service of readability, I often translated this in the plural as "humans," as I wanted to avoid using "man." I adopted the same approach when the text makes a general claim by using a word in the singular ("the worker," for example). Translating this in the plural ("workers") was necessary, as German nouns are gendered, and I sought to signal that the text is referring to workers in general. Where a specific individual or clearly gendered image seems to be evoked, I retained the singular word form. Anders also makes use of several words for "machine" and technological gadgets and devices: *das Gerät, der Apparat, die Maschine*. These have slightly different nuances in German. Besides being a reference to the world of machines in general, *das Gerät* evokes technological objects that have become part of everyday life and are not thought of as machines, but simply as things that one makes use of. Similarly, *der Apparat* is often a reference to a specific electronic device—a radio set, for example. In "On Promethean Shame," I mainly opted to translate all these terms as "machine," as this helps facilitate the sense of an encounter between human and machine that the text discusses. In the remainder, I mainly switched to "device," as this term seemed more suited to the specific context of radio, television, and nuclear weapons—that is, the portrayal of the world of specific devices. Finally, the heavy pathos of the moralizing register of Anders's narrative voice was often toned down, and I at times also transposed the indirect reported speech into the first person (the scene about our "*postliterary illiteracy*" being a case in point). I have also tried to capture the sense of humor that runs through the text, which at times might have rendered what is merely implied or potentially humorous in the German more explicit in the translation.[11]

The question of unity brings me to a short comment about the unusual composition of *The Obsolescence*, which might at times seem a bit "*zusammengebastelt*"—that is, "cobbled together" (to borrow another word from Anders's correspondence with Marcuse).[12] Although one should perhaps not read too much into this, as Anders offered this assessment in 1978, over twenty years after the first publication of *The Obsolescence*, it is somewhat contextualized by the text's publication history.[13] Prior to appearing in *The Obsolescence*, parts of various essays were published in other contexts. In fact, it even seems that the three main essays that make up *The Obsolescence* may have at some stage been intended for publication as three separate books (they would certainly be long enough). This has left some marks on the text—for instance, in the form of repetitions in the argument that hint at earlier serialized publication. (This is especially prominent in "The World as Phantom and Matrix.") It is perhaps also noticeable in the somewhat provisional introduction, which seems to forget that the book also includes a fourth essay, the comparatively short piece on Samuel Beckett's *Waiting for Godot*, which is itself a republication of an essay first published in 1954. As is the case with the German editions, these potential inconsistencies were retained, as they reflect Anders's precarious life during exile and his eventual return to Europe, a precarity from which *The Obsolescence* emerged.

There were a number of additional editorial decisions, which were made in collaboration with Christian Dries. We have tried to retain Anders's reliance on Latin, Greek, and French phrases and expressions, and also highlight the at times frequent code-switching into English. Wherever translations of these phrases could not be incorporated into the main text, they have been provided in the notes. While the German edition leaves most direct quotes unreferenced (Anders rarely provided bibliographical details), we have identified sources (or probable sources) where possible. Where the text assumes knowledge of German poetry or literature, we have at times also reproduced short passages—for instance, from Johann Wolfgang von Goethe's *Faust*. On a select few occasions, we also provided some further contextualizing materials where specific textual motifs or key ideas lead back to Anders's earlier writings and his poetic and literary praxis. The source text of this translation is the most recent (2018) German edition of *The Obsolescence*, which includes Anders's original dedication and the 1979 preface to the fifth German edition.[14]

This leaves the title. The German title reads *Die Antiquiertheit des Menschen*, which would literally translate into something like "The Antiquatedness of the Human Being." With *The Obsolescence of the Human*, we opted for a version that echoes the translation of the German title that is now adopted in the overwhelming majority of the already substantial literature engaging with Anders in English. *Der Blick vom Mond* (1970), Anders's philosophical diaries about space travel, characterizes *Die Antiquiertheit* as an attempt to portray "the obsolescence of the model of the human that has prevailed until today" (*Die Obsoletheit des bisher gültigen Menschenmodells*),[15] and the German adjective *obsolet* also features prominently in *The Obsolescence* itself. It appears most memorably, perhaps, in "On Promethean Shame," where it is deployed in a Molussian industrial song that effectively captures the mood of awkward outdatedness that the German *Antiquiertheit* conveys: "Each day out of machines arise / ever more beautiful machines. / Only we remain malformed, / only we're born obsolete."

—Christopher John Müller

The Obsolescence of the Human, Volume 1

In the year 1906, exactly half a century ago, my father, William Stern, published the first volume of his work Person and Thing. *At the time, he was twenty years younger and many generations more optimistic than I am today. It was not easy to rob him of his hope that the struggle against the influence of impersonal psychology could restore the concept of the "person." His personal kindness, along with the optimism of his era, for a long time prevented him from realizing that it is not science, but the way humans actually treat humans, that turns "persons" into "things." Dishonored overnight, driven into exile by the enemies of humanity, he was not spared the sorrow of better insight into a worsened world.*

It is in memory of him, who planted the notion of human dignity firmly and irremovably in his son, that these sad pages about the destruction of the human were written.

Preface to the Fifth German Edition

This volume, completed more than a quarter century ago, doesn't merely seem not to have become outdated; it seems to have grown in relevance today. Yet this doesn't speak for the quality of my original analyses; it speaks emphatically against the state of the world and human situation they discuss. This situation hasn't substantially changed since 1956, and indeed, it couldn't have changed. The accounts I wrote back then were diagnoses, not prognoses. Sadly, in view of the threats to the natural world that have become so obvious during the last twenty-five years, this book's three main theses—that we cannot live up to the perfection of our own products; that we can produce more than we can imagine and take responsibility for; and that we believe it permissible, no, compulsory, to do what we are able to do—have become even more topical and urgent. I wish to emphasize, therefore, that I didn't possess any "prophetic" powers back then. The situation, rather, was that 99 percent of the world population was already unable to see; in fact, they were being methodically rendered blind (as I indicated by introducing the term "apocalypse blindness").

In the years since these pages were written, my attitude toward nuclear armament (essay 4) hasn't changed. I have, in fact, intensified my efforts in this direction, as is demonstrated by my essays on the nuclear situation (*Endzeit und Zeitenende*), my Hiroshima diary (*Der Mann auf der Brücke*), and my correspondence with the Hiroshima pilot Claude Eatherly (*Burning Conscience*).[1] The long delay of the second volume of *The Obsolescence of the Human* is partly explained by how inappropriate it felt to only theorize about the apocalyptic danger in academic texts.[2] The bomb doesn't just hang over the roofs of universities. The time between the publication of the first and second volumes was thus mainly filled with activities directed against nuclear armament and the Vietnam War. Yet I still unreservedly endorse my original essay on the bomb. In fact, because nuclear power plants have obstructed our view of nuclear war, I believe it to be even more important today, as nuclear power has made us, the apocalypse-blind, even more imperceptive toward the apocalypse.

In the twenty-eight years since it was written, the topicality of "Being without Time," the essay on Beckett's *Godot* (essay 3), has also increased. This is no surprise, since it portrayed the world—or, rather, the worldlessness—of the unemployed, a misery that today, half a century later, is once again beginning to reach global dimensions.

Yet I am no longer fully happy with the totally pessimistic evaluation of mass-media in "The World as Phantom and Matrix" (essay 2). Humans are "rendered passive" by TV, or so it notes; they are "trained" to systematically confuse being and semblance. And the world, in turn, is transforming into a copy of the images taken of it, since world events now mostly seek to accommodate the needs of television. Although these theses have no doubt become truer, they nevertheless require some additions, even though certain "political leaders" are today adopting my warnings, twenty-five years after they were written. And these additions would be of an encouraging nature. For what transpires today is that there are also some situations in which television images deliver a reality to us that we would otherwise not be part of at all. And these images can profoundly affect us and motivate us to engage in historically significant actions. Although seen images aren't as good as seen reality, they are better than nothing. The images of the Vietnam War that were daily channeled into American homes really did "open" the eyes of those who were staring at the screen, and they caused a protest movement that very substantially contributed to the discontinuation of the genocide taking place back then.

When I wrote this plea for the continued existence of a human world—no, sadly, I have to say, for the continued existence of the world—many of my would-be readers hadn't yet seen the blinding light of our bleak world. In reading these pages, they will recognize that the revolutionary, no, the catastrophic situation they were born into—a situation in which humanity has the capacity to destroy itself, a situation they are sadly far too used to living in—had already set in before they were born. They will recognize that this definitely nonhonorable capacity is older than them, and that the duties they have today were already the duties of their parents and their grandparents.

It is my heartfelt wish for them and their own children that none of my prognoses will prove to be true.

—Günther Anders, Vienna, October 1979

Introduction

> For their last meal, those who are condemned to death have the freedom to choose whether they'd prefer their beans served sweet or sour.
>
> —From a newspaper report

. . . Because their options are decided by others.

We, too, can freely choose whether we'd like our *todays* served as a bomb blast or a bobsleigh race. We have this freedom because we are also in the hands of others—and so are our choices. We all must make our choices *as* consumers of radio and television—as beings sentenced to feed on world-phantoms instead of being allowed to experience the world. And maybe we no longer want any other options, and we perhaps cannot even imagine them—not even other kinds of choices. All this has already been decided.

As I presented this idea at a conference, I was interrupted by a heckler. Everyone is free to switch off their radio or television, he objected—free, even, not to buy any such devices and give their full attention to the "real world" and this alone. This I disputed. Those who go on strike are no less in the hands of others than those who consume. Regardless of whether we play along, we're part of the game because we're the ones *being* played. No matter what we do or abstain from doing, our private strike does not alter the fact that we are part of a humanity that prefers world-phantoms to world, and the consumption of phantoms to world-experience. This humanity is now our actual world—the social world we must deal with every day—and it is not possible to go on strike against that.

But the so-called real world—the world in which events unfold—has also changed through its *phantomization*. To a large extent, it is now already optimized for transmission—that is, it has been designed to ensure that its phantom-version is well received (and, of course, profitable). The claim that "*everyone*" is free to own or not to own, to use or not to use such devices, is a complete delusion. The friendly mention of "human freedom" doesn't change the fact that there is a *compulsion to consume.* And it is not exactly proof of freedom that in the very country in which individual liberty is writ so large, there are also products that are referred to as "must-have goods." And all this talk of "musts"[1] is fully justified. Even just one single missing "must-have" device is enough to put the whole life apparatus at risk—an apparatus that is itself defined and secured by other devices and products. Whoever exercises their "freedom" to reject a specific device rejects them all and, in doing so, rejects their own life too. "*One*" could do this, really? Who might this "*one*" be?

What is true for radios and televisions also applies to all other devices, *mutatis mutandis.* They certainly aren't "means to an end" [*Mittel*] anymore. A "means" is by its very nature something secondary; it is something added to a freely chosen end and used *ex post* to "mediate" [*vermitteln*] this end.

Our devices aren't "means" but "*predeterminations*"—the decisions that are made about us *before* it is our turn to decide. And strictly speaking, they are "*the*" predetermination.

Yes, "*the predetermination.*" In the singular. There is no such thing as an individual device. The whole is the true.[2] Every single device is itself only a device *part,* only a screw, only a component in the system of devices. As a component, it partly serves to satisfy the needs of other devices, yet its mere existence also imposes a need for new devices onto yet more devices. To claim that this *macro-device,* this system of devices, is a "means" and, as such, ours to use for any ends we may choose would be totally absurd. This system of devices is our "*world,*" and "world" is not the same as "means"; it is something categorically different.—

There is today nothing more awkward, nothing that makes one as instantaneously unwelcome, as *the suspicion* that one is *a critic of machines.* And there is nowhere on our planet where the risk of this suspicion is any less acute than anywhere else. In this respect, Detroit and Beijing, Wuppertal and Stalingrad, are simply the same. This is also true for all social groups. Regardless of the class, interest group, social system, or sphere of political philosophy one is in, one is immediately branded a ridiculous Luddite whenever one takes the liberty of speaking up about the "degrading effects" of some device. The moment one begins to speak, one automatically signs one's own intellectual, social, and journalistic death sentence. In light of this automatic humiliation, it's no surprise that most critics hold their tongue; it has already become a question of civil courage to conduct a critique of technology. I can't afford (or so the critic reasons) to have everyone (from Lieschen Müller to the computing machine)[3] tell me that I am the only one who is trying to thwart the march of world history, the only one who is obsolete, the only reactionary far and wide. And so, the critic's mouth stays shut.

Yes, so as not to be thought of as reactionary.— Which is precisely what happened to the present author at the conference just mentioned above.

Picture this: I was discussing what I call "*postliterary illiteracy*," providing an account of today's *global flood of images*. This flood, I suggested, stems from the fact that today's humans are endlessly invited to stare at pictures of the world and, in this way, are seemingly also invited to participate in the "whole" world (or rather, in what is meant to be the "whole" for them). This invitation is being extended by means of the full arsenal of reproduction technology—glossy magazines, films, and television broadcasts—and it is issued everywhere. And the less humans are granted access to major world decisions and insights into the interrelations that shape the world, the more generous the invitation becomes. They are "having their eyes stuffed," as I cited a Molussian[4] fairy tale as saying: The less people have a voice, the more they are given to look at and see. With the help of this systematic flood of images, I continued, an "*iconomania*" has been instilled in them that has already assumed the unpleasant traits associated with "voyeurism": When images proliferate and overrun the world, they are inherently at risk of turning into stultification-devices. For in contrast to text, images *as* images don't make any wider contexts visible; they only show world-snippets that are torn from the whole. In showing the world, I concluded, images conceal it.— Now, it is important to point out that this line of argument was simply portraying the situation without adding any suggestions for therapeutic remedies. Yet, as I came to an end, it happened: I was called a "Romantic reactionary," and this by a representative of the *juste milieu*. This classification puzzled me—I was not exactly used to being called reactionary. But my confusion was very short-lived; in the discussion that ensued, the heckler immediately revealed what he had meant: "Those who explicitly shine a light onto such effects and phenomena," he explained, "are *engaging in criticism*. Such critics disrupt the development of industry as well as the sale of products; or they are at least naive enough to try. *And since industry and markets must obviously continue to develop* (don't you agree?), *criticism is eo ipso a sabotage of progress and, thus, reactionary.*"

I could not complain that his account lacked clarity. The evidence it provided for the strong resurrection of the idea of progress felt especially instructive to me. In 1945, in the immediate aftermath of the catastrophe, there was a prevailing impression that this idea was in rapid decline. And although earlier restoration periods had treated the idea of progress as a thorn in the flesh, his words revealed that this very idea was now being used as the argument in support of the restoration prospering today.— As for being called a "Romantic," when I didn't let the matter go, my attacker was kind enough to clarify this label as well. My "Romanticism," apparently, stemmed from "*my evident and stubborn insistence on a humane conception of the human.*" The instant correlation of being "human" with being "stubborn," the unspoken assumption that categories other than "humane" could provide definitions of the human, and the fact, finally, that nobody participating in the discussion seemed taken aback by any of this, made the whole episode culminate in what felt like quite a horrific punch line.—

Similar things have, of course, happened to me elsewhere. But *the equation of "critical" and "reactionary"* still seems to be *more popular in Germany than in other countries.* In Germany, the principles of the past regime—a regime which had put an end to criticism by liquidating all critics—have partly survived in a rudimentary form and partly already exist again in a germinal state. One thing is clear: These principles aren't dead. That presenting "criticism" as a "reactionary stance" and branding critics as reactionary saboteurs belonged to the ideological tactics of National Socialism is well-known. The expression "*Nörgler*" [nag], which was widely used to this end, in fact cut both ways. By calling the critic a "nag," the "movement" also glorified itself as progressive, for if the act of criticism is rendered reactionary *per se,* the object of its criticism—the regime—must necessarily be progressive.[5] This, or at least something not all too dissimilar from this, is what my heckler meant to imply.—

But let's not exaggerate the amount of civil courage it takes to conduct a critique of technology today. It would be wrong to say that *we* are resurrecting the question of machine-breaking here again; the debate is well underway. What, for instance, is the debate over the abolition of the atomic bomb about if not the potential destruction of a device? This discussion has been taking place for over a decade now, but it carefully avoids any direct associations and any vocabulary that would make it all too obvious what is being discussed. Everyone fears the embarrassment of being accused by the other of machine-breaking: the scientist fears the layperson, the layperson the scientist; the engineer fears the politician, the politician the engineer; the left fears the right, the right the left; the West fears the East, the East the West.— Around the globe, there really is no taboo as widely agreed on as the taboo that "machine-breaking" represents—a consensus deserving of a much worthier cause.

But when I say that "the debate is well underway," I'm not trying to imply that it could quietly pick up the old arguments (for instance, from the age of "textile workers") and affiliate itself with the classic Luddite struggle once again. This time—and this is the fundamental difference between the question of machine-breaking in today's industrial revolution and the first—*this time, we aren't dealing with a dispute between representatives of two different modes of production.* It's not only the artisan (who today barely exists in the classical sense) who is at threat because of machines—an artisan uprising, occasioned by the fact that they would rather continue to manufacture their televisions and thermonuclear weapons at home, would be absurd. And neither is it only the factory worker, whose "alienation" has been recognized for a century, who is under threat—today *everyone is.* Everyone is now an active consumer, a virtual victim, and a user of machines and machine-*products.* I emphasize "*products*": for the central issue today is not *the conditions of production, the question of who produces, or even the scale of production.* At issue, rather, is *what kinds of products and devices are being produced.* And this introduces the second fundamental difference between the past and the present danger. During the First Industrial Revolution, products were not primarily being criticized because they were products; the struggle was almost exclusively concerned with the monopolistic nature of mechanical *production,* which could achieve prices that were ruining artisans and small-scale operations. But at issue today are the actual *products* being produced: the bomb, for instance, or the human of today, who is just as much a product—as humans now owe their characteristics, or at least their totally altered character, to television and the consumption of industrially produced world- and opinion-pictures.

The wider topic is thus no longer primarily about the replacement of an outdated mode of production by new ones, or about the competition between different types of labor. As it runs across all segments of society, it is no longer *politically divisive either,* and it affects many more people. One would almost have to make up the difference between an upper-class and a middle-class[6] television problem, but it would be outright absurd to try and distinguish a middle-class atom bomb problem from a proletarian one. And just as it affects all social classes, the issue also affects every country, every part of the globe—it knows no "Iron Curtain." Both in the East and the West the transformation and liquidation of the human by its own products is a burningly pressing question, regardless of whether the flames are seen, regardless of whether they are discussed—let alone extinguished. No matter how implausible this may sound or how outdated it may feel in the cold political climate that currently reigns, when compared to the atomic problem, the differences between the political "philosophies" of the two worlds (that falsely call themselves "free" while accurately brandishing each other "unfree") have long since become differences of the second degree. The psychological effects of technology care as little about these differences as technology itself does. Even if one doesn't believe in the "One World" thesis[7] drawn up the day before yesterday, this thesis is nevertheless corroborated by the souls being conditioned by technology both here and there—and no less, and in a no less macabre fashion, by the radium-contaminated atmosphere that doesn't care about national borders.

The phenomenon discussed in the three essays that form this volume is thus not confined to any specific part of the globe; it stands apart from all political systems or theories, and from all social plans or programs. It is specific to our epoch and is, as such, *epoch-making.* This book doesn't ask what Washington or Moscow make of technology, but what technology *has made, is making,* and *will make* of us before we have the chance to make anything of it. Napoleon declared politics our fate one hundred and fifty years ago, and one hundred years ago Marx declared the economy our fate. It is in this sense that technology has become *our fate* today. Although we might not be able to be the hand that guides it, this doesn't mean we should deny ourselves the opportunity to keep a close eye on what its fingers are doing.

With this I could actually jump right in, *in medias res.* But before I do so, some methodological remarks will be sent ahead—or to be more precise, two warnings. These warnings seem necessary to me because the texts in this book aren't simply literary essays, nor are they philosophical analyses in the usual academic style. Using the old expression "occasionalism," they might be called examples of "*occasional philosophy*"—they are "*Gelegenheitsphilosophie.*"[8] What I understand this term to mean must at first sight appear like something that ought not to exist—an *Unding*, like *a hybrid of journalism and philosophy*. It is a mode of philosophy that takes today's situation as its subject—things that are characteristic of today's world. But these are not merely its subject, for it is, in fact, the opaque and unsettling nature of these things that gives rise to this mode of philosophy in the first place. The hybrid nature of this undertaking results in an unusual style of writing.

Readers will especially notice, and probably be irritated by, the continual *changes of perspective*, which in Leibnizian terms switch between "*vérités de fait*" and "*vérités de raison*."[9] While following the discussion of highly topical phenomena (the "*occasions for philosophy*"), they will again and again notice how they are being thrown into fundamental questions that seem to have no bearing on the initial phenomena (since these questions concern "*basic problems of philosophy*"). While reading the first essay, for example, which considers a new form of shame (the shame humans feel in view of their "humiliatingly perfect" machines), they will suddenly be dragged into a detailed metaphysical exposition of the "human's nonidentity with itself."

Whenever this was somehow possible, I tried to relegate such philosophical expositions into the notes. This has made the notes section quite substantial, and it now also mainly provides contextualizing thoughts instead of mere side comments; it provides documentation instead of appendices. But because these studies claim to be philosophical, it wouldn't have been acceptable to relegate all philosophical elements "to the notes"; this would have created a shift of emphasis that would have distorted the truth.

If my essays were monographs in a discipline other than philosophy, readers would, of course, be rightly incensed about sudden digressions. *But "digression" means something different in philosophy than in any other discipline* or even in everyday life.— And for the philosophical rigorist, it even means the opposite.

To clarify the implication, we will let such a rigorist speak (he is a man who is categorically opposed to the kind of "occasional philosophy" attempted here). "If you are really claiming to be conducting philosophy," he argues, "then special problems or even issues of the day must remain off limits; people of faith can't believe in any arbitrary *creata* either, just as ontologists must avoid any merely ontic entities. In philosophy, there are no changing menus or daily specials; as far as it is concerned, news [*Zeiten*] and newspapers [*Zeitungen*] don't exist, and philosophical monographs are like square circles—they are *contradictiones in adjecto.* Any topical issue or isolated problem thus already amounts to a digression. By restricting yourself to the singular, or even the occasional, you are giving up 'the general,' you disown 'reason, cause, and ground,' you dim 'the whole'—or whatever else it is you think philosophy deals with—and with this, you've lost the right to call what you are doing philosophy."

"In your essay," he quickly adds (for he is trying to preempt delicate questions about what philosophizing could draw on and what its "busy musings" would consist of if anything specific were already an unworthy subject)—"in your essay on 'radio and television,' you've slipped in some basic remarks about the 'essence of images.' But what was your digression? Was it really this inclusion? Or didn't you begin to digress the moment you selected your so-called subject—the moment you allowed yourself to become involved with a purely empirical entity, such as television, instead of conducting 'philosophy of art' or, indeed, 'philosophy'? Your digression, because you have pressed forward into territory that besides being unworthy of philosophical inquiry is also inaccessible to it, can even be called a '*stretch.*' Don't misunderstand this expression; I am using it in a philosophical sense. It refers to the tie—the elastic band that is attached to the generality of the concept, the band you have 'overstretched.' Sure, it has some elasticity, but its elasticity is not unlimited—it can't be stretched so much that its loose end touches even the most contingent individual phenomenon, such as television. Your choice of topic was thus already your digression.— Yet it would be wrong to call your remarks about the essence of images a 'digression'—for here, you were attempting to go beyond the singular phenomenon and investigate 'its ground, its cause.' And a 'digression toward the cause' would be nonsensical. This, then, was no crime against philosophy; quite the opposite: It was an act of remorse—remorse before the tribunal of the idea, an attempt to make amends."— The rigorist has spoken.

And not just him. He is joined by warning voices reaching back two and a half millennia. What else had the passion of philosophy consisted in, the passion of the most disparate philosophers, if not the glorious turn away from the contingent toward "the actual"—the turn from "*mundus sensibilis*" toward "*mundus intelligibilis*"? Indeed, this very dualism—that is, the unspoken assumption that this division is legitimate—was *the* metaphysics of the West, which all philosophers shared. In fact, even the empiricists took it for granted. Had they not, they would never have sought to establish empiricism as a principle and a foundation. And it really isn't possible to ignore these voices. The simple fact that the writer of these lines repeatedly gives in and returns to most fundamental reflections while discussing our epoch (and the concrete and contingent phenomena that typify it) is in itself proof that he cannot simply escape these voices either. But neither can he simply heed their warning.

In fact, nobody can, and nobody ever has.

The intransigent ideal of philosophy is most peculiarly unrealistic; servile compliance would nip any philosophizing in the bud. It may be true that philosophers live "in the general," but it's just as true that doing philosophy, in contrast to mysticism or simply saying "Om," can never just mean that one stares at the idea of "the general," "the whole," or "the ground." Quite the opposite. A person doing philosophy must always set out for *something:* something specific, something distinct from this ground—that is, whatever it is they want to get to the bottom of. In short, it belongs to the reality of doing philosophy that this intransigent definition of the essence of philosophy is ignored; ignoring it is in fact one of the conditions for philosophy's existence and continued survival. This also means that people who practice philosophy are not only in the world as philosophers. They are never just enveloped by the horizon of the "as such," never just fascinated by the beauty of "the whole," trembling before the "foundational ground," or attracted by it. Besides practicing philosophy—no, first and foremost—philosophers also live as ordinary neighbors to the people on their left and their right; they exist as humans who were born, who have needs, and who are mortal; as humans who are pursued, lured, and encircled by a whirl of *haecceitates*, and it's precisely these *haecceitates* that drag or chase them into doing philosophy.— Yes, precisely these. Through the unpredictability of their existence [*Faktizität*], their unanticipated, inexplicable presence and mode of being, these *haecceitates* provide enigmas that are no less obscure—and no less unsettling—than any number of "as-suchs," "totalities," or "grounds" could ever be. Perhaps they are even more deeply unsettling, for aren't "totality" and "ground" usually treated as the kind of answer that makes all questions fall silent. "It is basically unnecessary to get to the bottom of things," as a dialogue of a Molussian Sophist tells us, "for if the ground were the reason for the way these things are, they would not be these things here—these singular entities that are torn off from the ground. Thus, there must be a reason these things are themselves rather than being part of the ground. Who knows? Perhaps there are even as many reasons as there are things."[10]

Should the Molussian Sophist's desperate reflection be true, the bankruptcy of philosophy would be confirmed (which, of course, doesn't make his reflection untrue). But this is not the place to pursue his argument further. We cite it here only to show—by means of an extreme example—how it is precisely the specific, singular, and occasional that can trouble those who do philosophy the most.

But this much is clear: Those who dismiss the singular as a merely contingent and empirical object unsuitable for philosophy sabotage their own philosophizing. They are like the fool who, before moving into his house, walled up the door, only to later be found frozen to death on the doorstep outside. His dying words, written on the doorpost, explained that he had walled up the door because "it was something ambiguous—a hole in the house, a thing that is half inside and half outside."[11]

But enough images. No philosopher, unless they are swapping philosophy for mysticism and only ever saying "Om," is in a position to say *what* belongs in their house and what doesn't. None is able to say to what specific degree they are allowed to be empirical and still move within philosophy's area of competence, and at what point they start to step beyond its permitted limits. And when they think they can specify this, they cannot justify why they have drawn the line at that particular point. It is not at all clear whether philosophy has ever genuinely attempted to confront this question. I certainly cannot recall anything like a history of philosophical etiquette—namely, a document recording the levels of specificity still considered *fit for polite philosophical society—that is, "philosophisch salonfähig."* Were an investigation of this kind to surface, its outcomes would probably be deeply embarrassing for philosophers. It would reveal that eligibility was only ever decided by convention—that is, that the criteria for inclusion were always nonphilosophical, if not even incompatible with philosophy. I strongly doubt that Hegel could have explained why, in his portrayal of world history, he deemed some events to be sufficiently *salonfähig* for his system, while rejecting others as empiricist *plebs.*— No one can pretend in good conscience that things have improved or that these boundaries have become any less blurred. There is no philosophical explanation as to why, for instance, in the officially accredited discipline of philosophical anthropology, we have the right to philosophize about "the human," although we'd be immediately accused of a lack of rigor were we to conduct a "philosophy of the mosquito" or a "philosophy of the child" (after all, the human is also but a specific empirical entity). There is, of course, no denying that professional philosophers usually think they know exactly how much specificity is permitted, how far they can go, what is still philosophically "proper." Yet, in most cases, their self-assuredness is just the unphilosophical confidence of those who are following a custom. The more confidently they present themselves in this respect, the more right one is to distrust them as philosophers. And they have often had to pay a high price for their confidence: The spectacle of the post-Hegelian epoch has not exactly been awe-inspiring. Afraid of committing a *faux pas,* philosophers left the truly new strides into the specific and the occasional to the Kierkegaards and the Nietzsches, the Feuerbachs and the Darwins, the Marxes and the Freuds, to the great scientists of individual disciplines, yes, even to the great novelists. These great amateurs of philosophy felt no inhibition about expanding the boundaries and relocating the prohibition signs—and they often didn't even call their philosophical practice "philosophy." And even though the professionals are now busily making themselves at home within the newly legitimate boundaries that the amateurs have advanced, and even though they are impressing us by officially announcing "elastic systems" or unveiling the methodological foundations of "methodlessness," this theater of catching up is not enough to compensate for past failings.

Readers will now understand how a follower of the truth, for whom these great open-minded figures of the past are role models, will take on specific empirical entities even if it remains unclear whether (or to what extent) his activity can be called "philosophy." It is quite irrelevant whether tomatoes are called a "vegetable" or a "fruit" as long as they are nourishing. Anyone who is genuinely philosophizing will persist even when unsure about whether they are *digressing,* even without knowing whether their work already exceeds the permitted degree of specificity. They won't even stop when it becomes impossible for them to tell whether the elastic band tied back to "fundamental principles" has already been "overstretched." At any rate, reflections on principles, about who might still classify this or that point as "philosophy," to them seem pointless. It's the things themselves that matter. Astronomers practice astronomy because they are interested in the stars, and not just for the sake of astronomy itself. And those who philosophize are the same: whether philosophy "exists" is the least important question, and philosophy's field of competence—the question of who owns a given hunting ground—is just as unimportant. The only thing that matters is what is brought back from one's excursions, whether it is something nourishing.—

Our warning about the continual changes of perspective has now been clarified: It is quite plausible that the more an "elastic band" is stretched, the more likely it is that it will "*snap back into shape.*" Because we will repeatedly stretch it with a lot of force, it will also continually snap back. Readers will notice the following pattern: In the middle of discussions of current phenomena, they will suddenly realize that they have been carried away and carried into the obscure realm of fundamental philosophical questions; and yet, before they have even had the chance to stretch their legs, they will already be back at the front line of current affairs.

Our second warning now closely follows from the first, and it concerns an "irregularity" that might seem even more objectionable. It does not so much concern our "too singular" objects of investigation but the way these are portrayed; for these portrayals—or at least some of them—will seem like "*exaggerations.*" And this is easily explained: They *are* "exaggerations."

As should already be clear from the fact that I am myself using this word, "exaggeration" is, of course, not intended in the usual sense here. It has methodological connotations.

There are phenomena for which exaggeration and magnification cannot be avoided; for unless they are distorted in this way, they would not be visible or identifiable at all. By withholding themselves from our naked eye, phenomena of this kind leave us with a simple *choice*: "*exaggerate or reject all insight.*"— Microscopy and telescopy are the most closely related methodologies here because they also seek the truth by means of exaggerated visualizations.

In what sense is "exaggeration" now also required for our topic? Why are the objects under investigation so obscure to the naked eye? To what extent do they fail to show themselves and thereby make "exaggerated" portrayals necessary?

The answer is provided, implicitly at least, by the subtitle of this book. It reads: "On the Soul in the Age of the Second Industrial Revolution," a title that encapsulates the overarching topic of the individual essays. A more precise version would read: "On the *Metamorphoses of the Soul* in the Age of the Second Industrial Revolution."

This revolution didn't just begin yesterday, of course. It has long been providing the soul with the material conditions for its metamorphoses, and it adds new ones daily. Does this also mean that the soul has been able to keep up with these fully modified conditions? And that it is still keeping up with them now as they continue to be modified further every day?

Nothing could be further from the truth.

No trait is in fact more characteristic of us humans today than *our inability to stay emotionally* "up to date,"[12] *up to date with what we are producing*. What this means is that we have become unable to match the speed at which our products change, a speed decided by us. And so, our machines and devices have run away from us, they have dashed ahead into the future—a future we call "the present"—and we are unable to catch up. By making use of our unlimited Promethean freedom to forever create new things—and because of the constant pressure to exercise it—we, as temporal beings, have brought ourselves into such disarray that we are now lagging behind what we have ourselves projected and produced. Slowly, and with a guilty conscience about our obsolescence [*Antiquiertheit*], we now continue on our path or simply loll about like bewildered dinosaurs amid our machines. If surrealistic formations are defined by the presence of two totally disparate, contradictory, and even mutually lethal elements that suddenly appear simultaneously together—no, even in a mutually paralyzing interdependence—then surrealism has never been more "classically" realized than in the configuration formed by a human standing in front of a "computing machine."

This growing *asynchronicity between the human and the world of their own products*—the daily widening gap between the two—is what this text calls the "*Promethean gradient*" or "*Promethean gap*" [*das prometheische Gefälle*].[13]

This "gradient" hasn't gone unnoticed, of course. It was sighted, for instance, as part of the Marxist doctrine of the "superstructure," especially in the discussion of the different speeds at which the "base" and "superstructure" develop. But it was only sighted. The gradient Marxism was concerned with was just one example among many, an example extracted from a much wider complex in which highly diverse gradient-phenomena can be differentiated. Besides the difference between the conditions of production and ("ideological") theories, there is also the gap between *making* and *imagining,* the gap between *doing* and *feeling,* the gap between *knowledge* and *conscience,* and, finally and most importantly, the gap between the manufactured *device* and the human *body* (a body ill fitted to the "body" [*Leib*] of the machine). Each of these "differentials" will come to play its part in the course of this study, and they all share the same structure: the "*progress*" of one capacity beyond the other, which now has to "*limp after it.*" Just as ideological theory lags behind the actual conditions, our imagination lags behind what we can make: We can make hydrogen bombs but are insufficiently able to imagine the consequences of what we produce. Our capacity to feel, likewise, limps behind our deeds: We can bomb hundreds of thousands into the ground but cannot sufficiently mourn them or feel remorse.— And so, right at the back, at the furthest distance, the human body is also trying to tag along. Still draped in its folksy rags, it plods on as the most embarrassed straggler, the very last in line, and equally out of step with all who have taken the lead.

Each of us thus consists of a loose column of individual entities, each with its own level of antiquatedness and obsolescence, each marching at its own pace. Even though this is just an image, its suggestive power should suffice to deal a final blow to the nineteenth century's ideal of "the harmonious personality," an ideal that was fragile anyway.

The asynchronicity of the various human "faculties"—the "Promethean gradient" arising from the asynchronicity of the human with its own products—is one of the central subjects of this book. This doesn't mean, however, that we endorse the decision that is usually deemed obvious as soon as this asynchronicity is noticed: the decision, namely, to present the speed at which production is changing as the speed everything else should match. There is, of course, no denying that our products are trying everything to force their own speed onto us humans,[14] or that we are ourselves frantically trying to live up to this demand. The question, though, is whether we are succeeding in our attempts and whether we are even right to try. It is entirely conceivable that our technological devices are transforming too fast—categorically too fast[15]—that the demands they place on us are exaggerated, if not impossible, and that their unreasonable expectations are actually propelling us into a collectively pathological condition. Or, to say the same thing from the perspective of the producers, it certainly doesn't seem impossible that *we*—who are actively producing these products—are in the process of establishing a world we are incapable of keeping up with, a world it is impossible for us to "grasp," because it categorically exceeds the capacity of our imagination, our emotions, and our sense of responsibility. Who knows? Perhaps we have already established such a world. After all, apart from its formally unlimited capacity to produce, the human nevertheless remains a more or less determined morphological type whose adaptability is more or less fixed. It is thus an entity that cannot be remodeled at will, not by other powers and not by its own—it has an elasticity that cannot be strained *ad libitum.* When stepping onto the stage of its own historical world as a changeable actor, the human quite evidently enjoys far fewer freedoms, and collides much sooner with unmovable limits, than it does as a "freelance" stage designer and prop-maker. Despite the incredibly colorful and always changing back and forth of history, the question of whether "the human has changed" has repeatedly been asked and is still being asked today—and this is neither by chance nor a sign of philosophical dilettantism. Today, as human production seems to have exploded all limits, and as this particular explosion has made the unchanged boundaries of all our other faculties even more visible, *a critique of the limits of the human* seems to have become *the desideratum of philosophy.* And not just a critique of the limits of reason, but of the limits of all human faculties: of the capacity to imagine, feel, bear responsibility, and so forth. Vague speculation about our finitude is today no longer enough, let alone speculation that doesn't even center on our material needs but exclusively on death (which is peculiarly much more metaphysically accepted, much more *salonfähig,* than hunger). The limits of the human must actually be delineated and defined.

What do any of these remarks have to do with my warning about my "exaggerated" portrayals?

Our own metamorphosis has been delayed by the Promethean gradient. Our souls are now lagging far behind the transformation level of our products—that is, our world. This now means, however, that many of the soul's traits are still off track and have not yet found any clear physiognomic lines; some only make themselves noticeable as indistinct sketches, and only a few have assumed a distinct shape and experienced a rearticulation for our age (i.e., the Second Industrial Revolution).[16] Yet such delayed transformations still represent comparatively "successful" cases. For there may also be transformation attempts that never yield an immediately recognizable outcome because they are utterly foiled by the fateful rigidity and limited capacity of our imagination or feeling. Because of this, and in the absence of any immediately recognizable shape, their contours can only be deduced indirectly from panicked or otherwise pathological behaviors that stem from *failing attempts to synchronize*. The case of a bomber pilot made its rounds in the American press a few years ago. During the war, this pilot had participated in the devastation of towns and countries alongside many others without feeling any moral qualms. But then, in the postwar era, he began to try to "fully intuit and emotionally comprehend what he had done."

He was thus attempting "to become the person who (in terms of his deeds) he already was." Yet despite his best efforts, he was unable to develop "a soul appropriate for our epoch." When his attempt to coordinate himself with himself failed, he withdrew into a monastery, fully distraught. He will not have been the only such victim of the times; there probably are, and will be, many more. But in its simplicity, his "I still don't get it," which circulated in the press in the first years of his retreat, remains the exemplary testimony to the desperate effort with which he, the spokesperson for today's humanity, was trying to catch up with himself.[17]—

This, then, is the situation: Because of the "Promethean gradient," the souls of this, our epoch, are in part still "in the making"[18]—*still unfinished*, but also, in part, *never finished* because they fail to assume any final shape at all. If one tries to portray such souls all the same (as we are attempting to here), one will inevitably run the risk of giving faces that as yet lack form and profile a physiognomy that is far too distinctive. There is a risk of enlarging the relief of their profile to a height it does not yet actually have, at least not when judged by photographs that capture them in the moment. One is in danger of passing off a caricature as an actual likeness; in short, there is a risk of *exaggerating*.

But if one rejects such exaggeration—if one does not extend the vectors along which our souls are struggling to move, if one does not show the ultimate outcome of the (often merely intended or half or not at all accomplished) metamorphoses of the soul—there is a real danger that one is actively making these traits and even their direction of travel unidentifiable. Such exaggeration is all the more legitimate because it is the *concrete tendency of our age to actually drive this metamorphosis forward by exaggerated means*—for instance, by means of "human engineering."[19]

Our "exaggerated" portrayal is thus only a facet of the exaggeration that is happening today; it is merely an *exaggerated portrayal of what is produced by means of exaggeration and augmentation.*—

With this, the connection between "the Promethean gradient" and "exaggeration" has been explained, and any suspicion that "exaggeration" is trying to achieve something sensational should have lost its edge. And so, we now jump *in medias res* into our first exaggerated image: the portrayal of "Promethean shame."

1

On Promethean Shame

If only I were one of you,
I'd no longer be ashamed;
I'd happily learn to enjoy
my eternal return.

I'd run on an appointed course,
attuned to the machines,
I'd do the same as yesterday,
and, tomorrow, the same again.

No one would know who I am,
no one would find out either,
who, in the dark beginning,
my parents were.

How, tiny in mother's lap,
like a fish I swam,
and as a lump of bloody clay,
my life on earth began.

Rather than, through the furnace's red heat,
and fully formed by things,
springing amongst your ranks,
with an armor-plated disposition.[1]

1. *First Encounter with Promethean Shame—the Prometheus of Today Asks: Who Am I, Anyhow?*

I begin with some diary entries from California.

March 11, 1942

Believe I have found the signs of an entirely new pudendum this morning: a form of shame that did not exist in the past. I will provisionally call it "*Promethean shame*" for myself. I understand this to mean the "*shame when confronted by the 'humiliatingly' high quality of fabricated things [selbstgemachte Dinge].*"

Joined a tour together with T. of an exhibition of technology that has opened here. T. behaved in a most peculiar manner—so strangely, in fact, that I ended up only looking at him instead of the appliances on show. As soon as one of the highly complicated pieces started to work, he lowered his eyes and fell silent.— Even more strikingly, he concealed his hands behind his back, as if he were ashamed to have brought these heavy, graceless, and obsolete instruments into the company of machines working with such accuracy and refinement.

But this "as if he were ashamed" is too tentative. The pattern of behavior left me in no doubt. The things, which he recognized as exemplars, as superior, and as representatives of a higher class of being, really did play the same part that persons of authority or of acknowledged "higher" classes used to play for his ancestors. To stand in his bodily clumsiness and his corporeal imprecision under the gaze of such perfect devices was really unbearable for him. He was truly ashamed.—

When I try to investigate this "Promethean shame" further, its basis—"*the basic flaw*" of the one who is feeling shame—appears to be the nature of his own origins. *T. is ashamed about having naturally grown instead of having been made.* He is ashamed because he owes his existence to the blind and uncalculated, highly archaic process of procreation and birth, which places him in stark contrast to the immaculate products which are carefully designed through and through. His shame thus consists in his "*natum esse,*"[2] in his lowly birth. He regards this as "low" (not unlike the chronicler of founders of religion) precisely *because* it was a birth. Once he is ashamed of his antiquated line of descent, the same naturally also holds for its faulty and inescapable result: *He is ashamed of himself.*[3]

Incidentally, throughout the entire tour, T. kept silent. He rediscovered his voice only long after we had left the exhibition. This seems to corroborate my shame hypothesis further because any form of self-expression would have contradicted shame: If shame shows itself at all, it does so precisely through self-concealment.

March 13

Reflections on T.'s behavior:

Promethean defiance is the refusal to owe anything, including oneself, to anyone else. Promethean pride consists of seeing everything, including oneself, as one's own achievement. Some vestiges of this stance so typical of the self-made man[4] of the nineteenth century are certainly still alive today. But I doubt they are still characteristic of us. It seems that feelings and attitudes of a different kind have taken their place—attitudes resulting from the peculiar fate of Prometheanism.

Prometheanism has experienced a truly dialectical reversal. Prometheus's triumph has in a sense been all too overwhelming. It has been so triumphant, in fact, that Prometheus, when confronted with his own achievements, is beginning to lose the pride that was a matter of course in the previous century. A sense of abject inferiority and self-contempt is replacing this. "Who *am* I anyhow?" the Prometheus of today asks himself, playing the jester at the court of his own machines. "Who am *I* anyhow?"

It is against this changed background that one needs to consider the desire of today's human to become a self-made man, to become a product: *Humans want to make themselves, not because they can no longer tolerate anything that was not made by them, but because they, too, no longer want to be something not-made. They feel indignant not because they were made by others (God, gods, or nature), but because they were not made at all and, as such, are inferior to all their fabricated things.*—

March 14

What's at stake here is clearly a variant of a classic error: namely, a "*confusion of creator and creation.*" This expression has its origin in the *Confessiones*. Augustine uses it to describe *the* religious error *par excellence:* the worship of a *creatum—that is,* of something that was created (the world or a worldly artifact) or the worship of a pictorial representation of a god in the guise of such a *creatum*—instead of reserving this worship for the *creator,* who alone is deserving of it.— The equivalence of the two errors is clear: When feeling "Promethean shame," humans also prefer what was made over the maker. Here, too, the creation is accorded a higher rank of being. With this, however, the analogy comes to an end because the role that humans play in "Promethean shame" has completely changed: Whereas for Augustine humans belong *eo ipso* among the animal kingdom, they now appear only in their *homo faber* quality—as beings, therefore, who fabricate things. "Confusion of creator with creation" in this instance means that the honor humans extend to their things is actually due to themselves—and themselves alone. The substitution is here not proof of undue pride, but of self-degradation. Humanity denies itself the credit it deserves. T. attests to this self-degradation with his behavior.

A legend from Molussia recounts how Bamba, the Molussian god of creation, immediately sought refuge in invisibility after he had formed the Molussian Mountains. It is said that "from that day onward he remained invisible because he was ashamed not to be one of the gleaming peaks he had created." T.'s behavior matches that of this Molussian god.

March 15

T. certainly also wants to "exist" [*"Da"* . . . *"sein"*].[5] But to exist as a "natural son" rather than as a legitimate product, to exist as someone who was fathered rather than fabricated, to be a human instead of being equal to other, precisely functioning, convertible, and reproducible machinery—that's what pains him; that's what embarrasses him. Probably he even deems all this an original sin (even though he does not use this term and does not express his disillusionment in words at all). In short, and in the words of an industrial song from Molussia:

Each day out of machines arise
ever more beautiful machines.
Only we remain malformed,
only we're born obsolete.

Too soon from obscure ground,
preformed and cast out,
at this late hour we humans stand
imprecisely in this world.

Alas, in view of machine-precision,
a head held high becomes us not.
Only things are granted confidence,
only machines are granted pride.[6]

2. *Three Reservations Are Discussed: Promethean Shame Is Absurd. It Is Invisible. It Is Trivial*

After I had read out these diary entries, the claim that a new—hitherto unknown—variety of shame exists or is at least emerging was met with the following three objections:

First Objection: "*We humans have made these machines ourselves. Our natural and legitimate attitude toward them is therefore pride.* The assumption that *homo faber* must blush to see what he himself has fabricated, or that the producer would feel shame at the sight of their products, is absurd."

Riposte: This challenge is rhetorical. The unifying words "we" and "our" express a purely verbal solidarity. That "we" are proud isn't true. Who is this "we"?

If anyone had the right to employ such a "we," then it would be the minority of researchers, inventors, and experts who truly master these mysterious realms. *We,* however, the remaining 99 percent of today's population, did not "make" the machines (the cybernetic ones, for example). They are not felt as "our" achievements but strike us rather as strange and disconcerting. One's own products are never disconcerting. Even if we had been involved in the making of these things, we wouldn't feel like their proud creators. Production processes today consist of so many single and separate steps that the workers who actually build the world of products and machines cannot see these fabrications as the fruits of "their own" labor. There is no longer any occasion for pride—be it for a single end product or the world of machines and products as a whole—because no end product shows any sign that it was invested with the specific qualities and achievements of individual workers. One can only be proud of achievements that bear traces of individuality that make processes of identification possible. For the workers, the world of machines is no more an object of pride than it is "their" property to own.— This, of course, is all the more true for all those who are not involved in production processes at all.

No, for the majority of people today, machines do not primarily exist as human products, and they are most certainly not viewed as testimony to some personal Promethean glory. This is the case despite the tremendous increase and ever more widely shared competence in the use of technology, and notwithstanding the common knowledge that products don't grow on trees. Products, rather, are simply "there." We encounter them primarily as necessary, desirable, superfluous, affordable, or unaffordable consumer goods that become "mine" only after I have bought them. As such, they are much more likely to be proof of one's own insufficiency than evidence of one's own power: In a highly industrialized country, the excess of displayed yet unobtainable products is simply overwhelming. The shopping mall is the permanent exhibition of what one *does not* have.—

As for the man who is for the first time confronted with a working computing machine, self-aggrandizement and pride are even more alien to him. An observer who erupts with the exclamation, "My goodness, aren't we great guys, to be capable of this!" when encountering such a machine is a clown, a figment of imagination. Quite the contrary! He rather murmurs with a shake of his head, "My god, it's incredible what *it*—the machine—can do!" At the same time, he feels highly ill at ease in his creaturely skin, for the machine half gives him the creeps and half puts him to shame.—

Second Objection: *"I have never been able to observe the onset of this Promethean shame."*

Riposte: Entirely possible. Only rarely can it be caught sight of in action. In most cases, one must deduce it indirectly from modes of behavior (provided that the one shamed by products does not detect this feeling themself). This has the following reasons that are related to the nature of "shame":

I. The best-known varieties of shame (e.g., the shame associated with the genital area) become acute in the interaction of humans and become visible—as barriers to communication—in encounters between human beings. "Promethean shame," however, is a form of shame that is felt in response to a thing. Because no human partner is present to make us feel shame, there is likewise for the most part no human partner to witness our shame.—

II. Shame does not "appear" at all when it sets in.[7] When shame "appears," it does not "step onto the stage of visibility"; it "takes a step back." Humans who are ashamed try to conceal their flaw and themselves. They are, however, unable to fully realize this desire to "sink into the ground and disappear completely." From this wavering shame, two peculiar, if you like, "dialectical" consequences ensue that explain the "invisibility of shame."

First consequence: Because the one who is ashamed *remains visible,* and because shame draws attention to its presence, a second form of shame immediately compounds the first—*the shame of being ashamed.* Shame accumulates ("iterates") automatically. It is nourished, so to speak, by its own flame, which becomes hotter and hotter the longer it burns.

Second consequence: In order to put an end to this self-accumulation of shame, which becomes increasingly unbearable with every passing moment, the one affected resorts to a trick: *Instead of concealing his flaw and himself, he conceals his shame,* and even his gestures of concealment. He adopts an attitude that is diametrically opposed to shame—brazenness, for example, or a stance that declares, "I couldn't care less." In order to conceal his shame, he thus practically rips the shirt from his own chest. With this motion, he often not only seeks to deceive the one before whom he is ashamed but also himself.— It is not only in individual (erotic) situations that this mechanism can be observed: In America's love life of today, there is no more powerful motif than the shame of still being considered puritanically coy. Result: the bikini.— In other words, in order to overcome his urge for self-concealment, the shamed person decides to return to the level of the most normal visibility. "Anyone who does not conceal themselves," or so this person calculates, "anyone who remains conspicuous is not suspected of shame (and thus, also not suspected of having a flaw)." They hence behave like someone who is wearing a torn item of clothing but who remains confident enough to treat this tear as nonexistent.— If shame remains imperceptible and invisible, this is so because it is "*concealed through visibility.*"—

Third Objection: *It is possible that T. was ashamed. In this case, however, this "Promethean shame" is not an exciting new phenomenon but rather the sign of a well-known state of affairs: It is merely a symptom of the "reification and objectification of the human," which has been discussed ad nauseam.*

Riposte: No. It is a sign of much more. What T. deems to be the occasion of his shame is precisely not that he has been "reified," but the other way around. He is ashamed because he has *not* been reified and because he is not a thing. That's also why the expression "thing-shame," which I had initially considered, is out of the question. *A new, second level in the history of the reification of human beings* has been reached with this response—this *shame at not being a thing.* Humans now acknowledge the superiority of things, bring themselves into line with them, and *welcome their own reification.* This also means that nonreification is condemned as a shortcoming. Or—if one believes that this stage has already been reached—*a third level* must be considered: one in which the response of the one who is reified (approval or rejection of reification) has already become second nature and has become so immediate that it is released not as a deliberate judgment but as a feeling. This is certainly something entirely new—and T. has clearly reached this stage: Intimidated by the superiority and supremacy of products, he has already *deserted to join* *their camp.*[8] Not only does he now share their point of view, not only has he fallen in line with the standards they set, but his *feelings* have also become theirs: He despises himself in the same way that things would despise him if they could. And when he feels shame, he does so precisely because of this contempt, which he deems well deserved.— Such self-degradation in front of fabricated things has not recurred since the end of idolatry.—

3. *An Example of the Reification of Self: Makeup*

Two expressions used in the above ripostes require clarification: "endorsement of one's own reification" and "desertion to the camp of machines."—

(a) An example of the endorsement of reification: "*makeup*"[9] as a reification of self.

For girls,[10] it is out of the question to go out in public without makeup. This does not only mean, however, that they are embarrassed to show themselves in a scruffy or unadorned state, as their mothers and grandmothers were. No, the pivotal point here is *when*—that is, in what state of adornment—they feel attractive. *When* are they deemed to be "well turned out"? *When* do they believe that they have no reason to be embarrassed? Answer: once they have transformed themselves into *things.* Once they have changed into crafted objects, into manufactured goods—as far as the raw material of their limbs and faces allows this to happen. To show oneself with "naked fingernails" is "impossible": Fingernails only become fit for society, fit for the office, yes, even fit for the kitchen, once they "match" the appliances these fingers have dealings with; only when they show the same dead, polished thing-"finish"[11] as things and when they come across as manufactured. In short, the fingernails have to *disown their organic past life.*— The same standard also holds for hair, legs, facial expressions, and indeed the body as a whole (although, in this last instance, nature is a bit recalcitrant). *Today, it is the unmodified body that is "naked," not the unclothed one:* the body, then, that does not sport any thing-like qualities and does not bear signs of reification. One is ashamed of this body that is "naked" in a new sense even when it is dressed—much more so than one is ashamed of a body that is "naked" in the conventional sense, as long as the body on show has been reified in a pleasing manner. Every bathing beach demonstrates this, not just fashionable ones.— To vary a famous word by Nietzsche: The body is something "that must be overcome."[12] Or rather: It *already has been* "overcome."[13]

4. *Examples of the "Desertion" of Humans to the Camp of the Machines—the First Calamity of Humankind: The Human Body Is "Stubborn and Rigid"—the "Stubbornness" of Their Bodies Turns Humans into Saboteurs of Their Own Achievements*

(b) Let us now turn to the expression "Humans are deserting to the camp of their machines."

What was meant by this metaphor?

An American air force instructor teaches his cadets that humans, in their natural state, are "*faulty constructions*"[14] when measured against the tasks they face.[15]

Regardless of whether the instructor was joking or serious, a better attestation of the "desertion" in question could not be thought up. Humans can only be considered constructions, "faulty" ones even, once they are classed as machines. Only when the category "machine" has become universally applicable and is deemed all-inclusive can that which was not constructed be reinterpreted as a "faulty construct."

It cannot be denied that humans are inferior to their technological devices in terms of strength, speed, and precision, in the same way as human thinking power fares badly in comparison to the thinking power of "computing machines." From his point of view (which he has borrowed from machines), the instructor is thus right. And this even more so because he no longer even has this type of inability to compete in mind.— He would probably already deem it embarrassing to occupy himself with this comparison. No, he is already making a much more modern comparison: He is not thinking of *humans as machines existing alongside other machines,* but of *humans as machines existing for the sake of machines.* He is considering humans as working parts of machinery that has already been built or as parts of already specified blueprints for future technology. As such he calls humans "faulty," meaning "fitting badly," "inappropriate," and "not measuring up"—that is, not made to suitable standards. This being "not made to suitable standards" could still mean that humans are merely unformed, unshaped raw material. But they are not even this. For humans *are* preformed, they *are* pressed into a shape, they *have* a form: just the wrong one. Not only are humans not made to specification, but because only "matter" can be reworked, they cannot be made to fit the specification—unless, that is, already-shaped humans are demoted to the status of mere material, are regarded as "scrap metal," are treated as such, and are melted down. With this image, however, we are already getting ahead of ourselves.—

From a machine point of view, the inability to be made to measure [*anmessen*] naturally appears to be a human "pretention" [*Anmaßung*]—yes, even a scandalous resistance and revolt. And in the mind of those who have turned the measure of machines into their own benchmark, a disgraceful failure.[16] Even if it was just meant as a clever remark, behind the instructor's phrase "humans are faulty constructions" lies the indignation of machines and the shame of humanity. "After all" (one would probably be allowed to let the instructor say), "we cannot expect our designs and machines to undergo alterations just because *we*, as human components with limiting requirements, are made so unreliably."—

Nothing would now seem more obvious than the following conciliatory response: "Regardless how great the superiority of machines is in strength, speed, and precision, they are and remain mere rigid and definite objects. Whereas we humans—yes, and this constitutes our dignity—are alive, malleable, adaptable, elastic, and hence free."— But this is precisely what the air force instructor would contest. "It is the other way round!" he would exclaim. "When I call us humans 'faulty constructions,' I do so precisely because we are rigid and unfree in comparison to things!"

What does he mean? How did he arrive at this conclusion?

Let's forget for a moment the distinction between a dead object and a living human being that serves as the premise for our pride.

Our world of fabricated products does not define itself as the sum of single, finalized pieces but rather as a process: the daily new production of things that are daily new. Hence, the world of products does not "define" itself at all. It is indefinite, open, malleable, daily intent on reforming itself, daily adaptable to new situations, on the verge of new tasks every day. Reshaped continually through trial and error, the world of products presents itself every day as a new and different world.

And we? And our body?

No sign of daily change. Our body is today still yesterday's body, still the body of our parents, still the body of our ancestors. The body of the human who builds rockets can hardly be distinguished from the body of the troglodyte. It is morphologically constant; morally speaking, it is unfree, stubborn, and unaccommodating. Seen from the perspective of machines, it is conservative, antiprogressive, obsolete, and unchangeable: a deadweight in the rise of machines. In brief, *we must invert the relation between freedom and unfreedom. Things are free, and human beings are unfree.*[17]

The air force instructor had this and nothing else in mind when calling humans "faulty constructions." His figure of speech was much more than a witty remark. For, if we assume that the future tasks that will fall to technology are close to his heart—and we have no reason to doubt this—then we can see his words voiced by a grave concern: *the anxiety that our body,* regardless of whether the rift between it and our products broadens or even merely remains constant, will *threaten all new projects* that new machines show to be necessary; that our body will end up wrecking one project after another.

"In principle," or so the instructor could argue, "if it only depended on machines, we could soar to such altitudes—but this 'in principle' is redundant, because *we* couldn't survive it anyhow because our body would go on strike, and so we can't do it after all. 'In principle,' we could travel to this or that planet—but *we* are prevented from doing so by our own limitations, *we* cannot survive the stress test." If the instructor has heard anything of Greek mythology he might conclude in this or a similar manner: "The Icarus of today falls from the skies not because the wax holding his wings together fails, but because he himself, Icarus, is a failure. If he could only jettison himself as ballast, his wings could conquer the skies."

Not the worst description of humanity today—and the instructor would certainly endorse this—would be the following: *"Humans are saboteurs of their own achievements."* For sure, they are not saboteurs in the sense that they would voluntarily harm their machines—in fact, despite all the weapons of destruction in existence no such idea has ever crossed their minds. Our contemporaries have never been more selfless toward any other entity than toward their machines. No, they are saboteurs because they, the "living ones," are rigid and "unfree." "Dead things," on the other hand, are dynamic and "free." Humans are saboteurs because as natural products, as born bodies, they are too emphatically defined to keep up with the daily changing world of machines—a world that makes a mockery of all self-determination. Deeply concerned, ashamed, and with a guilty conscience, the human of today thus observes his highly talented children, fearing that he will ruin their future because he can neither send them on their career path without support nor can he accompany them. And it often happens that he, the lame father, endows his fledgling children with servants designed to replace him because he is bedridden with his earthliness, unable to be part of their stellar rise, and must stay at home. As a result, crews of escorting personnel and supervisors in the form of machines are assembled (e.g., the monitoring instruments he builds into his space rockets and artificial satellites). Or, to continue with a different image: The lame father sees himself forced to give them a "boomerang-like" post to carry on their journey—mail made up of questionnaires[18] full of question marks and bearing the notice "Please return to sender." Mail that is tasked to return to Earth completed to tell the story of what he himself would have liked to have learned along the way, if only he had been able to take part in the journey.[19]

The greater the misery of humans who produce goods becomes, the less they are a match for their own creations; the more relentlessly and tirelessly, the more voraciously and frantically they multiply the civil service that their machines, their submachines, and their sub-submachines form into. By doing so, humans are naturally multiplying their own misery further: the larger, more intricate, and complicated the manmade bureaucracy of machines becomes, the more futile are their attempts to keep up with it. It may hence be right to conclude that human misery leads to an accumulation of machinery, which in turn leads to an accumulation of human misery.— Happy days, when the pastoral idyll of Hydra was still regarded a frightful myth!

5. Today's Human Tries to Escape This Calamity by Means of Physiological Synchronization with Machines, by the Means of "Human Engineering"—the Extreme Perversion of Supply and Demand

For the man of today, it is of course out of the question to accept the stubbornness and inadaptability of the human body and settle once and for all for human inferiority and retardation. He must, therefore, do something. His dream is, naturally, to become equal to his deities—the machines—or put more accurately, to be one and entirely part of them. *Ὁμολογουμένως ζῆν.*[20] As a Molussian hymn proclaims:

If only we could succeed,
to shed our burden and stand
like machine shafts,
neatly slotted into others,

as prostheses with prostheses
most intimately bound,
then our defect would be overcome,
our shame would be unknown—

if this could still be granted us,
if this favor still could be bestowed,
no sacrifice in this world
would be for us "too much to make."[21]

Yet, as already said, this Molussian image of blissful "*cosubstantiality with technological instruments*" remains wishful thinking. For humans, true integration into machinery is no more possible than real competition among equals. Machines are once and for all ontologically superior to beings who are born. This does not mean, however, that humans could wash their hands in their creaturely innocence and let the matter rest, or that they would even want to do so. No religion has ever accepted the argument that the human is a mere creature and not a god as a *carte blanche* to be morally indolent. Neither the industrial religion of today nor its followers would accept the argument that "humans are not products but merely creatures" as a valid excuse for being idly content with their own creaturely inadequacy. It is expected of humans that they prove their "piety toward things" by at least attempting self-reformation and an "*imitatio instrumentorum*."[22] The minimum requirement is that the man of today makes the token effort to "better" himself to such a degree that the "sabotage" he practices *nolens volens*[23] as a result of his "original sin" of being born is reduced to the smallest conceivable amount.—

This is being attempted through certain *metamorphoses of the self* called "*human engineering*"[24]—that is, "engineering work on human beings." We will describe a few examples in more detail. These attempted metamorphoses invariably begin as follows: The human engineer exposes his body to extreme and artificially created "*situations that test his corporeal limits*."[25] To be able to study corporeal reactions, conditions are created that are *just about* bearable. Needless to say, these reactions are not examined in order to find out what the human physis is. The objective is not to determine how this physis has "naturally" delimited itself in the form of a body, and then trace these limits cartographically; in any case, a body subjected to unnatural conditions would hardly disclose its own natural limits. What is true for modern physics hence also holds for our current example: The object of experimentation is modified by the parameters and mode of observation the experiment entails. No, it is the other way around: The aim is to see if the human physique shows weaknesses and catch such *"soft parts" in flagrante delicto.* I call those parts "soft" that can still be modeled and shaped because they are currently still undefined, amorphous, changeable, and ambiguous. Parts and areas, therefore, that might still allow the body to be adapted to the demands of machines.

The "human engineer" does not want to know what the human physis *is.* He rather wants to find the threshold within which it *could* "just about" *still be* (without biting the dust during the stress test). *He does not want to know how human physis measures up in its organically grown form,* but rather *which* alien *demands it could still "measure up to"; not what is fixed and certain about human physis,* but *which of its limits are not fixed and can still just about be altered.* He is hence only interested in artificially induced corporeal limit situations because he aims to surpass these limits. If he has found a situation that is "just about bearable" and, through dressage of the self, managed to become accustomed to it, he fraudulently moves the limits to a new threshold. A new "limit situation" is staked out that he must become accustomed to by training his body.

Like a pioneer, he endeavors to push his limits further forward, ever further away from himself, "transcending" himself more and more.[26]—Although he does not step across into the supranatural, he does enter the no-longer-natural. By leaving his innate limitations behind, he enters the realm of the hybrid and artificial. In brief, *the aim of the experiments is to make the human physis* undergo a metamorphosis. With the exception of magic and medicine, this physis has always been regarded as "fate." The aim is, hence, to *strip physis of its fate, of all that is fatal.*—This also involves the removal of all that is "embarrassing" and shameful (as the word *fatal* hints at).[27] Who knows whether the eagerness with which these great-grandchildren of the Puritans dedicate themselves to this masochistic transformation of the body is not in part explained by the secret current of residual energy generated by the Puritan hatred of the body, energy that they do not recognize as such and which cannot be used otherwise in today's world.

The "theory of the body" has thus been replaced by a specific "*praxis*," by a "physiotechnics" (if this analogy to "psychotechnics"[28] is permitted). Not, however, a "physiotechnics" of the kind we already know from medicine, but rather a new revolutionary technology intent on toppling and abolishing the presiding "regime" of physis as such. The aim of this technology is to shape from the "existing state" of the body radically new ones. Following a famous pattern, its maxim could be described thus: *"It is not enough to interpret the body in various ways; one must also change it."* And this needs to be done again each day and each time to fit the specifications of the machine in question.—

If a machine requires a body capable of operating in icy conditions, one exposes one's body to the freezing cold in an icing-up installation that is specifically built so one can train in these icy conditions and transform oneself into a being that is their match.

If another machine needs endurance of abnormally low pressure, one builds low-pressure chambers in order to force a metamorphosis into a low-pressure being.

If a third requires resilience to centrifugal forces, one builds a centrifuge that pulls one's innermost parts outward until one has let oneself be transformed into a being that can live in such ex-centricity.

The first step in this process invariably involves the determination of the precritical point, the environment of the "just about bearable." The second step is to train and exercise in the said environment, and the third step is to push the threshold further "*behind the back of physis*" once the exercise has succeeded. Human ingenuity really knows no limits here. In point of fact, there are no catastrophic conditions that could not be fabricated, or rather modeled in true-to-life (this actually means true-to-the-machine) training- and martyr-simulations. No ingenuity or luxury is spared in the construction of these models, for the human engineer has staked everything on them. With their help, he hopes to be able to pass the high school diploma for machines [*Geräte-Abitur*] and, in doing so, be absolved of his shame by the machines he admires.

What the body is to become is in each case determined by what the machine demands. The relationship between "*supply and demand*" has here experienced a peculiar perversion, one that far exceeds the conventional perversion of this relationship that is already normal today.

The "conventional perversion" of supply and demand, as I understand it, has two aspects:

(1) *The inversion of the chronological sequence*—that is, the fact that it is not usually demand that leads to supply, but the other way around, that supply leads to demand. (2) The fact that *demand is also produced* as a "*secondary product*" that is essential for the sale of the actual product being supplied.[29] One thing, however, is still left unaltered in this normal perversion, or so it seems at least—namely, the division of roles between demand and supply: *Humans are here still being* treated as "*subjects of desire and demand,*" even if, as consumers, they are terrorized or flattered into their needs.

The destruction of this semblance is the new element that perverts the relationship between supply and demand further. The *machine* now openly and shamelessly presents itself as the "*subject of desire and demand.*" It demands to be offered what it needs, and expects that humans exert themselves to come up with ever better offers (because humans in their present state are not acceptable to machines). Humans are thus required to supply everything the machine needs to function as well as it "could" function. This "could" is not merely a subjunctive clause, suggesting that it does not commit us to anything. No, this "could" describes what is "due" to technology, and the meaning of the word *due* already shimmers across into being an obligation.

Into being an obligation. This means that moral demands are now also transferred from the human to the machine. What is "due" now also becomes what one "ought to be." The maxim "become who you are" is acknowledged to be the valid maxim of machines. The purpose humans now pursue in life is limited to guaranteeing the success of this machine-maxim through the offering, provision, and preparation of their own bodies.

Only one generation ago, there were many phrases (in pedagogy and the branches of psychology concerned with aptitude) that sought to express the sentiment "give the capable a free hand." Such phrases convey the view that the "potential and talents" of humans are sacrosanct in themselves. They imply that developing and utilizing these talents is a moral duty and that neglecting or suppressing them is immoral. What was reserved for humans yesterday is now reserved for the machine: Its "potentials" are nurtured. It is now a duty to "give it a free hand"; the stifling of its "potential" has become an immoral deed. *Machines are the "gifted talents" of today.* Even the inherent "potential" of the H-bomb, the wunderkind among machines, is held to be sacrosanct. As manifold as the excuses for the existence of the bomb may be, a good part of the vehemence with which protests against its existence are met can be explained by indignation at the fact that there are humans who take the liberty to oppose the maxim "become who you are," who want to nip the "organic development" of the machine's potential in the bud and block its "free development."— Today, even the most dreadful machine can be successfully justified and defended if its critics are suspected of being Luddites. And since nothing is easier, this always succeeds.

6. Human Engineering: Initiation Rites of the Robotic Age— "Dehumanization" Does Not Alarm the Dehumanized, for It Does Not Belong to Their Area of Responsibility

The reader who comes across the details of these human engineering experiments for the first time is plagued by an irritating feeling of *déjà vu*—or, rather, *déjà lu.*[30] The irritation lasts until they suddenly remember what it so strikingly reminds them of: namely, ethnological accounts of *the rites of passage* that tended to precede the admission of adolescents into the community of adults in the life of a clan. This association is certainly not unwarranted. For what the human test subject hopes to achieve by subjecting their body voluntarily to torture-like experiments is really a kind of investiture. The test subject hopes to be admitted into the community of "adults" and be ranked among those who "count." Only it is machines that now count as "grown-up."— *The experiments that make up "human engineering" are* in fact *the initiation rites of the robotic age.* The human guinea pigs are candidates and eventually neophytes who are proud to have left their "childhood" behind. They are proud to have undergone "the education of humankind" that is demanded today.

However, if "machines" are considered the "adults," then "to move on from one's childhood" and "the education of humankind" means "to leave the human condition behind and stop being human." For the moralist, who cannot let the idea of the human be swept aside, the trajectory of this development is the total catastrophe. What humans are hoping to achieve with their experiments really is the *climax of all possible dehumanization.* I say "climax" because for a century the most extreme image of dehumanization, degradation, and deprivation of liberty conceivable was that "*workers are being exploited by their working conditions.*" In the light of what has now actually occurred, this image has become the epitome of naivete and harmlessness. That the *passivum* "being exploited by" still brings shocking images to mind for contemporary man—images of suffering, unfreedom, and misery—is really out of the question. *What he deems his plight today* is, inversely, the limitedness of human "usefulness and application," the *worrying possibility* that there might be *definite boundaries* to human passivity and usefulness—in short, *to human unfreedom.* The expression "climax of dehumanization" is certainly not an exaggeration: Contemporary man doesn't shy away from any amount of blood, sweat, and tears and uses all his spontaneity and ingenuity to extinguish the spontaneity and humanness of his performance and achievements. He has mobilized everything so he can implement his own reification and become passive, for he hopes to explode the boundary of unfreedom that threatens him until he finally obtains the *summum bonum* of total usefulness.

It is part of the nature of machines that they conduct one single, specialized task, that they are fully consumed by the performance of this task and find in it the sole justification of their existence. Should machines display additional features beyond their intrinsic purpose—for example, the heat generation of the light bulb—then these are merely "put up with" if they cannot be exploited further or eliminated. This aspect of machines now also defines the entity that is to be produced through "*human engineering.*" This being is to obtain one specific ability—for example, the capability to execute this or that maneuver at a certain altitude, at such and such a speed, in specific low-pressure conditions, within a certain response time.— No matter how "superhuman" such ability may be in comparison with what the body is actually capable of, the desired outcome is something *subhuman.* The "actual human being" (if it cannot be eliminated completely) here merely becomes an unavoidable appendix to the machine, for the desired outcome of human engineering is a machine-function, a proficiency in an isolated "skill."

Of course, those who are part of "human engineering" see nothing morally questionable in their activity: it is deemed neither "outlandish," nor "impossible," nor even dehumanizing.

What strikes them as "outlandish" is merely the fact that, as humans, they form such "impossible" exceptions in a world that evidently belongs to machines and devices. And it is precisely this fact that makes humans conceive of themselves as "hopeless," just as an inept machine is called "hopeless." As for the charge that "human engineering" is dehumanizing, this is completely incomprehensible to those who are involved in it. After all, to be like a machine is what they desire and the task they have set themselves. For them, the robots clanking about in "cartoons,"[31] disguised as puppetlike humans for fun, are not undignified or terrifying figures; they embody their dreams and their duty to conform and synchronize.[32] Why the satisfaction of their yearning and the pious devotion toward things, why such pleasant and honest activities might be considered debasing, is destined to remain inexplicable for them. If there is anything at all that is bothering them about "dehumanization," it is at most the fact that now and then old codgers appear who think of attaching the ugly epithet "dehumanizing" to their activity. But at most this. Usually they do not even notice such old codgers.— Those who are invested in human engineering are not disturbed by the question of *what* (apart from the special skill they hone to the point of virtuosity) *should become of them as human beings* (for, the codgers suggest, they were once "human," too, and, therefore, should not deny that there is still something humanlike attached to them as an irremovable deadweight). Human engineers are not unsettled by such questions. Such questions are usually not even noticed since they do not belong to the area of their specialized experimental tasks. If such questions are acknowledged, it is only in order to respond that such matters belong to a different area of responsibility and that someone else will have to take care of them. For human engineers would certainly feel that occupying themselves with such matters would "interfere with their own work," making them neglect their duties, and that this would be immoral and ultimately even *indiscreet*. If one translates the unspoken moral code of specialists into words, "indiscretion" means that one has more interests than one's position requires, that one has mistakenly brought along a surplus of knowledge and interests not required for the job one is about to begin.[33] In short, in response to the question what should become of him as a human being, the human engineer would answer that's "*his* business"[34] while pointing his thumb back over his shoulder in no specific direction, for he has no clue to which authority, person, or department this "*his*"[35] refers; yes, he doesn't even know whether such a department exists. In point of fact, the expression "*his business*," which once meant "everyone has the right to become happy or unhappy in their own manner," today means the following: "If it is someone else's business to have an opinion about me, I will take special care not to involve myself in his business (that is, in my business that has been placed in his hands)." If this way of putting it seems too pointed, the following analogous example should be considered: today's relationship between doctor and patient. Patients are often not only prevented from knowing about their own state of health—about their proximity to death—they even voluntarily take no notice of it. For the patient's state of health is the doctor's business.

Our human engineer would hence not get at all agitated if, instead of asking the question of what will become of him as a human being, one were to shout the ready-made answer in his face: "You are turning yourself into human dross; you are making yourself a mere appendix of your artificially acquired specialist skill." Either he would not register this attack at all (for he has immunized himself against everything that does not belong to his area of responsibility), or he would retort, "You are barking up the wrong tree!" At best he would just shrug his shoulders, for he feels utterly confident in his cause. When it is a question of providing machines with what belongs to machines, he is certain that no sacrifice and no self-sacrifice can be called "an outrageous human sacrifice."[36] There is no chance that he could become a suitable target for attempts at moral intimidation.—

7. The Attitude of the Transformed Prometheus: "Hubristic Humility"

Never before, not even in their most mystical practices or magical transformations of the self, did humans attempt an operation that constituted such a radical "*rejectio fati*."[37] Compared to the aims of human engineering, the building of the Tower of Babel looks like an undertaking as meek as a lamb: The limit humans transgressed was here merely one pertaining to their workmanship and the permitted dimensions of human products. Their transgression was thus quite "innocent" because criteria that would have helped them discover and conform to the maximum permitted height had not been defined. Since humans can produce things at their own discretion, such criteria are not at their disposal, unless *humanity itself* is taken as a measure. That is, unless they define the critical point as the moment when humanity, "*smaller than itself*," can no longer keep up "with itself"—that is, the moment when human beings are no longer a match for what they have produced.[38] This moment is now.—

This much, at least, is clear: If humans disregard this critical point and begin to also alter their body, this marks a fundamentally new, truly unprecedented and vile chapter.

This, however, should not be misunderstood. It is not the alteration as such that we deem "vile." Anyone who hears in our arguments a hidden, "metaphysically conservative" voice would misunderstand their meaning.[39] Nothing is further from my mind than assuming the position of a "metaphysical moralist" who regards the existent (as it is or as he imagines it to be) as "good" and the "preordained order of things." I am not an ethicist who puts the morality of humans into the framework of what "is as it is and must so be." Nor am I someone who infers what is allowed and forbidden from the given order of things. The chances of developing such a "metaphysical morality" were forfeited a long time ago. It would be a futile exercise to want to retrospectively find a common denominator for questions of what is just [*quaestiones iuris*] and what is fact [*quaestiones facti*],[40] including those pertaining to metaphysical facts. To level the charge that "reason" was split into these two parts by philosophy, in particular by Kantian philosophy, which established this split once and for all, is ridiculous; since Paulsen, synthesis enthusiasts, who regard everything to be organized like a box of toy bricks, have mounted this attack on philosophy time and again. Because of the natural sciences, the world has transformed itself for us into a thing that is "beyond good and evil." The result is that moral problems—and not only theoretical "problems" but above all *our own "moral" and "immoral" deeds*—are now, regardless of whether we like it, *drifting without roots in the ocean of morally indifferent being.* They are drifting, so to speak, as a "*metaphysical bunch of cut flowers*" that concerns nothing and nobody apart from us humans, and about whose inconsequentiality we are not allowed to harbor even the slightest illusions.[41]— I say "without roots" because the opportunity that Kant saw in the fact that we could turn our non- or only partial affinity to morality into the metaphysically positive fact of "freedom" has also been lost.— Today, the life of a moralist is certainly no fun.

Far be it from me, then, to regard the form [*eidos*] or morphological fixedness of existing "species" (the "human" species included) as "sacrosanct" or "good" just because they are as they are, or to retrospectively sanction them (or condemn their transformation as a "cosmic prohibition"). All the more so because nature itself is clearly open to mutation and doesn't seem to place any great value on the permanence of species; in brief, the existence of eternally immutable species has not been established.— Even *if* there were such permanence, even *if* it were proven that the various species have endured unaltered to this day, this persistence would still not be morally conclusive, precisely because *quaestiones iuris* are independent of *quaestiones facti*. Or, put otherwise, because what exists does not have to be.—

No, the alteration of our body is not fundamentally new and vile because we are abandoning our "morphological destiny" or transcending the assigned natural limitation of our capabilities. It is despicable, rather, because we are undergoing this self-transformation for the sake of machines and because we employ these as blueprints and models for our own modification. We hence relinquish ourselves as our measure as humans, and with this gesture, we either restrict or give up our freedom. It is for this reason that the expression "*hybrid*" does not seem a suitable designation for what we are creating, no matter how adventurous our experiments and their aims may be. The connotations of breeding that the word *hybrid* evokes might be justifiable, for we are creating new hybrid beings through the crossbreeding of fabricators and products; but is this also true for the ordinary sense of the word *hubristic*—namely, "arrogance" and "pride"?[42] After all, acts of self-denial are not hubristic. And "*fitting oneself to a pregiven measure*" [*Anmessung*] cannot be equated to "*presumptuous self-evaluation*" [*Anmaßung*].[43] These acts are symptoms of curtailment—yes, even of self-humiliation.

Even so, the impression is not easily averted that this act of curtailing oneself in the attempt to "measure up to a machine" is also presumptuous. It would, of course, be unreasonable to dispute that the attempt to turn us humans into machinelike beings is monstrous. Were a "human engineer" to ask us what we now took him for, presumptuous or unassuming, hubristic or humble, we wouldn't be able to give him a clear answer. For the phenomena "of hubristically overestimating oneself" and "of making oneself into a hybrid by fitting oneself to a machine's measure" are phenomena that are like twins who cannot be separated. The "human engineer" is, in fact, both "overconfident" *and* "self-deprecating," both hubristic *and* humble. His attitude is one of "*arrogant self-degradation*" and "*hubristic humility.*"

Of course, these phrases sound mystifying. It is, however, only the constellation of words that is baffling, not the actual idea that they formulate. Just how old this idea actually is can be seen, for example, in the prohibition of suicide shared by many religions. This prohibition is not justified on the grounds that the suicide victim is aiming for a titanic or godlike existence—to suggest this would be absurd—no, it is the other way around. The person attempting suicide limits their own significance in such an absolute manner that in the eyes of religion such a judgment is the sole prerogative of the highest authorities. What is condemned is hence *"arrogant self-degradation."*—

The choice of this example is not a coincidence, since the self-degradation at issue in our discussion of human engineering is "hubristic" in a very similar way to suicide. It does not aim at total bodily annihilation, "merely" the annihilation of the human *as* a human. Yet this "merely" is debatable, since it is more "hubristic" than suicide. Why? *Because the hubristic and hybrid intervention in the human that the human engineer performs does not just "make use of what already exists" but invents something new.*— What does this mean?

Whereas the suicidal person limits themself to taking an inevitable event (dying) out of the hands of higher powers to bring it about preemptively, the intervention the human engineer dares to undertake is a first, a *novum:* It would never happen to him "by itself." It is a wrong that he would never be expected to face and suffer from the powers that be. His intervention is not merely the appropriation (and "requisition") of an existing injury but rather the presumptuous production of a new one. To the injuries he must expect from above—hardship, illness, aging, and death—he now masochistically adds a further one: self-reification.— This is how one might imagine a theological account reflecting on this in the year 2000: "Because the daemon or Marcionite God[44] who condemned humans to a machinelike existence or turned them into machines did not exist, humanity invented one. Yes, humans were even presumptuous enough to cast themselves in the role of this additional deity. They assumed this role, however, only to damage themselves in ways that they could not suffer from any other gods.— Humans turned themselves into masters only to find a new way to become slaves."—

It may be true that humans are doing all this in the blind hope, or at least on the pretext, that this self-reification will reduce or even abolish all other forms of human frailty and degradation. But what counts are not reasons, but only what is actually done: In order to combat the limitation of their liberty, humans limit themselves in a new way.—

We have usually associated the figure of *Prometheus* with "hubris." For the last 175 years (from Goethe via Shelley and Ibsen to Sartre's *Mouches*), our fathers treated *Prometheus* allegorically, and until recently, we did the same. If we ask ourselves whether this figure still has validity and allegorical significance for the representation of our contemporaries who conduct "human engineering," we arrive at the following ambivalent answers.

Our contemporaries are certainly still "Prometheans," but strangely perverted ones. They also sneer, "I should honor you—for what?"[45] But with this response they are refusing to honor themselves.

They also have presumptuously self-aggrandizing ideas of entitlement—but these are so aggrandizing that they begin to feel inadequate themselves.

They also suffer lacerations—but not because Zeus punishes their high-flying ambitions, but because they chastise themselves on account of their own "backwardness" and the "shame of having been born."—

Our contemporaries would, of course, recognize themselves even less in these mythological parables than in our account of human engineering in the previous sections. For them, their conduct is utterly nonmythical, or at least something that they can accommodate within the framework of the "Myth of Progress"[46] (although they do not realize that this is a myth). They automatically see "progress" in each new step they take, even when this step takes them into low-pressure or simulated frost chambers. They are reassured because things are moving forward and they are proud of this novelty.

After this digression, let's return to our more narrowly defined topic.

We had established the following: When humans suffer from feelings of inferiority in the face of their *machines*, they do so primarily because they are forced to realize—in their attempt to measure up to their machines—that they are a "poor quality" raw material when it comes to being molded into this or that machine-part. Instead of actually being raw material, humans are morphologically "unblessed" because they are already pregiven. Moreover, every pregiven form turns out to be "wrong" and "already pressed into the wrong shape" because the actual form machines require varies in each instance. In "human engineering," people attempt to "melt down" this "misshapen form" in order to retroactively source the material from which the shape required at any particular time can still be made.—

The fact of "having already been formed in the wrong shape" is the main defect the human engineer seeks to address and the main reason for "Promethean shame." It is only the main source, however, for he has many deficiencies that "generate shame." It would be impossible to compile a list of all these shameful shortcomings because this would correspond to a list of the virtues of all existing and all potential machines. In the following, we will therefore limit ourselves to a discussion of the defect that seems to be of most consequence after that of "being misshapen": the fact that *human beings "perish easily."*

8. The Second Inferiority of Human Beings: They Perish Easily and Are Excluded from the Process of Industrial "Reincarnation." The Human "Malaise of Being Unique"

Although humans are less malleable than their products, humans are also shorter-lived, more mortal. It is impossible for humans to compete with the durability—yes, even "immortality"—they can give to their products if they want to.

Of course, the things we make are not "immortal" in the strict sense of the word. The durability and shelf life of our preserved fruits, our frozen scrambled eggs, vinyl records, or light bulbs is limited. In most cases, however, it is *we humans* who have allocated mortality to our products and limited their life expectancy (so as to secure a steady or even increasing demand for them, for instance). Only our own mortality is not of our making. It alone has not been calculated. Because of this, it is a cause for shame.

With some restrictions, the use of the word *immortal* to describe our products, even the most obsolete ones, could be justified. This is because a new variety of immortality has come into existence: *industrial reincarnation,* meaning *the existence of serial products.* Certainly, viewed in the singular, the life, utility, and performance of each individual item produced (*this particular* screw, *this* washing machine, *this* vinyl record, *this* light bulb) has a limited time span. But is this also true if it was produced as part of a series? Doesn't the new light bulb that replaces the burned-out one continue the latter's life? Does it not become the old bulb?[47] Doesn't each broken or lost item continue to exist in the model, make, and blueprint according to which it was made? Cannot every item console itself with the hope of being there anew, once its sister piece has taken its place? Has it not become "eternal" on account of its replaceability—that is, because of the technology that makes this possible? O death, where is thy sting?[48]

As strange as this may sound, it is only the vocabulary deployed here that is strange. The phenomenon itself is neither uncommon nor new.— Example:

In contrast to the great fire that destroyed the Library of Alexandria, no page was actually burned when thousands of pages were reduced to ashes in the course of Hitler's book burnings in 1933 because every page had hundreds or thousands of identical siblings. As ignominious as the intentions of the arsonist were—and as ominous the gesture of his hand that betrayed the fact that it would soon commit more than paper to the flames—at this stage, his acts of annihilation were still totally farcical. Amid the jeering crowd dancing around the pyres of books, there also danced unseen a feather-light group of hecklers, whom the flames could not touch. Even while being scattered to the winds, this group of the "*book prototypes*" proclaimed, "Go ahead, burn our copies! Burn them! Us you cannot burn."— And today these supposedly incinerated books live again and do so in thousands upon thousands of copies.—

The world we live in today is certainly no "Kingdom of Ideas." But it is surely undeniable that it is more "*Platonist*" than any human-world has ever been before. This is precisely because our world is made up of things that are for the most part mass-produced to a specific norm. These products owe their existence to *ideas:* They have come into the world as imitations or imprints of prototype models, blueprints, or master copies. *None of the things* produced in this way now makes, or can make, the arrogant claim of being "itself." None is *individual and singular in anything other than a numerical sense,* for the ideas they incorporate also provided the blueprint for their sister pieces. As long as items produced to an unchanged model are kept in stock, as long as a new individual item can step in for a damaged one (for this specific burned-out light bulb here, for example), it is merely a question of money whether one deems the thing one possesses mortal or immortal. *For the man who is flush with cash, each thing can reincarnate itself in a new one.* And this prospect of reincarnation only ends when the "idea" of the thing also dies—that is, when its make is dropped in favor of another one. The thousand copies produced from the blueprint of that particular make, of course, then also gradually disappear.

Of course, this industrial Platonism, this "immortality through reincarnation," is not an accomplishment of the products themselves. They owe this virtue to us, and this needs no further elaboration. This admission, however, doesn't provide a counterargument to what has been said so far. For the object of this investigation is precisely the fact that we feel inferior to our products, *despite* the fact that we produce them ourselves. Here, too, what counts is only our own disadvantage: only the fact that we cannot share the virtues that we bestow on our products. In this case, the disadvantage in question is that none of us has been granted the ability to exist in the form of multiple (simultaneous or successive) copies. In contrast to the light bulb or vinyl record, none of us has the chance to outlive themself in a new copy. In short, we must continue to live our lifetime in obsolete singularity and uniqueness. For those who recognize the virtues of the machine-world as exemplary, this is a flaw and as such a reason for shame.

That's right: a reason for shame. I am not using this expression as a metaphor here any more than I did last time. To show that the emotion here really is "shame," I will share a particular experience:

About ten years ago, I visited someone who was terminally ill in a Californian hospital.

In response to my "How are you?"[49] he made a gesture that seemed to include not only his immediate fellow hospital patients but humanity as a whole. He murmured, "There is not a lot happening with us—any one of us." He met my query about what he meant with a shrug of his shoulders, as if to indicate that the answer was self-evident. Then he added a rhetorical question to counter mine: "Well . . . can they preserve us?"[50]—The word *they* referred to the doctors; "preserves" are bottled fruits—he wanted to say, "Can they perhaps turn us into preserves?"

I said no.

"And," he continued, "spare men they haven't got either."[51]

"Spare men?" I asked, not understanding.

He elaborated, "Well, don't we have spare things for everything?"[52]

Now I understood. He used "spare men" as an analogy with "spare tires" or "spare bulbs."[53] Like another light bulb, so to speak, that one could screw into the fitting in his place once he had burned out. What he meant to say was, "They do not keep spare men in stock for us."

His last words, however, were: "Isn't it a shame?"[54]

He was thus suffering from a double feeling of inferiority: (1) that he couldn't be preserved like a fruit, and (2) that he couldn't be replaced like a light bulb. For he was quite simply—and here the disgrace was undeniable—a categorically perishable individual item.—

It was not easy to console this man. Who on earth is mentally prepared to see a fellow human being suffer because he can be neither preserved like fruit nor replaced? The words of comfort I improvised were certainly inadequate. And, of course, his distress, in shame and in death, was in no way less than that of others in their final hours. It cannot be denied, either, that he also had a particular experience of mortality, and that this was even premised on an image of "eternity" [*aeternum*] or at least "being everlasting" [*sempiternum*].[55] Only, this image was no longer modeled on God, the starry heaven, the realm of ideas, or humankind (and while I sat by his bedside holding his hand, the considerable differences that no doubt exist between these terms seemed to dissolve). Instead, he looked to the world of preserved peaches and branded products that have become immortal through reincarnation. This was thus the eternity for which he had respect; this the ideal before which he was ashamed. In this state of shame, the light of his life went out. And because he could not ascend to the Mount Olympus of manufactured goods, but plunged into the Hades of raw material, he ended up more destitute than all the fabricated things he had valued during his lifetime.—

There is nothing more rebellious and cunning than us, the people of today, when we think we are being cheated out of even the merest shade of unfreedom. So, of course, we naturally do not submit without resistance to this *numerus clausus*[56] and our exclusion from the class of reproducible branded goods. We are no more willing to put up with our "faulty construction" than with our irreplaceability and existence as unique beings.

"Why," someone could object, "why do we still need this? Aren't we all what we eat and 'what we do,' and since millions of us already consume and do the same things, hasn't our indistinguishability—and with this our interchangeability—reached its peak? Are we not also only 'individuals' in a numerical sense? What reason could we still have to complain about a *numerus clausus*?"

Viewed from the perspective of institutions, the economy, the leisure industry, politics, and warfare—all of which make use of us as instruments of labor, consumers, or victims of some other kind—this replaceability is already an undeniable fact. It is, however, not merely a bare fact, but one that is daily confirmed by public opinion and a state of affairs that is also expressly approved of by scientists. For even social psychologists and moralists have got nothing better to do than idealize "adaptedness" and "inconspicuousness"—that is, to present everyone who still insists on a rudiment of selfhood or a trace of personal difference as a "crank"[57] or pathological weirdo.[58]

It cannot be denied that our transformation into reproducible mass-products has already succeeded when seen from the perspective of these institutions (which make use of individuals). "*Spare men*"[59] are always available (because everyone is merely a part and an actor in a labor process). However, this conclusion is only true from this institutional perspective. It is not shared by individuals themselves. Our already mentioned patient, for example, testifies to this. He did not complain about his replaceability; he grieved over his irreplaceability.

Each of us, every single individual, shares the fate of this patient. As replaceable as my contribution at work may be, as guaranteed the gapless continuation of my function, I myself, my identity as XY, cannot be continued by a substitute. Even if the "I am me" that my substitute will utter corresponds verbatim with mine, it will now refer to another I—namely, his. My own self will stay behind unreplaced and unreplaceable.

Anyone who doesn't pay careful attention here can easily miss the crucial point. After all, it was the credo of all ideas of humanity that the singular, as singular—that each individual, as an individual—is irreplaceable (and "irreplaceably valuable"). It could sound, therefore, as if we are here speaking about a basic characteristic of humanity or what remains of the experience of being human.

The opposite is true. What is decisive here is *how* the individual confronts their own irreplaceability—*what* they feel it to be. What matters is that one deems one's singularity an undeserved handicap—an attribute that makes one recoil, even though one cannot deny being identical with it.— In short, this individuality is experienced as a defect—a *flaw* of which one is ashamed. This "*malaise of being singular and unique*" is decisive here.

If one gives this "malaise" the trivial wording of a pop song—"I'm unique, I won't come again"—it is immediately clear that it is closely related to the fear of dying. And this makes sense: We said above that mass-products have achieved a "deathless" state due to their replaceability. If humans are excluded from this serial existence and replaceability, they are also excluded from their "deathlessness." The experience and knowledge that we humans are not mass-produced hence acts as a *memento mori.*[60]

9. *Humans Attempt to Escape This Second Calamity through "Iconomania"*

There is one phenomenon that now offers overwhelming proof of how ubiquitous the feeling of disadvantage—this "malaise of being unique"—has become: "*iconomania*," the current addiction to images.

It is certainly indisputable that this addiction to images is a first in the history of humankind; as regards intensity and scale, this phenomenon has left all other current addictions way behind. Indeed, it is key to any theory of our age. Since only a new term can indicate that it is a concept of philosophical significance, we will coin it "*iconomania*."

There is a utopian description of our world in words we have put into the mouth of an extraterrestrial reporter. He states that the human-world appears to be, first and foremost, "the vehicle and occasion for potential images." "At times," he recounts, "I was even under the impression that the purpose and function of the human-world is limited to this. At any rate, images play such a tremendously prominent role that, when I imagine their world emptied of its billions of photographs, films, TV-phantoms, and posters, pure nothingness remains. I have not met a single human being who did not voluntarily present several pictorial duplicates of themselves and their loved ones, nor anyone who did not carry such images on their body or at least possess them."—

One of the most important reasons for this hypertrophic production of images is that images allowed people to seize the opportunity to create "spare pieces"[61] of themselves and, with this gesture, belie their unbearable singularity. The excessive production of images is a large-scale countermeasure against humanity's "I am unique." Whereas people are usually excluded from serial mass-production, they transform themselves into "*reproduced products*" when they are photographed. *At least in effigies, they can also attain a multiple, at times even thousandfold, existence.* Even if a single human "only" lives as the model for their own photographic replica, "they" somehow still exist in them.

It follows as a matter of course that *movie stars,* who are seen by more eyes than us common mortals and are the most successful among us in assuming this multiple existence, become our role models. *We honor them with wreaths of laurels because they have made successful inroads into the sphere of serial mass-products—a sphere we recognize as ontologically superior to our own human one.* We idolize them because they have in the most triumphant manner realized our dream of existing as things, our dream to be upstarts in the world of products. There is in fact no fundamental ontological difference between the star actress, disseminated by thousands of photographic replicas, and nail varnish that circulates in countless examples. It is entirely logical that stars[62] and products form an alliance and support each other in advertisements (the star by recommending the merchandise, the merchandise by indulging images of the star): They are "birds of a feather that flock together." Not only are they equally well-known, they have also overcome their mortality in the same way: Both can continue to prove themselves after their death in the form of their reproductions. Like nail varnish that can continue its path and keep on shining once the original prototype has ceased to exist, the smile of the star can continue to shine after the original has gone the way "of all flesh." In a certain sense, the star is even "immortal while still alive" (the "immortal Garbo,"[63] for example) and not bound by the way of all flesh: She is younger than she is because the majority of her pictures feature the eternalized version of her actual self (i.e., the only version that counts commercially), the one that portrays the divine and wrinkle-free appearance from her youth. The way of her actual flesh is a process that no longer counts and that can at best only be shamefully concealed. "Hollywood privacy," an entirely new type of "privacy," is in the process of taking shape. It seeks to hide everything that could become detrimental to the validity of images.—

Nevertheless, in comparison with the truly multiple existence that our mass-products enjoy, our own multiplication through images is destined to remain an "as if." Despite our intense iconomaniac activity, the satisfaction we can derive from it is only compensatory. The difference between actual examples and mere copies, the fact that we have to settle for multiplying ourselves in the form of photographs whereas products are allowed to disseminate themselves across the world as truly identical pieces, cannot be eliminated. Humans can still not completely extinguish the shame they experience in front of their "superior things" through the images they have of themselves.

So, does this mean—one will ask—that in their most deeply held dreams and desires, humans see themselves as actual mass-products? That they positively desire to live as mass-products?

It is not possible to conclusively answer this question in the affirmative, for the same reason that it is impossible to assert that humans positively long to be forever—in the sense of "*sempiternitas*," and not "*aeternitas.*" It would be very easy to see evidence in support of this assertion in our fear of death and our precautions against dying. But this reasoning would be misguided. Certainly, it's true that we *do not want to die.* It's not true, however, that we positively want to continue to live on and on, that we want to become countless millions of years old. We are not even capable of "imagining" such a forever on and on. This means that the certainty is exclusively negative and precautionary, not a positive desire. We cannot pursue here why it is that no positive mental image can match this most fundamental desire to negate death, why our imagination remains *blind,* and why we are in a sense "too mortal" to even positively "*think*" that we will never die. But that's the way it is.

Our desire for multiple existence also remains "blind" in the same way. It's true that each of us wants to divest ourselves of our "unique existence." It is not true, however, or at least it is a "philosophical exaggeration," to suggest that everyone shares the desire of the Californian patient—the desire to divide himself up into "spare pieces"[64] so he can populate the world in the form of many or even countless examples. This is why our "iconomaniac compromise"—which consists of taking part in the mass-existence of products through our images while remaining ourselves—might be the best solution after all.

10. *An Illustration of Humiliation Sourced from History: MacArthur as a Precedent*

Following the analyses offered so far, in which we consistently employed anonymous phenomena of our age to illustrate the symptoms of "Promethean shame" and the lengths to which we go to combat this feeling, we want to portray at least one case in which this *humiliation* is visible *as an identifiable historical event.* It is evident that the outperforming of humans by machines that puts them to shame must be noticeable as *instances* of *degradation* when humans are in the public eye and on the highest rung of the social ladder. MacArthur was a human being of this kind.

For many years it has been an open secret known around the globe that at the beginning of the Korean War, General Douglas MacArthur proposed measures that arguably could have triggered a Third World War. It is also common knowledge that the decision as to whether such an outcome should be risked was taken out of his hands. Those who removed this responsibility from him, however, did not do so in order to take it on themselves or pass it onto other people with political, economic, or moral expertise. Rather, they removed the decision from MacArthur to hand it over to a *machine* (for the "final word" in this matter had to be objective—and today only statements made by objects are deemed "objective").— In short, they "transferred the responsibility and answerability"[65] to an "*electric brain*,"[66] which they considered to be the highest and final authority in this matter. The decision was thus not merely removed from the specific person MacArthur, but from him as a human being. The machine-brain was pitted against MacArthur's not because there were specific reasons to question his intelligence, but because of the simple fact that MacArthur merely had a human brain.

It is, however, misleading to say that this responsibility "*was* being taken away" from him as a human being. After all, the power wresting the decision from MacArthur was not some superhuman authority—no "Moira," "Tyche,"[67] "God," or "history," but *humanity itself.* In a sense, humanity used its right hand to rob its left, offering up the loot—its own conscience and freedom to decide—on the altar of machines. With this, humanity proved that it had submitted itself to this manmade calculating robot, was willing to accept this machine as a substitute for its own conscience, and acknowledge it as an oracle machine and even as the machinic eye of Providence.

This is to say, therefore, that *humanity harmed itself* by subordinating the general to this machine. One should not misunderstand me here. This is anything but an expression of solidarity with MacArthur. That he was placed center stage as the *dramatis persona* in this exemplary moment was a mere whim of history; any other person in the public eye would have served history just as well. What we are alluding to, then, is merely that anyone who transfers responsibility from a human being to a machine also transfers their own responsibility. And with MacArthur, humanity practiced this form of self-degradation for the first time in a representative manner. For the first time, it publicly said to itself: *"Because we aren't as good at calculating* [rechnen] *as our machine we are unreliable and incapable of taking responsibility* [unzurechnungsfähig], *so let's not 'make decisions'* [rechnen] *anymore."*[68] *For the first time, humanity didn't feel any shame at publicly shaming itself.*

This oracle machine was "fed" with all the data about the American and the enemy's economy—"to feed" is the technical term for supplying machines with the data they require to reach decisions.[69] However, it is an exaggeration to say "with all" the data. For a quintessential aspect of machines is that they have an "*idée fixe*"—that is to say, they impose determinants that are limited artificially by what they can do: The fact is that the perspective from which the data is computed is regulated in advance and remains fixed in this manner once and for all. And so, they "fed" the machine exclusively with the type of data that did not offer any resistance to quantification. Datasets, therefore, merely touched on benefits and damaging effects, as well as the profitability and unprofitability of waging the war under consideration. Of course, for reasons of methodological unambiguity and clarity, this procedure had automatic consequences—for example, the annihilation of human lives or the devastation of countries could only be considered and evaluated as figures of profit or loss. Questions like "Is this a just or unjust war?" were not even put to the "electric brain." In fact, they would have even been embarrassed to dish up such fare, for it was foreseeable that the machine, in its incorruptible object-objectivity, would have refused such sentimental and subjective fodder. Or, if they had nonetheless violently stuffed such questions down the oracle's throat, it would have not digested them and would have responded with an electric constipation.— In short, out of consideration for the machine's exalted allergy toward the subjective, they renounced moral questions in advance. Following the model of "logical empiricism," they only classed questions as "sensible" to which the machine could provide an unambiguous answer. They rejected all other types of question as nonsensical (and this even though they tried to convince themselves that the machine had been consulted in the first place only because of acute moral scruples).

Two things are thus sidelined, no longer "at work," and no longer "count" [*rechnen*] when recourse to such a machine has been found:

1. The competence of humans to resolve their problems themselves, because in comparison to the machine their ability to calculate and work things out equals zero.— And:
2. The problems themselves if they cannot be evaluated and computed by a machine.

As is well-known, it only takes a ludicrously short time for the data to be processed by a mechanical digestive system. They had hardly finished feeding the apparatus when it excreted its oracle. And since they were willing to accept this excretion as a verdict, they knew at this point whether they should dare the undertaking and announce a holy war, or write it off as loss-making business and scrap the whole idea as immoral.

After a few seconds of profound electric reflection or digestion, the "brain" excreted a decision that by chance was more humane than the decisive strike the human MacArthur had proposed. The "brain"[70] resoundingly proclaimed that MacArthur's approach would be "loss-making business"; yes, it even portrayed the potential war as disastrous for the American economy. All this was certainly very fortunate—possibly even the best fortune humankind has ever had, for the oracle sounding from the mouth of the machine was already being carried into the atomic age. However, the process, as such, by which this decision was reached was at the same time the most epoch-making defeat that humanity could have inflicted on itself. For never before had humanity degraded itself to such a degree that it entrusted judgment about the course of history, perhaps even about whether it may be or not be, to a thing. It is irrelevant that the machine's verdict on this occasion was a veto and a reprieve, it still constituted *a death sentence, precisely because the source of possible reprieve had been transferred to a thing.* The status of humanity was not decided by the machine's positive or negative answer but rather by the fact that this question had actually been submitted to the thing and the answer was abided by. If one realizes that thousands of humans—and we are possibly among them—owe gratitude for their chance of not being dead yet (which is what counts as "life" today) to the "no" excreted by a machine, shame would make one want to sink into the ground.[71]

But let's return to MacArthur.

His proposition has been consigned to the oracle machine for evaluation and arbitration. Even before the machine has voiced its support or its veto, the "*right of decision is taken away from*" the general. MacArthur has to accept what the machine utters as the final word. Even if it had given the go-ahead, he would, of course, still have lost "the right to decide": for what would be left for him to do after the machine's announcement could only be done because of its electrical permission. The humiliation has hence begun.

But as we know, the proclaimed verdict was a *veto.*

With this, the humiliation is complete. For a man with MacArthur's ambition and imperiousness, the situation is untenable. The consequence cannot be avoided: Although he is not demoted in the military and technological sense, his demotion is nevertheless a fact. His military career has come to an end. He becomes a civilian.

This was the first consequence. It is logical and holds no surprises.

But the story of MacArthur's humiliation is not over yet. There was a second consequence, which, if I am not mistaken, constituted a *countermeasure.* And this countermeasure is anything but ordinary; it is, rather, highly astounding. For this story reached a climax that one would more likely expect from a novelist than from reality: MacArthur became the president of a corporation making office machines.[72]

It is admittedly not impossible that this move was a coincidence—that, among the countless job offers MacArthur received upon his return to civilian life, this was the most attractive. But this seems unlikely. We believe, therefore, that we are at liberty to portray this climax, at least on a trial basis, as the product of an *action* that *means* something. We will do so with every reservation and hence leave the question open as to whether we are staying within the boundaries of fact or have already leaped into the realm of "Science Fiction"—or, rather, "Philosophy Fiction."[73] Let's spell this out.

If we were to portray MacArthur's choice of career in a novel, we could only explain his motivation as follows: Unable to come to terms with the humiliation the machine had inflicted on his authority, MacArthur was looking for ways and means to rid himself of the blemish. Under these circumstances, the executive position in a company that manufactures business machines was the only position that would allow him to gain compensation. It was the only position in which he could show the machines that his voice and verdict still had weight and prove to them that they, too, were still under his dominion. It was only here that he could master and humiliate the arrogant machines—at least in business and administrative matters. It was only here that he could quench his thirst for revenge. And that's why he chose that particular profession.

This motivation could be translated into the dialectical language of Hegel without any difficulties. In a version of *The Phenomenology of Spirit* written for today, one would then read: "After the former 'slave' (the machine) had become the new 'master' by demoting the former one (MacArthur) to the rank of a 'slave,' this 'slave' attempts to become once again the 'master of the master' (the machine)." This is the story of the humiliation—at least the hypothetical one.[74]

11. *Shame as a Disruption of Identity—the Expression "Ontic Endowment"—the "I" Is Ashamed of Being "Itself"; the "It" Is Ashamed of Being an "I"*

In conclusion, I must engage with the strongest objection that was raised against my thesis of "Promethean shame." Contesting this most fundamental argument requires such an unusually extensive "digression" that it was not advisable to bring it into the course of the investigation so far. The objection in question tries to demonstrate that the whole theory is unfounded and claims that the word *shame,* as used in the expression "Promethean shame," is *only a metaphor.* "Real shame," the objection explains, is always felt in front of an authority acting with a superior power, whose surveilling gaze is feared. Now, in the case of "Promethean shame" (if there were such a thing), the authority would be the world of machines and therefore consist of sightless things. Because these cannot see the imagined human flaw and consequently cannot evaluate and condemn it in any way, the whole thesis is delusional. The word *shame* amounts to nothing more than a metaphor and, while one may well allow the passive verb form that suggests that humans *are being "humiliated and embarrassed"* by the sophistication and multiplicity of manufactured things, this is not the case with the self-reflexive verb form, suggesting that humans are actively "ashamed of themselves."

If we are to meet this fundamental point of contention, we first have to understand the nature of "shame as such."

What, then, *is* shame?

I. a) A reflexive act (to be ashamed *of oneself*)—that is, it involves a *reference to self.*
b) Shame, however, is a reference to self that *fails.*
c) This reflexive act does not fail only occasionally (as it does in other self-reflexive acts, such as trying to remember oneself); it *fails categorically and fundamentally.*
d) This failure is categorical *because those who are ashamed encounter themselves at once as identical and as not self-identical* ("It's me, yet it isn't me after all").
e) As a consequence of this, the *act never comes to an end:* Because the one who is ashamed cannot "get over" the

contradictory encounter, shame is never over and done with either. (In this and the traits *f* and *g*, shame resembles astonishment.)

f) So much so that it actually forfeits its quality of being a reflexive act and degenerates into a state.
g) It does not, however, become a state of equilibrium, a stationary "mood," but an oscillating state of *irritation and of disorientation;* it turns into a condition that seems to perpetually begin afresh, even if one already deems oneself to be in the thick of it. In brief, *shame is an interference in processes of self-identification;* it is a condition of "being distraught and confused" [*Verstörtheit*].

II. In contrast to most of the harmless "acts" that usually are the objects of psychological and phenomenological investigation, shame essentially entails a "*double intentionality*": It is not solely directed toward its (normal) "intentional object" (the perceived "flaw") but is also always simultaneously turned toward an *"authority" before* which the one who is ashamed feels shame. It entails a "*coram.*"[75]

III. This authority is an *unsolicited authority,* often even one that is unwanted; it is hence not actually "intended" but is rather evaded. The intentional turn toward this authority is in fact a turn away from it. The intentional reference to it is a rejection—that is, it is "*negatively intentional.*"[76]

The traits listed in section I will be the main subject of our investigation. For this reason, I will quickly make some comments about points II and III first.—

Neither "*coram*" nor "negative intentionality" has been given any specific importance in phenomenology.

Ad II: "*Coram*"[77] is not a trait specific to shame, even though it is particularly evident in this case; it is a trait that belongs to all social acts. (Example: "to boast in front of someone.") "*Coram*" is probably an element that can be discovered in every act, however hermit-like it may be. Even if one is oblivious to this, even if the act takes place in private, it is doubtful that there is an act that does not make reference to a particular social world [*Mitwelt*] on which it counts and before which it takes place or wants to pass unobserved. The conception of consciousness in the classical phenomenology of Husserl always had to closely border on solipsism because this phenomenology, without being clear about its own principle of selection, almost exclusively restricted itself to the analysis of acts that did not openly imply this "*coram*."

Ad III: "*Negative intentionality*" remained unknown because the "ego" was assumed to be *eo ipso* "free," or at least to be "*active*." The fact that the word *freedom* was seldom used (e.g., in Husserl) doesn't say much to contradict this. No different from other philosophies of subjectivity, "transcendental phenomenology" is marked by a classical lapse. In its pursuit of knowledge of the "ego," it only captured the tip of the ego, the "cognitive, perceiving I," and mistook this for the whole ego or as representative of it.[78] It is undeniable that the "cognitive, perceiving I" pursues an object and is therefore "positively intentional." But that the ego is only ever "pursuing something" is not true. Besides, the ego also withdraws into itself (in the manner of a snail), and intentionality only gains its meaning in coordination with this countermovement; the ego does not only see, it is also seen; it does not only intend, it is also the object of intention—for example, when it is pursued or chased. It is highly questionable whether the object someone is fleeing from can be called an "intentional object" in anything other than a grammatical sense. After all, not everything that is grammatically an accusative is automatically also an "intentional object."— Shame now also belongs to this class of two-way intentional acts because it turns its back on the authority before which it feels ashamed. Shame neither intends to see nor to avoid seeing the higher authority. What it wants is not to be seen. This relationship differs so fundamentally from what is usually called "intentionality" that one ought to actually invent a new term to describe it.—

If one sums up the traits mentioned so far, shame is *a reflexive act that leads to a state of confusion. This reflexive act fails because, in the process, humans experience themselves in front of an authority from which they turn away. They experience themselves as something "they are not," in a manner that inescapably impresses on them that "they are this after all."*

After this academic-phenomenological description, which still leaves the fact *that* shame exists unexplained, it will be good to jump to a concrete example.

Example: The hunchback is ashamed of his hump. (More correctly: He is ashamed to be the one with the hump.)

In a certain sense, the hump appears to him as something contingent—that is, not as what he "is," but only as something that he "has."— *What one "has,"* however—"has" in the way that one "has" one's body—*one "is,"* and this inescapably so. Hence, the hunchback is inescapably also "the one with the hump," with whom he now has to identify, even though he cannot do so; even though he "can't help" his physiology. Because he cannot get over this contradiction, he cannot get over his shame: It does not abate.

As for the "authority" in front of which he is ashamed, it consists of a *tribunal that decides how humans actually "ought to be."* He, too, belongs to this tribunal. He, too, approves the human canon this tribunal represents—no less so, in fact, than the members of the tribunal with a more fortunate build. Because he accepts the judgment of the tribunal, but then again, as the misshapen one, he cannot let this judgment sink in (and he cannot want to let it sink in), he turns his back on the tribunal—and also turns away from himself. Here, too, a "crisis of identity" is evident.

Moralists will find this basic example repulsive. "If he can't help the fact that he has a hump," they will exclaim with moralistic blinkers, "why does he need to be ashamed of it?"— The response to this interjection is not that the hunchback has not also thought of this ingenious line of argument. (Quite the opposite; as long as he can remember, he has been telling himself, "I can't help it—my body is not my fault," and he never tires of trying to extinguish his feeling of shame in this way.) No, the answer to the moralist's interjection is that these great efforts remain futile because he does not care about whether he "needs" to feel ashamed.— In short, he is ashamed precisely because of his inability to do anything about it. *The hunchback is ashamed of his physiology not despite the fact that it is not his fault, but because of it.*

As it happens, no saying points to what shame is as unambiguously as *"I cannot help it, it is not my fault." Everything that I "cannot help," everything that I "cannot do anything about," is beyond the reach of my freedom and self-determination; as such, it belongs to the realm of fate and is "fatal," a measure of "impotence," in the broadest sense.* Shame has its source in the contradiction that exists between the pretention of freedom and the "fatal," between ability and inability. Shame is ashamed of inability.— What does this mean?

The aspiration to freedom by its very nature knows no scale and is without measure: to be partially free is not enough. The self [*das Ich*][79] does not only want to be itself now and again; the individual does not only want to be individual as an attribute. The self, rather, desires to be absolutely free, individual through and through, nothing but itself.

This overstretched aspiration is "pathological."[80] It cannot be sustained in the long run. There is always a point at which the self comes up against the limits of its own freedom, individuality, and self-confidence. In such moments, the self finds "itself as something" that it is, too, even though what it finds is not identical to what it is as a self or as an individual. This is the moment when the self discovers that it also exists as an "it."

With the word *it*, I do not only mean what Freud called the "id" [*das Es*], but something much more general—everything that is not of the self. The "it" designates everything preindividual, whatever this may be—everything that the self is part of without being able to do anything about it. The "it" denotes what is "*given*," what the self *also* has to be, insofar as it is. That's why we address it as the "*ontic gift*."[81]

Shame is generated in the instant "something given" is discovered. I say explicitly "in the instant." Shame does not limp behind, setting in after the event. Shame is not to be understood as a reaction to the realization of one's inabilities. Nor is it a statement [*Stellungnahme*] about this state of affairs (as sadness is, for example). Shame is this inability; shame is this failure as such. This makes shame incomparably more fatal than sadness.— When, for example, an ascetic finds himself confronted with the fact that he "has" a body (and "to have" here also means "having to have," "not to be able to do anything about this circumstance," and thus passively "being had"), and when he is confronted and perplexed by this fact—which is as unacceptable as it is undeniable—then this bewilderment and perplexity is his shame. When the "self" demanding freedom has to realize that it was not posited by itself (in Fichte's sense), but that it rather "became," that it would not exist as a "free self" at all if it were not at the same time arrested by the "endowment" of an unfree, conditional, and creaturely humanity—then this helplessness and this failure is the shame of the self. "To be ashamed," hence, means "*Not being able to do anything about the fact that it is not one's fault.*" Put another way, the very thing on which the moralist stakes his claim for an absolution from shame is precisely its generative moment—its basic motif.—

What holds true for the hunchback's hump, for the body, and for the origin also applies *mutatis mutandis* to the "bad deed": Because one is primarily ashamed of *something that is* [*eines Seins*], of *having* something, and not because of what one has done, the latter cause of shame is not nearly as fundamental as the hunched back in our example. Ultimately, the perpetrator is not ashamed of his misdeed, but of *being* "the one who" is identical to the one who committed the bad deed. In brief, he is ashamed of *something that is.*[82]

Shame erupts because one is simultaneously "oneself" *and* something else. At the same time, it is in a sense also an attempt to shed this "other quality," the "endowment." We have already seen that this attempt is futile.[83] In fact, shame is so perplexing and without method that it leaves the person feeling shame no other escape route than the defeatist one of going under together with the shameful "endowment" (to which they are chained). The figures of speech "to want the ground to open and swallow one up" or "to sink into the ground" are not metaphors. They are accurate descriptions that cannot be replaced with more "exact" ones. And yet, we know that this obsession with disappearing is futile because the ground petrifies under the feet of shame. And now, shame begins to accumulate frenetically as "shame about being ashamed."—

Once one has grasped the basic condition of the one who is ashamed and has understood that this disorientation and despair are brought on by the recognition that freedom and individuation are limited, it is evident why *sexual shame* has time and again ended up being regarded as *the quintessential* form of shame. Sex and that which pertains to the genus are the preindividual sphere *par excellence.* Its realm lies beyond choice and freedom, and as such, it provides the preeminent "it," which does not belong to the individual *qua* individual.

For the individual, the claim that their sex does not belong to them remains hollow: for a bond exists between them, and in two senses. On the one hand, the individual "belongs" to a sex (and is subject to "longings"). On the other hand, sex can only actualize itself in an individual. The bond is thus so close that it doesn't make any difference if one asserts "the individual is an attribute of its sex" or "sex is an attribute of the individual." The truth lies in this ambiguity—the simultaneous validity and nullity of these two statements.

This ambiguity—or this "dialectic" fact, if you will—is just the reason why one's sex is a pudendum, why it is something of which one is ashamed. Inasmuch as individuals are sexed, beings who "belong" to a sex and who "long," they are not masters of themselves and hence not "themselves" and not free.— In a most ambiguous manner, they are *at once themselves and not themselves,* and thus the formula according to which we have begun to conceive shame is confirmed.

Viewed from this perspective, shame is thus not any random psychological state (insofar as any such states exist at all). It is rather a *metaphysicum,* the embodiment of the dialectic of "the particular thing" [*res*] and "the universal," a dialectic that has long been the object of contestation in universalism. The "*res*" (here: the individual) is ashamed of embodying the "universal" (or at least the general—the individual is ashamed of being its attribute, of being it). It is not without reason that the one "thing" that all have in common, the "shared feature" incorporated in every single body, is called the "*private parts*" [*die Scham*].

If one restricts oneself to the example of sexual shame, our definition of shame is confirmed: Shame is a disorientation of human beings who are forced to find an "it" instead of an "I," and laboriously and futilely try to identify themselves with themselves completely. But does this hold for other forms of shame? Will they confirm our definition just as well?

Let's put this to the test.

The shame of the shy child is surely no less elemental than sexual shame. When a child hides behind its mother's skirts, does it do so to conceal its existence as an "it"? Does the child hide because it feels cheated out of its entitlement to only be an "I" (or exclusively a "self")?

No.

Does it even make a claim to such an entitlement?

Likewise no. It seems pointless to attempt to establish even the most artificial link between this example and our previous definition. Our classification of shame as a "disturbance in processes of identification" thus seems to contain a mistake.

But where is this mistake?— Let's engage in a short reflection.

"Identification" requires two parties: the *one who* wants to identify—we will call them the "*identificator*"—and the "*identificandum,*" the thing *with which* they attempt to identify.

So far, we have obviously limited ourselves to cases in which the "self" was assigned the role of "*identificator,*" whereas the general and impersonal realm of the "it" was the "*identificandum*": *The "self" was ashamed of the "it."*

But is this the only possible distribution of roles? Is an inversion of this dynamic not equally possible? Is it not conceivable that the "it" could end up in a situation in which it was forced to identify with the "self" being imposed on it? And would this imposition of identification not lead to disorientation? Precisely to the disorientation we call "shame"? In brief, is it possible for the *"it" to be ashamed of the "self"?*

This question sounds strange.— The child we will summon now will immediately convince us of how little this is the case.

The existence of the child is still embedded in the restrictive ground of the family, limited to a mere togetherness, a "being-with-others" [*Mitsein*]. The child has not yet developed into an "I" and has not yet individuated into a "self." The thought of "*taking itself out of*" the ground in which and with which it lives has not yet entered the child's mind.

I say "*take out of*" [*herausnehmen*]. Initially, this seems to mean nothing other than "to remove something from something else and to make it an exception [*Ausnahme*] in doing so." But because the turn of phrase "to take oneself out of" [*sich herausnehmen*] already shimmers with the connotation "unrestrained impudence," we can expect a certain relationship between "being-oneself" and shame, if one can trust the wisdom of language.[84]

Let's assume that this child is addressed by a stranger who asks the child's name, for example. This customary opening question can stand in for any other question that might be asked, as it requires that the one addressed "*take*" the step "*out*" of the group by saying their name and identifying themself as themself.

The status of the child has been transformed by a single blow: Instead of still only existing "with-others" as a mummy's child, instead of remaining indistinctly in the background, it now has to stand its ground as an "exception," as a single being who is no longer part of the ground it was taken from. The shock of this social birth is violent, so strong that it can perhaps even be understood to be a variant of what Freud calls the "trauma of birth."[85]

The child is, of course, not entirely ignorant of this new state. Although it had not yet "taken the liberty to separate itself" from the ground in which it was embedded, it had already been exposed to the danger of "being-itself" from the moment it was born—the birth itself was the first "abandonment": Every child is an "abandoned" child. In its hours alone, particularly when "left to itself" in a state of deprivation, it had to experience for the first time existing as a self: It had to come to terms with being without being-with-others. Since anxiety is the engine of insight, the child had already begun to recognize—or at least to intuit—that the blissful indistinctness of existing exclusively with others would one day end for good, and that a colder, more exposed mode of being would become its destiny.

But *who* had this intuition? *Who* was the subject at the heart of it, recoiling anxiously?

The self?

Is it conceivable that it is the self that is recoiling, when the thing making it recoil is the looming anxiety and dread of being a self?

Inconceivable.

And as strange as this may sound, the only conceivable answer to the question of whom the recoiling, intuiting subject could be is the being not yet fully formed into a "self"—namely, what we have called the "it."

And this catastrophe now unfolds. The very instant the stranger addresses the child, the threat becomes real. Even though the child is still an "it," it sees itself challenged and called forth as an "I," and because the child cannot yet meet this challenge, too much is demanded of it. That the child already knows and recognizes "*I* am being addressed" or even "It *is* me" doesn't even have to be disputed. It is equally certain, however, that the child also knows "It *is* not yet me"; "*I* am not yet a self." In brief, *the situation is a concurrence of being-a-self and not-being-a-self.* This paradox is no less indissoluble, no less indigestible, and no less unsettling than the situations we have recognized as shame-inducing above—situations to which there is "never any closure."

The "it" now really has no other options than the disturbed ones described: Overwhelmed by the anxiety of having to emerge as a "self" in public, the child only knows the desire to sink into the ground and become invisible. In the manner of a kangaroo that jumps back into the pouch from which it came, the child attempts to regain the alibi of merely existing-with-others and hides in its mother's skirts.

Schelling said, and this expression was already invoked above, that "the arrogance of man rises up [*sträubt sich*] against this origin from the ground." We may now also add an accompanying sentence: "The origin is reluctant to take the leap into freedom and into the exposedness of existing as a self." In this refusal and reluctance there is also "shame"—and probably not just "also": There is every indication that the shame of being "oneself" is the more common and originary form of shame than the feeling of shame about not being oneself.[86]

That's right: the more general and originary form. Why, after all, and because of what, does one usually feel shame?

A speaker who has to confront a crowd (instead of remaining comfortably concealed within it): Is this person not ashamed of appearing *as themselves*? Is the speaker not ashamed of being abandoned to "the public" under a barrage of looks?— And what about someone wearing unusual clothes or a foreigner in a foreign land? Do they not run the gauntlet because "they take the liberty of standing out" and because they are "exceptions"? They attract attention because they *are being themselves,* and as a consequence, they contrast with others and stand out. And does not everything indicate that the shame of breaking norms is the prototype of *moral shame*? The disobedient who "take the liberty" of standing outside the normal and the norm, are they not ashamed when *exposed as being themselves*?— There is here a stark contrast to our earlier examples. These now begin to lose some of their relevance.— Do those who are ashamed not generally conceal themselves because they perceive their *selfhood* as a flaw? When they "voluntarily stand in the corner," or when they pretend not to hear, do they not all attempt to extinguish their shameful conspicuousness by denying that they exist as a self in order to reassume their old status prior to their lapse?[87]

This digression has delayed us for a long time. But since it seems to have given us some insights into what shame is, it has left us better prepared to face the objection that originally gave rise to it.

The objection was that the expression "Promethean shame" was purely metaphorical and therefore would not allow us to see true feelings of shame in the experiences the expression describes. The reservation was that the "authority" provoking these experiences could not be a truly shame-inducing authority, since it consists merely of blind machines. The thought that humans could be eccentric enough to accept eyeless objects as true authorities, or even that they would be afraid of these objects as "awe-inspiring powers" [*tremenda*] that can make them shudder and tremble, is simply absurd.— This was the objection.

Clearly, we are now left with two tasks: (1) We are obliged to examine the accuracy of this objection—an examination that in itself has two components: (a) the question of metaphor; and (b) the question of "eyelessness." (2) We need to examine—and this is the main task—whether

"Promethean shame" represents "a disturbance in processes of identification." Since we have defined shame as such an interference, only this will allow us to decide whether "Promethean shame" can be classified as a "true form of shame."— With this, we have sketched the path of our concluding analysis.

12. *Dismissal of Objection (a)—No Expression Is Purely "Metaphorical"*

We will begin by responding to the objection that we are letting ourselves be led astray by a metaphor.

What is the point of this objection? What does the claim that a description of a mental and emotional reality is "exclusively metaphorical" mean? Is there even such a thing?

Are human beings that we call "bright" only "bright" in a figurative sense? Is a "heavy heart" [*Schwermut*] only an allusion because the true literal sense of "heavy" [*schwer*] is deemed to be an exclusive attribute of physical objects? Why exclusively of these? Where is it written that the word *heavy* was originally deployed to describe physical qualities?

When psychologists discredit the diagnostic value of such expressions, admitting in a voluntary gesture of self-limitation that they know for a fact that the employment of such metaphorical expressions leads to very little that is rigorous or worthy of being taken seriously, they do so primarily because they are intimidated by the rigor and authority they accredit to their bigger sisters—the natural sciences—and thus not on account of real methodological concerns. And here they are in the wrong. For, stripped of their linguistic dimension, the mind and feelings are entirely without essence: The way psychic reality represents itself linguistically is part of this reality itself. When something "weighs on" the soul, when it describes itself as "weighed down," when it recognizes itself in this word and is able to convey this to fellow human beings, this is evidence that the "heart" is truly felt to be "heavy." The idea of seeing expressions such as "heavy" merely as a label that was peeled away from another (physical) source and stuck onto "heavyheartedness" is rooted in the unspoken presupposition that the soul invents its forms of expression in the same arbitrary manner that industry comes up with names for branded goods. In fact, the allusion to "weight" is a real component of the emotional disposition felt. And even if one were to admit that the term "weight" is a label that has been removed and stuck on elsewhere, the very fact that the soul has reached for precisely this label and not another to characterize its own condition would signal that it actually recognizes itself in it.

Put otherwise, the suspicion directed at metaphorical expressions is based on an erroneous notion that is nevertheless taken to be self-evident. This is that different realms of sensation and experience are autonomous and hermetically sealed from one another, and that competent people—namely, scientists—are not permitted to engage in small-scale border crossings between these realms. To prove that such a psychology is absurd and the expressions imported from allegedly foreign areas are readily understood is not really necessary. On the contrary, the very fact that expressions imported from allegedly foreign provinces are so readily understood demonstrates just how widely open the borders are. In everyday life, no scientist can get by even for an instant without having recourse to the very metaphors he discredits. And he cannot contest that metaphors are an excellent resource for clearly communicating to others things that resist "nonmetaphorical" formulation. This state of affairs should make us suspicious, but not of metaphors—every beginner who has heard anything of the rigor of science has this suspicion anyway. No, it should make us *suspicious of the legitimacy of our suspicion of metaphors*. To complain that we cannot access the reality of the psyche because of our "merely metaphorical" language is misguided; *on the contrary, precisely in recognizing metaphor, we grasp one of the most essential traits of the soul.*

When people who are ill moan to us about a "dull" [*stumpf*] pain, we understand exactly what they feel and what they want to say. An imaginary doctor who dismissed this expression as "unscientific" or "merely metaphorical" would be a character for a farce about our scientific age—especially if he were to accuse the patient of using an expression that belongs to another realm of meaning and being guilty of *μετάβασις εἰς ἄλλο γένος*[88] on account of this. That the sick person is "transferring" (*μεταφέρει*) the meaning of the word *blunt* [*stumpf*][89] from another physical realm to their pain is not at all true. The expression "blunt"—or, rather, the quality this expression points to—does not only inhabit one specific realm. Or, put positively, the expression denotes something "pre-specific." It portrays a relationship between the world and the subject that does not yet have anything to do with the division of the subject into different sensual realms. The specific actualization of this quality in a particular sensual realm is always secondary.

Only with this secondary motion does "the blunt" become something that is blunt to the touch, or blunt acoustically, osmotically, or emotionally. If this were not the case, we would not be able to comprehend the words used at all (and the language of poetry would remain obscured from us). We can only apprehend what a "dull pain" [*stumpfer Schmerz*] is, and even what a "blunt knife" [*stumpfes Messer*] is, because we comprehend what "bluntness" is in a pre-specific manner.[90]— And what is true for "bluntness" also holds for all other experiences that are actually or allegedly characterized with "metaphorical" expressions, such as "exaltation," "relief," "feeling overwhelmed," "depression," and so on; and, of course, this is also true for shame.

13. *Dismissal of Objection (b)—the World Is Not without Eyes*

So much for the subject of metaphors.

My opponent had offered the following reason for his accusation that my use of the word *shame* was metaphorical: One can only speak of real shame when those who are ashamed know themselves to be under the watching eyes and surveillance of an "authority." The world of machines *does not have eyes;* no human being is eccentric enough to assert that machines can see them. It is therefore out of the question that machines could function as "authorities" of this kind.

This seems plausible, but only because we are biased by theories and no longer know what is plausible to someone who isn't. To those who are unbiased, nothing is more persuasive than the notion that they are being looked at by things. Such humans are anything but theoreticians of epistemology and cognition [*Erkenntnistheoretiker*] who conceive of themselves merely as observers (of the world). No, they know with the same level of certainty that they are also seen and observed (by the world).— And this doesn't just mean that they know they are seen by fellow humans and animals, but by the whole visible world. Normally, at least originally, humans deem "visibility" a reciprocal relationship: *Everything that they can see also sees them.*

It is unnecessary to emphasize that we are not asserting that humans are actually seen and observed by a chair, table, or picture; what we are suggesting is only that, with an unprejudiced "worldview," humans consider themselves "viewed by the world." They do not make this into a theory, precisely because this is a "pretheoretical worldview" that dissipates the very instant a theoretical attitude is assumed. That is also why the term *worldview* might be inappropriate here. At any rate, the use of the word is irrelevant in view of the fact that we *behave in a way that suggests that "we feel seen."* This is more conclusive than any theory that could be laboriously deduced in retrospect. This claim could be supported by countless bits of empirical evidence. For instance, by the material compiled—albeit with a completely different aim—by psychoanalysis. This provides examples of how the inhibitions of shame are still at work in completely undisturbed isolation in a natural environment. Robinson didn't walk around naked, either.

We also readily allow poets to tell us that the mountain peak is "facing us" with menace or that the moon is "looking down on us kindly." And we do not reject such information as eccentric but understand what the poets mean to say with it. This shows that we still take "seeing things" for granted. We, too, see something in their "*look and appearance*" that also makes things "*look at us of their own accord.*" In the idiom of poets, we thus rediscover our earliest and most familiar way of viewing the world. What one calls this "mutual visibility" is quite irrelevant, be it "animism" or "anthropomorphism." It probably has its roots in our tendency to attribute to the world that it is "looking out for us" [*auf uns schaue*] (in the Austrian sense of "taking care of us") when it seems to "favor" us, while appearing to "have it in for us" [*auf uns absehe*] when it makes us feel threatened.

The following experiment—which is not at all that easy to conduct—underscores how far from being true it is that someone has to be "eccentric" in order to feel looked at by the world. On the contrary, a human being has to do violence to their sense of immediacy and naturalness [*Unbefangenheit*] to produce the feeling of *not* being looked at. If we attempt to consciously imagine that things *cannot* see us and do so with ruthless and uncompromising insistence, the outcome affects us in a most peculiar manner. The idea that the chair on which we usually sit, the table we write at every day—yes, that even our mirror—has never ever seen us because it does not know us; the thought that the picture hanging on the right above our bed is plunged into eternal night, that it knows nothing about the companion picture to its left, nothing of its own beauty, and finally also nothing about us, even though we look at it every day; the thought, in short, that the picture is as blind as a stone and that we have to spend all our days surrounded by totally blind things, living as beings who see while remaining unseen—this idea is so adventurously alienating that it makes one think one can feel "air from other planets."[91] Only with great difficulty can we fend off the feeling of having just conjured up the image of the ghostly living conditions on a darkened other world.—

No, it is entirely unwarranted to believe that we cannot acknowledge things as authorities on account of their "blindness." We are far too blind toward the blindness of things to sustain such a belief, as long as we stay uninhibited enough to accept this. It is pure theory to believe that humans reason with themselves that "things do not see me—*ergo,* it isn't necessary to be ashamed in front of them." The question is whether such an "*ergo*" must be presumed at all. If this is the case, it would probably be entirely different and would very likely attest that our mechanism of deduction unfolds in an anticlockwise direction: "I am ashamed in front of something which is superior—*ergo,* this thing can see me."[92]

The components of the objection are hence highly questionable. Rather than pursuing this question any further, it is more important to examine whether "Promethean shame" fulfills the condition that we've established as being the essence of shame. We need to check whether it reveals itself as a "*disturbance in processes of identification,*" as a faltering process of identification between the "self and the it" or the "it and the self."

It was not without good reason that the term "it" was left utterly unqualified when it was introduced earlier. We limited ourselves to using the term to describe everything the "self" is *nolens volens "party to"* (as it is forced to realize in the disconcerting encounter with itself). The "it" hence names *what* the "self" *is also,* without this "it" being a *part* of the self. It was not permissible to define the concept more narrowly because the most diverse entities emerged as such an "it": the body, sex, and species, "*what one does*" [*das "man"*].[93] We were able to content ourselves with not defining the "it" further because we were exclusively interested in the disturbance of identification that sets in when the "self" encounters itself as an "it"—that is, we were interested in the character of its "ontological endowment." For, we recognized that the essence of shame is disturbance or confusion of the "self." The phenomena the "it" covers are divergent to such a degree that we do not wish to insinuate that the "it" corresponds to something ontologically homogeneous. To do so would be meaningless.—

We cannot be surprised, therefore, that the "it" that we are discussing now is fundamentally different from the quasi-"natural" "it" phenomena (the body, sex and genus, the family, etc.) that were previously foregrounded in our analysis of the basic workings of shame. The whole point of our analysis is in fact rooted in this fundamental difference. In "Promethean shame," it is the machine or technological device that is encountered as the "it." We are confronted with this "it" while operating machines, a relationship in which humans have to function "with" and "alongside" technological objects, "as" part of the working machine, and as such they no longer encounter themselves as a "self." To demarcate the "it" that manifests in engagement with machines from the "it" of our previous discussion, we will call it the "*it of the technological device*" [*das Apparat-Es*].

We can, in a sense, picture the human being—and this is now truly only an image—as clamped between two brackets, constrained by two forces that both challenge the "self": On the one side, the human is constrained by the "natural it" (by the body, sex, species, and so forth); and on the other side, by the "artificial" (bureaucratic and technological) "it of the technological device." The space left open for the "self" is already very narrow today, and because the "it of the technological device" encroaches on this space ever further, this space is getting smaller, while the danger that the "self" is crushed by these two colossal nonindividual forces is daily increasing. With every day that passes, this hope of being crushed seems more justified—yes, hope, because millions hope that this catastrophe will happen and a technocratic totalitarianism will set in. When this outcome is finally achieved, tomorrow or the day after, the final triumph will exclusively belong to the machine. In its greed to devour everything and especially that which is alien to it, the machine will not only have incorporated the "self" but also the other "it": the body.

Today, technological devices are already trying their best in this respect. The saying that the machine is "getting at us and our bodies" is not merely an image, in fact, it is actually "getting into us." It even attempts to overpower our sexuality in order to include it in its domain. Here is an example.

14. The Orgy of Identification as a Template of the Interference and Disturbance in Processes of Identification—Jazz as an Industrial Dionysus Cult

Jazz music, which is still today often called "Black," does not owe its existence exclusively—if at all—to "visceral memories of desert and jungle drum"; it is (at least equally) "machine-music," a type of music, therefore, that initiates forms of dance that are fitted to the humans of the Industrial Revolution.[94] It is not only the "torpor of an archaic existence" or the "lust of urgent sexuality" that has become a voice in jazz, but always also the tenacity of the accurately working machinery of the steel mill, which impassively and neatly cuts the glissando of animality into ever uniform pieces.

It might at first appear as if the two adversarial, impersonal forces—the "foundational force of the body, sex, and species" and "the force unleashed by the thing"—are unified in an exhilarated ridicule of the "self." It seems as if these two forces flay and whip each other up, working together in mutually reinforcing yet counterpoised motions to grind up the "self" that is caught between a mechanical system that has become orgiastic and an organism that has become mechanical. But in this alliance, the machine has not yet achieved its ultimate goal: to liquidate the sexed body [*das Geschlecht*][95] as such. The alliance with the sexual is not entered into so that the machine can work in unison with it, but in order to transform the trapped sexual energy of the body into energy of its own kind: The machine is attempting to turn the dance into a *process of transformation* and the dancers into transformers, whose duty it is to convert animalic into mechanical energy.

Of course, we do not mean to say that the vital energy generated while dancing has use value for machines. But despite this, the machine—in its autocratic aspirations and because it does not tolerate forms of energy from which it cannot benefit—releases the superfluous vital energy from the frenzy of dancing and so exhausts it. And this is done in such a way that *the exhaustion of energy as such takes on the form of machinelike movements*. The types of music that mobilize such forms of dance contain all the hallmarks of an automatic and thing-like motion: Their *fury of repetition*, neutralizing any sense of time and shattering it, is the fury of the constant meter of the working machine.[96] *Syncopation*, which is the underlying musical principle of jazz, should not be understood as a "purely musical phenomenon" of placing a rhythmic accent "off-beat." It should rather be seen as the symbol of the ceaseless and never slackening persistence with which the machine cuts in to interrupt the rhythm of the body. In a sense, then, *syncopation* is the interference with which the machine pounds its victory into every single beat of the music, and the music does so as the continual *refutation of the body* and its birthright of "counting" also and having its own meter.

What the dancer dances is not only the apotheosis of the machine. Because the body coexecutes its own refutation to prove its conformity with the machine, the dance is simultaneously also a ceremony of abdication and synchronization. It is *an enthusiastic pantomime with which the body reenacts its own utter defeat.*—

The expression "industrial religion" we coined above in the discussion of "human engineering" is corroborated here: The orgies these dances represent, for example, in the dance halls of Harlem, have nothing to do with "pleasure." They are much more and much less than this: They are *ecstatic sacrificial dances that are cultically celebrated in honor of the Baal of the machine.*

The ecstasy is real. The dancers are genuinely "beside themselves"—not, however, in order to be one with chthonian forces,[97] but rather to become one with the God of the machine: an *Industrial Dionysus Cult.*—What holds true of syncopation also applies to the "final cadences" of these pieces of music: the victory of the machine-god is drubbed into the body. Expressed in musical terminology, these cadences are always "false cadences" [*Trugschlüsse*], and like syncopation, these are mechanical. They function as "*brakes*"[98] and represent pure "*deceleration.*" If there were a typology of what meanings "final cadences" can be made to signify—in late Romantic music, for example, the "final cadence" would signify a "peaceful and gentle dying off" or "redemption"—this "brake ending" would be absolutely new. Such "brakes" are "*interruptiones.*" These endings are like *coitus interruptus* (because machines are always stopped at an arbitrary point in time, "just when they are right at it"). This expression explains what this really means for the human dancers who have matched their bodily motions to the music. But not clearly enough, for it isn't just sexual arousal that is being interrupted, but life as such. When music and dance are slowed down or abruptly switched off like a machine, the dancer is "beaten to death" as a whole. They seem to have become a victim of murder or an industrial accident. In brief, *when seen as a cult, the interruption signifies sudden violent death.* It is, however, precisely this death that is a central component of the cult, since there is nothing more machinelike than being arbitrarily shut down. Being able to end in this way is proof that the transformation into a machine has been accomplished. "*I am being switched off*"—the one who is suddenly being put out of service says and feels—"*therefore I must be part of a machine.*" Being stopped in this way is even enjoyed. One can be certain that the music machine will immediately recommence from the beginning for a new shift, which promises new life in the form of a renewed defeat and a new hope of being murdered again. As such, the interruption represents the culmination of the union with the new deity.

That the dancers are actually "finished off" by this rite and fully forfeit their sense of self can be seen in a striking phenomenon: During the orgy, the dancers *lose their faces.* What I mean by this is not the often-observed fact that even today's faces are shaped by identical role models and become like similar mass-products. Such serially produced faces only differ from one another in the way two pieces of fabric with an identical pattern also differ from one another: they can only be told apart by individual flaws in the weave. No, what I mean to say is that even this mass-produced face is lost. This loss can manifest itself in several ways. The face, for example, can become slack and unkempt in the course of the orgy. It may cease to be the center of the human being's field of expression and the mirror of individuality. The face *either* mutates into a mere *body part,* the naked and uncontrolled appearance of which no longer surpasses the likeness of a shoulder or backside; *or* it turns into a blank poker face that is *cut off from the orgy* and so remains in the dark about what is going on downstairs. Here, the face is merely "carried along" as an "ontic endowment" because it cannot be handed in at the cloakroom like a jacket before the onset of the ritual. During the orgy, the face can also *glaze over,* by which I mean that it clearly ceases to see anything or to be conscious of being visible. It would not be at all surprising if during the intoxication of their mechanization, one of the dancers had the idea of pulling something over their head in order to conceal their face (which in any case no longer has any significance)—for these orgies are not without humor. It would be equally unsurprising if a new variety of shame were to emerge in this setting: *facial shame.* I don't mean the shame felt on account of having a repulsive or unattractive face, but rather the *shame* the dancer might feel *about having a face in the first place.* Analogous to the ascetic shame about having a body discussed above, facial shame is ashamed of the fact of being condemned to carry around this stigma of selfhood as a compulsory ontic endowment.

Even though these thoughts may be "philosophical exaggerations"—overstated portrayals [*Entstellungen*] pointing in the direction of the truth—it cannot be denied that the face has become a residue and an obsolete fragment. And it is certainly no coincidence that this loss of the face has started in the very epoch in which the face is increasingly also neglected as a subject in the visual arts, and when commercial drawing counts as chic only if the face is omitted altogether when humans are being depicted.[99]

But let's once more turn to the "music machine" that sets this orgy in motion.

When bandleaders attempt to perform this music in the form of "serious music"—that is, as a concert in which the music is merely to be listened to—they are falling victim to a misunderstanding. This, in turn, stems from an appetite to provide the music with the social respectability of acknowledged "cultural values." This aspiration is not misguided on account of the music being too "light" or merely being "pop." The opposite is the case. The music is terribly serious—too serious for the concert hall. What I mean is that it affects humans too deeply and violently and in a manner that alters their "ethos" (in the musical sense of the ancient Greeks)[100] much more radically than solemn concert music can today. Classical music, once it has reached its final symphonic apotheosis, leaves its listeners—who remain a mere audience—with an "experience" that evaporates straight away because it has nothing to do with life outside the concert hall. There is nothing more inconsequential than the effect of serious music. There is nothing more serious, on the other hand—that is, nothing more dangerous or more destructive—than the effect of the type of music so often dismissed as merely being "superficial." This music constitutes a real intervention, a violent metamorphosis that has an awful lot to do with the world and life beyond the concert hall. The mode of being it works its listeners up to is that of the machine, a mode of being, therefore, which rules from morning to evening, and the music now makes people conform totally to this mode of being.

Because every ritual demands *full bodily participation,* this is also required here. That is also why those who stage this music "*in concerto*" merely as art, in front of a mere audience, do not create audiences that are newly receptive for art or self-improvement; rather, they betray the cultic community. Its members, instead of being able to participate in the mysteries of the cult, are condemned to be content with the role of being mere listeners. They have to be mere voyeurs of turning into a machine, so to speak. That such audiences, unwilling to be fobbed off with this role, begin to revolt in outrage about the fraud—as was recently the case in Vienna during a jazz performance "in concert"—is truly anything but surprising.—

What we have just portrayed is a situation in which an ecstatic rite acts as a means of violence to force an identification with the machine. Because no real machine surfaced in our discussion, the situation described is in a sense delusional—it represents a "pleasant" or macabre "illusion." Unless, of course, one views the music itself as a "machine" that was created by humans as the special apparatus with which they can accomplish their own machinization.

To explain why this excursus is connected to our main investigation is hardly necessary. It is obvious that a separate and specific initiation rite never would have been invented, never would have needed to be invented, if humans were usually able to identify successfully with the world of devices, appliances, and machines. They wouldn't have needed such an orgy to savor this identity. The fact that the rite exists can thus be seen as a symptom—or at least as an indicator of a symptom—that such identification usually fails.

But we do not yet want to accept this indication as constituting proof. We will rather "test" it further by scrutinizing the "natural" relationship between humans and machines—that is, the relationship that realizes itself in the workplace—and we will examine whether it involves a confrontation of the self with itself in which disturbances in processes of identification become acute. And finally—and with this we return to our original question—we will ask if this interference is "shame."

15. *The Relapse—the Self Encounters Itself as a Failing Self—the Example of Failing Labor as the Prime Test-Case of "Promethean Shame"*

What can it mean to have a "disruption of identity" [*Identitätsstörung*] in the face of machines?

This question, by analogy with the twofold principle outlined above—the "I is ashamed of the it," "the it is ashamed of the I"—can once again be answered in two ways:

Either, instead of encountering themselves when using machines, humans find something that already conforms to the world of machines: They discover themselves as a part of a machine.

Or humans have already willingly integrated themselves with a machine (or the world of machines as a whole). Despite this, because they are unable to "loyally hold the line" and achieve a full conversion and cosubstantiality with machines, they now encounter themselves instead of finding a machine-part.

In the first case, the "I" encounters itself as an "it"; in the latter instance, the "it" meets itself as an "I."

Discussing the first of these possibilities any further is unnecessary because it is already obsolete. The type of human being *Chaplin* portrayed in *Modern Times*—who compulsively enacts involuntary, machinelike movements and continues to do so even when he is no longer operating a machine—does not exist. There is no human being who realizes with feelings of alienation and horror that they have already become a part of a machine. No, such Chaplinesque humans do not exist. The portrayal is distorted. It's the other way around. "Modern man"[101] is disconcerted, rather, because a resilient residue of self still remains even when he "actually" ought to be nothing—and "actually" wants to be nothing—other than a "cog" in a machine. What shocks him, for example, is that he cannot keep up with the conveyor belt he is standing at because it is moving too fast or because his body cannot adjust itself to the combinations of movements it is asked to perform to work in unison with the machine. Alternatively, he may be lost in thought or may simply desperately want to scratch himself because of an itch. Nothing as scandalous as this can happen to any other part of a machine.—

But let's begin at the beginning, with the process, therefore, by which workers are meant to become familiar with the way a machine works. Let's see if this, too, already leads to something like an ambiguity in the self-identity of workers.

Anyone who has been confronted with a new job at a conveyor belt knows what effort it takes to transform this first confrontation into a synchronization with the working rhythm of the machine. They know how hard it is to keep up with the machine and they are familiar with the fear of not being able to do so. Because it is mundane, the antimony that this initial training causes usually remains invisible. If one considers, however, that in order to be able to work *without difficulty*, workers have *to exert* themselves in a most concentrated manner in order to take on the pace and rhythm of the machine—if one contemplates that workers are required to *exercise the strictest self-control* in order to start *an automatism*, that they have to *pull themselves together* so as to not *function as themselves*—one has to admit that this task is paradoxical. The ordinary expressions "adaptation" and "training" only provide names for this exercise.— For the paradoxical demands these processes entail—workers are asked to extinguish what they are *as* acting individuals, and *transform* their actions into *mere processes*, heteronomous and automatic ones at that—are utterly concealed by words like "adaptation" and "training." All the more so, because these workers, having acquired this automatism by means of successful transformation, are then expected to scrupulously control and maintain it. And the suggestion that all actions that entail adjustment to an instrument or implement require a level of "adaptation," whether using a hammer or playing the violin, and that such "adaptation" cannot therefore be considered degrading to human dignity, cannot offer any evidence against this task's paradoxical nature either. For sure, the violin player has to become attuned to their instrument. They have to adjust the strokes of their bow as the instrument and the movement of the music require, and they must even make sure that this becomes "natural and intuitive." In comparison with the induction of factory workers, however, this task is still perfectly human. It is free from contradiction inasmuch as the violin player is allowed to be unambiguously active while learning to master the instrument, thus transforming the violin into a part of their body (which increases its field of expression), incorporating it as a new organ into the organism. The labor of adaptation that workers are asked to perform at the workplace, on the other hand, is the absolute inversion of this task. They are asked to turn themselves into an organ of the machine. Their task is to *let themselves be ingested* into the movement of the machine; they must put themselves in a position in which they are *incorporated* by the processes of the machine.— In brief, *they are asked to actively take into their own hands and carry out the process by which they become passive machine-parts.* The paradox of this expectation is indisputable.

In other words, since workers are required to mobilize all their powers of concentration to try to displace their focus away from themselves onto the machine, they must *simultaneously* be "*themselves*" and not be "*themselves.*" We are already familiar with this formula. We employed it above to describe the ambiguous identity of the person experiencing shame. It is, of course, no coincidence that it already appears here in the description of the workers' first familiarization with the workings of a machine. In fact, we are already in the close vicinity of shame, our problematic. We will reach its core when we give an account of the self-encounter in which this ambiguity becomes acute.

This question is only difficult to answer if we believe that we have to look for an explicitly theoretical act of reflection. It would certainly be pointless to try to locate such an act. The worker's *encounter with themself happens in the course of using and operating a machine.* It is itself a moment of work. Or, if one understands "moment" as a reference to time, this encounter is a particular instant in the labor process.

Let's assume that a worker has become successfully attuned to a machine. As long as the work progresses smoothly and without friction between human and machine, the worker isn't "self-aware." In fact, as long as the worker cofunctions in accordance with the rules and as a "convert," who has become a "cog" in the machine, their self is "not present" at all and has no being, at least not as a self.[102] Workers only "assume a self" when their conformity is insufficient or their work suddenly goes wrong. It is only then that the self "becomes conscious" and encounters itself—namely, as something objectionable: as *a failure.*

This means that the self only attracts its own attention because it falls out (of the "it of the technological device" and its conformist existence with the machine), because it has become visible as a counterforce, as a resistant part that works against the machine. The contours of the worker's individuality only become visible because—to use the words of a famous formula—it is a *"negation,"* an antithesis. Put even more clearly, *the encounter with the self is not the cause of a "disturbance of identity"* [Identitätsstörung] *but the other way around. This encounter only occurs because there is interference* [Störung].[103]

But *who* now encounters *whom* in this case?

The answer certainly is that the self encounters the self. But is this response adequate?

It would be totally wrong if this were taken to mean that "the self that is affected" is an equal partner or even identical with "the self that mobilizes this affect."

"The self that mobilizes this affect" is the exemplar, the "it-and-thinglike" machine-self, the cofunctioning and conformist self that has been transformed into a part of the machine and so has already left being a self behind. At least, it had hoped to have left this behind. What surfaces as "the affected self," on the other hand, is the old self, the old-fashioned being, lagging behind, which might have once had a *raison d'être* way back in ancient times, before the birth of the machine. Now, however, in an age in which adaptation to the "it of the technological device" has become obligatory, this old self has lost its right to exist.

Or, if one prefers not to speak of two selves but would rather see this as two different states of the self, the encounter with the self only sets in because the conformist self "*gives way*" [*nachlässt*] and so "*relapses*" into its old condition—for example, because it dwells on its thoughts or has a headache and therefore forfeits its previously achieved state of perfect conformity. If the expressions "giving way and relapsing" [*Nachlässigkeit und Rückfälligkeit*] sound moralistic,[104] this is unintended. The association of "encountering the self" and "to fall" with which we are familiar (for example, from phenomena such as having a "guilty conscience") is a fact with the widest possible validity. And if it appears here, this isn't because we are trying to force an analogy with moral shame, but simply because it, too, becomes visible while observing the present situation.—

This portrayal of the one "affected," however, is still incomplete. We said that it is the "old self" that is "affected" in the encounter. Is it really only this?

Let's once again assume that the self encounters itself as a failure. Measured by the intransigent yardstick of the machine, it is, of course, entirely inconsequential *who* disrupted its function, be it the self or its body. The body and the self are both equally incongruent with the machine; in fact, for the machine, this difference doesn't exist at all. The machine, figuratively speaking, would shrug its shoulders if a worker had the idea of justifying their failure by blaming their clumsy hands, which they can do nothing about.

But the culprit doesn't seek excuses of this kind at all. For when, in the instant of failure, he suddenly encounters himself as himself, as the old Adam, rather than as a machine-part, it is clear that he already judges himself by the *machine's* measure, already views himself from *its* perspective. It is as if he were a loyal party member who declares—yes, even feels—himself guilty in the interests of the party, which must be defended at all costs against any deviation from its line. The distinction between body and self—between these old antagonists who once played such a fundamental role (because the self was ashamed of the body)—has even lost all interest for him. In fact, this difference has become so irrelevant that the two seem to represent one unified formation. Instead of the rift between the body on one side and the self on the other, there is now only the rift between the machine (or, rather, the conformist self that is representative of the machine) and the old residue, which is made up of body and self in an undifferentiated manner.

These are thus the partners in this encounter, and these are the antagonists.

But how do they relate to one another?

With this question, we have reached the critical point of our investigation. We no longer deem it possible to describe this encounter as anything other than *shame* (in the proper sense of the word).

If the example with which we opened our investigation—T.'s attendance at an industrial exhibition—still invited the suspicion that we are only using the expression "shame" metaphorically, this is understandable. The situation of merely looking at technological devices is not the one in which shame actually erupts. The perfection of the product is here only theoretically compared with the imperfection of the body.

But we have now left this situation behind and replaced it with the situation that the appliances, devices, and machines demand from us: the *operation of machines.*[105] If it were merely a question of certain "psychological difficulties" that can result from observing machines, as it was in T.'s case, we could simply go back to our daily business and dismiss this difficulty as an emotional luxury. But the difficulty of operating a machine, the efforts required to learn how to do so, the constant threat of failure, and then, finally, the actual failure are all essential aspects of operating a machine, and they are emotional processes that are not just metaphorical. The effect of this actual failure is not limited to highlighting the different levels of perfection between humans and machines—that is, between the producer and the product. The failure rather "ejects" [*wirft heraus*] the one who fails, throwing them back onto themself so that their old residue of self now suddenly stands there without a world—inept, "rejected and disproven" [*verworfen*]—no longer knowing what to do with itself.[106]

For someone who fails thus, it's simply not possible to identify with the "unadaptable fellow," the reject, he now finds himself to be. It is impossible to be content with that "residue standing there." As the conformist he was up to this moment (and actually still is), he "is" not a "residue" and "is" neither this body nor this self.— At the same time, however, it is impossible for him *not* to identify with this residue: for he cannot deny that which he was "ontologically endowed" with. And as we know from earlier, the fact that this endowment was not his fault does not offer any absolution; rather, it precisely occasions shame and disgrace: for precisely the fate that he cannot do anything about is "fatal." And so, he is ashamed.

As difficult and complicated as this may sound in the language of theory, this does of course not put an end to the intensity of the actual condition. The one who is ashamed has no idea about theoretical difficulties. He knows nothing at all. For, to be ashamed means to know nothing. To know nothing, in turn, means to be completely at a loss, to not know where to turn. And this state of being at a loss is now precisely his condition. Having fallen out of the state of "being it"—a state he had still inconspicuously belonged to just a moment before, and which he had recognized as exemplary until he was jettisoned—he now stands there publicly exposed, in full view of that which he has squandered. And instead of this, he now has to be a "self"—that is, nothing more than a "deficient mode" of being a machine. And despite having a definite name, a definite body, or maybe even portraying the definite flaw of an individual peculiarity, he now stands there as nothing but a scandalous nondevice and conspicuous nobody. And although he no longer has any business in the realm in which the machine reigns, he nevertheless has no option but to remain, for not being there also exceeds his powers. Thus, he has no other wish than to sink into the ground; he has nothing beside this utterly unaccomplishable desire—only shame.

Anyone who has never blundered while handling a machine; anyone who has never incredulously looked back in disbelief at the assembly line that wordlessly keeps on moving; anyone who has never found themself cast onto the shore of their old self and never experienced what it means to suddenly find there *oneself,* precisely oneself; anyone who has never looked down at their hands in an alienated and bemused manner, at these clumsy instruments here, because their obsolescence and incorrigible incompetence has caused the fall from grace—any such person does not know the shame that is the shame of today, the shame that erupts daily a thousand times over. Anyone who denies the existence of this form of shame does so, because the admission that we've come such a gloriously long way only to now feel shame in front of things, would itself make them blush with shame.—

An earlier version of this translation of "On Promethean Shame" is part of *Prometheanism: Technology, Digital Culture and Human Obsolescence* (London: Rowman and Littlefield, 2016), 29–95. It is republished here with slight modifications.

2

The World as Phantom and Matrix

Philosophical Reflections on Radio and Television

> Because the king didn't like how his son, straying from the controlled roads, was traveling cross-country to form his own opinion of the world, he gave him a horse and cart. "Now you no longer need to go on foot" is what the king said. "Now you're no longer allowed to" is what his words meant; "now you no longer know how to do so" was their effect.
>
> —From *Children's Stories*

I. The Home-Delivered World

1. *No Means Is Just a Means*

The first reaction to the critique of radio and television presented here will be that such generalizations aren't permitted—that everything depends on what we "*make*" of these installations, on whether they are used for good or bad, for human or inhuman, social or antisocial ends.

This optimistic argument is well-known (insofar as a pattern of speech can actually be called an argument); its origins lie in the First Industrial Revolution, and it lives on in all camps, unexamined as ever.

Its applicability is more than questionable, though. The assumption that we have freedom of use and power over technology is a mere illusion, as is the belief that certain parts of our world are nothing but "means" (to which "good ends" can be attached at will). Radio and television installations are facts in and of themselves; they mold and influence us [*prägen*][1] regardless of the ends we use them for. And this circumstance cannot simply be undone by verbally downgrading them into "means." The argument thus applies a schematic to our lives that has nothing to do with reality. Our technology-suffused existence doesn't break up into individual, neatly demarcated sections, each marked either with the road sign "means" or "end." This schematic is valid only in the context of goal-directed actions or isolated mechanical procedures, but not where "the whole" is at stake—not in politics or philosophy. Anyone who views our entire existence through the lens of these two categories applies the model of purposeful action to our life as a whole; yes, they even treat it as if it were a technological process. This view demonstrates the very barbarism that sparks such outrage whenever it is spelled out in the saying "the end justifies the means." Yet this fashionable outrage is as inept as its (only rarely explicit) endorsement: They may not be saying as much, but those who reject this motto nevertheless agree with the two categories it relies on; they, too, concede that it is legitimate to apply them to existence as a whole. But humanity, true humanity, only begins where it no longer makes any sense to make any such distinction, where means and ends are so imbued by lifestyle and custom that it becomes impossible to recognize (or even ask) whether some specific part of life or the world is a "means" or an "end." Humanity only begins,

> where the way to the fountain is as good
> as the drink it supplies.[2]

We can of course use television to participate in a church service. But whether we like it or not, the very fact that we are *not* participating in the service and are *only consuming its image* "molds" or "transforms" us just as much as the church service does. And this picture-book effect is not simply at odds with the "intended" effect of the service; it is its exact opposite. The transmitted content is thus not the only thing that influences us and undoes influence, that molds and unmolds us; the "means"—the devices themselves—shape us too. Our devices are no mere objects of use: Their predefined structure and functionality determine how they can be used; they also define the style of our engagement and of our life. In short, they define *us*.

These pages were specifically written for readers who are radio listeners and watch television; that is to say, they were written for consumers. Academic philosophers and people employed in the radio and television industry were only a secondary consideration: Philosophers will feel alienated by my subject, professionals by my treatment of it.— But I am of course not writing for every consumer—only for those who have suddenly found themselves taken aback during a program or after switching it off; only for those who have asked themselves, "Hang on, what am I doing here exactly?" "What exactly is being done to me?" Those who are familiar with these feelings might find some answers here.—

2. *Mass-Consumption Is a Sum of Solo Performances—Each Consumer Is an Unpaid Homeworker Employed to Produce the Mass-Human*

In the days before radios were installed in their flats to make culture flow like water from the tap, the Schmids and Müllers, the Smiths and Millers,[3] would stream into cinemas together, where they would collectively (i.e., *as* a mass) consume the stereotypical mass-products created for them. It is tempting to see a certain stylistic harmony in this situation, a congruence of mass-production and mass-consumption. But this would be a mistake. Nothing is more at odds with the intentions of mass-production than the creation of a setting in which the same copy of a product (or reproduction) is enjoyed by several consumers at once, let alone by a large group. For mass-producers, it is irrelevant whether collective consumption represents a "genuinely collective experience" or remains the sum of many individual experiences—the concentrated mass is of no significance to them. They are not hoping to create an opportunity for everyone to consume the same thing together; they are looking for a mass that can be broken up into the largest possible number of *buyers,* a situation in which everyone buys the same thing on the basis of identical needs (which must also be produced).

This ideal has already been achieved in countless industries, or almost, at least. It seems doubtful to me, however, that the film industry can achieve it too. In keeping with the theater tradition, it still serves up its product as *a show that many consume together.* This is no doubt an antiquated remnant. It is no surprise, therefore, that not even cinema's gigantic level of development could prevent the radio and television industries from competing with it. Besides offering a product to consume, these newer industries enjoyed the added opportunity of providing the equipment required for its consumption. This equipment, in stark contrast to cinema, could be offered to almost everyone. And because it could be used to deliver the product directly into the living room (which cinema couldn't), it is just as unsurprising that almost everyone snapped it up.

And so it happened that the Schmids and the Smiths, the Müllers and the Millers, soon sat at home on many an evening—evenings they would have once spent at the cinema together. They stayed at home "to receive" radio plays and "receive" the world. The situation that was so unambiguous in the cinema—the consumption of mass-products by a mass—had hence been abolished. This, however, didn't lead to a decline in mass-production; quite the opposite. The production of mass-products for mass-humans—yes, the production of mass-humans as such—was ramped up further every day. Millions of listeners were supplied with the same ear fodder, and each of them was treated as a mass-human, an "indefinite article," and thereby assured of their own personal character [*Eigenschaft*]—or rather, their lack thereof [*Eigenschaftslosigkeit*]. Because the mass-production of receptor devices had made collective consumption superfluous, the Schmids and the Smiths now consumed mass-products as a family or even alone. And the lonelier they felt, the more insatiably they consumed—and with this a new type of human, *the mass-hermit,* was born. Millions of specimens of such mass-hermits now sit at home alone—although detached from each other, they are all reclusive in the same way. Yet they don't go into solitude to renounce the world, but because of the unbearable, godforsaken thought of missing even one of the world-nuggets served up in effigy at home.

Everybody knows how industry, mainly for strategic reasons, has abandoned the principle of centralization (which was undisputed a mere lifetime ago) and adopted "decentralization" instead. What is less well-known, however, is that this new principle already applies to the production of the mass-human as well. Yes, I did just say *its production,* even though our focus was on decentralized *consumption.* This switch is justified because production and consumption coincide in a most peculiar way. If humans, in a nonmaterialistic sense, "are what they eat," then mass-humans are produced by making humans feed on mass-products. This also means that consumption turns consumers into collaborators in the production of the mass-human; that is, they become collaborators in the reshaping of self this requires. Consumption and production coincide; when consumption is "decentralized," production is decentralized, too; it happens wherever mass-products are consumed—in front of every radio and every television set.

We are now all in a sense employed as *home-based workers,* but we are homeworkers of a most unusual kind. Our labor, the transformation of self into a mass-human, consists of *leisure*—it is performed through the consumption of mass-products. While the workers of the classic cottage industry produced products at home to secure a few consumer goods and a minimal amount of leisure, today's homeworkers consume as many leisure products as possible to participate in the production of the mass-human. This process is fully paradoxical: instead of being paid a salary for our efforts, we must pay to be able to work. We must buy the means of production we utilize (the receptor devices, and in many countries also the transmitted programs) to let ourselves be transformed into mass-humans. In other words, we are paying to sell ourselves. Because "unfreedom" has also turned into a product, we must now also buy our own "unfreedom," including the unfreedom we are actively helping to produce.—

But even if one objects to the strange move of seeing consumers of mass-products as employees in the production of mass-humans, there is no denying that actual massification via mass-gatherings is no longer required to create the type of human desired today. Le Bon's observations about mass-situations and the changes these effect in humans have become outdated: Today, individuality is unmolded at home, where rationality is obliterated too. The orchestrated mass (Hitler-style) has become redundant; to turn humans into nobodies (and make them proud to be nobodies), there is no longer any need for them to be drowned in a flood of bodies and cemented into a monumental edifice, a monumental mass. No approach is more effective at unmolding and disempowering the human-as-human than one that appears to preserve the free expression of personality and the right to be an individual. When "conditioning"[4] happens separately for every individual—when it happens within the cage of the self, in solitude, in millions of solitudes—then the procedure is doubly successful. And since the treatment presents itself as "fun"[5] and doesn't disclose to its victims that it demands sacrifices—since it leaves their delusional sense of privacy (or at least of private space) intact—it remains utterly discreet. The old proverb *eigner Herd Goldes wert,* which stated that "a man's own home is as precious as gold," has become true once again, albeit in an entirely new sense.[6] It no longer speaks of the homeowners who are lapping up the conditioning soup but of the cooks and suppliers—of the owners of the homeowners—who are dishing up ready-made soup that is mislabeled as a hearty, home-cooked meal.

3. *Radio and Screen Become Family Tables in Reverse; the Family Turns into a Miniature Audience*

It hardly needs to be mentioned that this new form of mass-consumption is usually not introduced by its real name. On the contrary, it has in fact been presented as the conduit for a renaissance of privacy and family life. This was perhaps predictable, if only for its predictable hypocrisy: New inventions especially like to channel any old ideals that might otherwise affect sales and stifle demand. "The French family," or so we can read in Vienna's *Presse* (December 24, 1954), "has discovered that television . . . is an excellent means to keep young people away from expensive pastimes; to tie children to the home . . . and give family gatherings a new appeal."[7] The opportunity this mode of consumption actually creates is the exact opposite of what is portrayed: it makes it possible to fully dissolve the family while simultaneously preserving the appearance of cozy family life (or even generating it). Yet the family is dissolved regardless of appearances. It is now the broadcast *external world* (be it fictional or real) that rules at home. Its reign is so boundless that it invalidates the reality of the home and gives it a phantomlike quality, invalidating the reality of not just the four walls and furniture but communal life itself. When distant things step too close, closeness becomes distant or blurred. When the phantom becomes real, reality attains a phantomlike quality. The home is reduced to a "container," and its sole function is to house the screen for the outside world. A *Weltpresse* report from London (October 6, 1954) reads, "Welfare officers retrieved two neglected children, aged one and three, from an East London flat. The children were playing in a room furnished only with a few broken chairs; in the corner there was a brand-new ostentatious television set. The only food items found in the flat were one pound of margarine, a can of condensed milk and a single slice of bread."[8] And with this, the last remnants of homeliness, of communal life and atmosphere to have survived in standardized countries, have been liquidated. The moment the television enters the home, the phantoms establish their reign, and do so without a fight or even a semblance of a struggle: The apparatus arrives, provides vision, and has already won. As soon as it enters, something begins to rustle in the walls; the walls become transparent, the cement holding the family together begins to crumble, communal privacy is ruined, and the end of the home has begun.

It was already possible to observe many decades ago how the piece of social furniture most symptomatic of the family—the solid table at the heart of the living room—was becoming obsolete and beginning to lose its gravitational pull.— Even back then, this table designed to gather the family around itself was often not reinstalled when homes were refurbished. Only today, however, with the advent of television, has a true successor been found. Only now has the table found a replacement in a piece of furniture that can match its symbolic power and persuasive force. This, of course, doesn't mean that the television set has become the center of the family. On the contrary, the apparatus both screens and embodies the family's decentralization—its excentricity; it is a negative family table, *a family table in reverse.* Whereas the table had made the family centripetal—it had forced family members to continue weaving the fabric of family life by making the shuttles of looks, interests, and words run back and forth—the screen now gives it a centrifugal orientation. It doesn't provide a shared focal point and center; it is the family's communal vanishing point. The family members, in a very real sense, no longer face one another; the chairs in front of the screen sit side by side. If the family members are to see (or even look at) one another, their eyes must accidentally glide off the screen. And as they watch alongside one another, conversations only happen by chance (if they even feel like talking and have the conversational skills to do so). No longer together, the family is now only in each other's company, only side by side—a mere audience. One cannot speak of a single fabric collectively woven, of a world formed or shared by them, for the family members are merely heading toward the same point of escape; they at best go there together but never as one. In fact, they do not actually share the realm of unreality—or the world they fledglingly fly out toward—with anyone (for they themselves are not really part of this world either). Or if they do share it, they share it only with all those other "soloists of mass-consumption," the millions who, at the same time, are staring at the screen just like them. *The family has been restructured into a miniature audience, the living room has become a miniature auditorium, and the cinema has been turned into the model for the home.* If there is one thing the family still genuinely experiences together and not only simultaneously and side by side, it is the hopeful anticipation of the day on which the apparatus will have finally been paid off and their communal existence will be over once and for all. The unconscious goal of their last shared experience is hence its extinction.

4. Because Our Devices Speak in Our Place, They Turn Us into Obedient and Spellbound Beings Who Have No Say

In front of the screen, or so we just observed, conversations happen only by chance (if television viewers even feel a need to talk and have the conversational skills).

This is even true of radio listeners. They likewise speak only by mistake, and with every passing day their desire and ability to speak further declines. This, of course, doesn't mean that they are genuinely falling silent, but simply that their talkativeness has assumed a passive quality. The fable in the epigraph relates how the king's words "now you no longer need to go on foot" end up meaning "now you no longer know how to do so." In analogy to this, "now you no longer need to speak" turns into "now you cannot do so anymore." *Because our devices speak in our place, they also take our language away from us;* they rob us of our linguistic competence, our opportunities to speak, and even the joy we find in language—just as the gramophone and radio have robbed us of the music we used to play together at home.

The lovers strolling along the banks of the Hudson, the Thames, or the Danube in the company of a prattling "portable" don't talk to each other; they are listening to a third person instead—to the public and mostly anonymous voice of the program. They are taking this voice out for a walk, as if it were a little dog—or rather, it has made them go out on walkies. This voice has transformed them into a miniature audience obedient to its every command, and so it is now three, not two, who promenade. Their stroll can hence not be called an intimate *tête-à-tête* since intimacy was suspended from the start. If shared intimacy does develop, it is not because the lovers are excited by each other, but because they have received instruction, stimulation, and even their arousal from the program's husky, sultry, or crackling voice. Night or day, at all hours, this voice now tells them what to do and feel—for what else does "program" mean? Since they follow these instructions in the presence of a third, they are in an acoustically indiscreet setting. Regardless of how entertaining their obedience may feel to them, one cannot say the lovers are still entertaining and conversing with *each other* [*sich unterhalten*]. They are *being* entertained, *being* engaged in conversation—and it is the third member in their union that is doing so; it alone has a voice. And it engages them not just in the sense of "*converser*" or "*amuser*," but also in the sense of "*soutenier*": It gives them the confidence and "support" they cannot give each other (as soon as they are left to their own devices, they don't know what to do).[9] The fact that they also make love to the radio (and not just when it plays music that induces "swooning")[10] doesn't need to be hushed up as a shameful secret, for we live in a world in which everybody considers this an obvious thing to do. The radio, desired and welcome in any situation, is our equivalent to those torch-bearing chaperones, those small earthen lamps[11] the ancients used to call on to witness their love's passions. Only, today's witness is a mechanized "public utility," and its flame not only provides light but is also meant to heat things up. Discreet silence is no longer expected either—quite the opposite. The witness is to go on and on, so its songs and words drown out the *horror vacui* that might even grip the lovers in the middle of the act. This "background"[12] has become so indispensable that it is now also added to the recorded Magnetophon tapes, the "voicepondences"[13] that lovers have now started to send each other since around 1954. So, when some smitten guy compiles his illiterate love letter today, he is careful to record it onto an acoustic (i.e., a musical) bedding track. This is the only way he can alleviate his concern that "just his voice" might be too naked a present for the adored recipient. When the tape is received, it is again this third voice—the voice of a suitor who has in a sense turned into a thing—that is actually meant to speak or please.

But romantic situations are only one example, albeit the most blatant. The couple let themselves be entertained and supported in all other situations, regardless of what they are doing. If, by some mistake, they happen to talk to each other, the voice of the radio speaks behind them in the background as the main protagonist and mood-setter, giving them the comfortable and reassuring feeling that it will keep on talking even when they are done—even when they are dead.

Since speech is now always guaranteed and is constantly being trickled into their ears in a ready-spoken form, our lovers aren't really *ζῶα λόγον ἔχοντα*[14] anymore (just as they have ceased to be *hominis fabris* when eating bread): To attain their intellectual nourishment, they form their own words as little as they "make" their own bread. For them, words are no longer something to be spoken but are now something only to be listened to; speaking is no longer something one does but something one receives. That they "have" *logos* in a very different way from what Aristotle's definition had meant to imply is thus evident. It is also obvious that they are turning into "*infants*" (in the etymological sense of the word), beings who do not speak and do not have a say.[15] Regardless of the cultural or political context in which it happens, this progression toward *ἄνευ λόγου εἶναι*[16] must always end the same way: in the creation of a type of human who no longer has a say, since it no longer speaks; a type who does nothing but listen and is merely a "*serf*" [*Höriger*], always obedient, always at the call of another.[17] The first effects of this restriction to only ever listen are already manifest:

All cultured or sophisticated forms of language [*Kultursprachen*] are becoming coarser, poorer, and less playful.[18] But this coarsening and impoverishment is not just affecting language, but also lived experience and thus the human itself. If speech lacks subtlety and richness, the "interior life" of humans lacks subtlety and richness, too: Language isn't just a human mode of expression—humans are also a product of the way they speak. In short, *humanity finds expression in the way humans express themselves, and the more they fail to do so, the more unexpressed it remains.*[19]

5. *Events Come to Us; We Don't Go to Them*

The treatment of humans now consists of a home delivery that emulates the direct supply of gas and electricity into our homes. And cultural goods are not the only things being supplied. It is not just music and radio plays; real-world events are delivered too—*especially* real-world events, in fact; or at least those that are selected, dry-cleaned, and prepared as "reality" for us or presented in its place. To be "in the picture" about the world outside, one must go home, where world events are waiting to shoot out like water from the tap the moment they are "ordered into view." How else, in the chaos of reality outside, would it be possible to pick out something real with more than mere local significance? The outside world conceals the outside world. It only becomes visible when we step inside and the door closes after us. The universe only shows us its reflection once we have become windowless monads. Lynceus, as we know, delights in looking out into the world from the parapet of Faust's tower.[20] Yet, before the world can fall to us, we need to swear our allegiance so fully to the tower that we only ever sit inside instead of keeping watch on top. Only now, within the tower walls, do we delight in our "happy eyes."[21] The small poem that was meant to cure our fathers of their wanderlust—"Just look, the good things are so close" ("So why long for distant lands?")—would today need to be reworded: "Just look, the distant is so close," if not even "Only what is distant is still close."[22] And with this, we have arrived at our topic. Besides creating the mass-hermit and transforming the family into a miniature audience, *the truly revolutionary effect of radio and television lies in their capacity to bring the mountain to the prophet and the world to the human. Events*—be they football matches, church services, nuclear explosions—now *come to us;* and they do not just reach us as reports or in the form of news—we are visited by the real thing.[23]

This third revolution is the actual object of our study, which is almost exclusively concerned with the peculiar changes humans undergo when they turn into beings that receive the world as a delivery, and the no less peculiar consequences this has for the conceptualization of the world and for the world itself. To show that real philosophical issues are at stake, some effects of this transformation will be sent ahead in a preliminary and still scarcely systematic list of statements:

1. If the world comes to us instead of us going to it, we are no longer "in the world," but are solely its idle consumers who live as if in a land of plenty.
2. If the world comes to us, but only in image form, it is half present and half absent—in a word, phantomlike.
3. If we can summon the world into our presence whenever we want, we possess godlike power (we might not be able to govern it, but we can switch it on and off).
4. If the world addresses us but we cannot address it, we are silenced and issued with a gagging order, and hence condemned to being unfree.
5. If we can perceive the world but do so without any option to ever intervene, we are turned into eavesdroppers and voyeurs.
6. If an event can be sent elsewhere in the form of a "transmission" so it can appear wherever one wants, it has become a mobile, almost omnipresent good. Its spatial position is forfeited as its *principium individuationis*.
7. If an event is mobile and can appear virtually in any number of copies, it belongs to the class of objects called serial-products; if this serial-product is transmitted in return for payment, the event is a commodity.
8. If an event only attains social significance in its transmitted form, *only* as an image, the difference between being and semblance, between reality and image, has been annulled.
9. If an event is socially more significant in its reproduced form than in its original occurrence, the original must adjust itself to suit its reproduction. The event must turn into a matrix, a mere template for its own reproduction.
10. If world-experience is primarily obtained by feeding on such serial-products, the concept "world" has been abolished (provided "world" is still understood to mean the "world we

> are *in*"). Once this feed predominates, the world is bargained away, and, on account of the attitude these transmissions make us assume, becomes "idealistic."

By now it should be sufficiently clear that our study does not lack philosophical problems, and we will discuss each point in turn—except for the last. It makes use of the term *idealistic* in such a strange way that we will clarify this last point straightaway.

That the world, for us as consumers of radio and television, is no longer encountered as an outside world we are *in*, but as something that is *ours* instead, was already signaled in the first point on the list. And the world really is *relocated* in a most peculiar way. It may not be "in our consciousness" (as crude forms of idealism suggest) or even "in our brain," but because it has been moved *from the outside to the inside*, because it now resides in my room as an image for consumption, as a mere *eidos*, this relocation nevertheless bears a most striking resemblance to classic idealism: The world has become mine, my representational idea [*Vorstellung*]. And as long as one is willing to accept the double meaning of the word "*Vorstellung*," it hasn't just turned into "my representational idea" (Schopenhauer)[24] but also into a "*Vorstellung*" in the theatrical sense: It has become a performance, a "performance put on for me." With this "for me," we have isolated the idealistic element. In its broadest sense, the term "idealistic" applies to *any attitude that transforms the world into something that is mine, ours*; something that belongs to us or is available to us. It hence applies to attitudes that turn *the world into a possession:* into "my representational idea / a performance put on for me" or a "posited product" (Fichte). If it is surprising to find the term "idealistic" in the context of a discussion of radio and television, this is only because idealism usually refers to speculative, merely theoretical ownership ("my *idea* of the world"), whereas it now describes a situation in which this theoretical ownership *has been actualized with the help of technology: The world really has been turned into something I have command over.* The very fact that it claims the entire world as its property in itself reveals that this assertion is the product of an immoderate claim to freedom.—Hegel used the term "idealism" in this its widest sense. In his *Philosophy of Right*, he didn't shy away from calling the feeding animal an "idealist" because it claims the world as its "own" as it ingests and incorporates [*einverleiben*] it as its prey.[25] Fichte was an idealist because he deemed the world *his* product, something "posited" by him, the product of an act of his *ego*. All forms of idealism thus share the basic assumption that the world is there *for* humans, either in the form of a gift or as something that is freely produced; they don't, therefore, conceive of humans as belonging to the world or as world-parts; they treat them as its opposite pole. And the interpretation of this gift, this "datum," as a given percept and "sense datum" is just one of the many variants of idealism, and hardly its most significant one.[26]

If it is true that all forms of idealism convert the world into a possession—into a dominion (*Genesis*), a perceptual image (sensualism), a consumer good (Hegel's animal), a posited or created product (Fichte), a property (Stirner)—we can also use this expression in good faith here, for in our usage, all possible nuances of possession converge.

Radio and television may open up windows onto the world, but regardless of how widely they fling these windows open, they also turn the world-consumer into an "idealist."

"Idealist"? As we have only just established how the external world triumphs over the internal one, this claim will inevitably sound strange and contradictory. To me too. The very fact that both claims are possible seems indicative of an antinomy in the relation of human and world, an antinomy that cannot be resolved just like that. If it could be, there would be no need for this study, which was set in motion by this very contradiction. Viewed as a whole, therefore, this study is but an attempt to unravel this contradiction.

6. Because Everything Is Delivered to Us, We No Longer Venture Out; We Remain Inexperienced

In a world that comes to us, there is no specific need to venture out into it; "*experience*,"[27] as we called it until yesterday, has become redundant.

Until recently, expressions such as "to come into the world" and "to experience" provided unusually productive metaphors for the school of thought called philosophical anthropology.[28] These expressions introduced the human as a being weak in instincts, one that could get to know the world only after the fact, *a posteriori*. To "come into the world" meant that one needed to *come to it*—to experience it firsthand and get to know it, until one gained sufficient experience, until one arrived. Life was deemed a journey of discovery, and the great educational novels were right to portray nothing but the paths, detours, and adventures that needed to be endured until one finally arrived at the world, although one had already been in it all along.— But now that the world comes to us—now that it, or rather its effigy, is let inside so we don't need to get involved with it—journeying out and gaining experience are no longer necessary; and because things that are redundant waste away, venturing out is no longer even possible.[29] It is becoming increasingly difficult to meet "experienced" people, and esteem for those who are old and experienced is clearly in rapid decline. Since we are more like aviators than pedestrians, we no longer need paths to travel on; consequently, our knowledge of the ways of the world, of the paths we journeyed on and which gave us our experience, is rapidly disintegrating—and so are the paths themselves. The world is becoming pathless. Instead of "covering the distance" to the world ourselves, the world is made to cover this distance for us (as if it had been "put aside for us"); rather than traveling to things ourselves, these are now paraded in front of us.—

I admit, this sketch of today's humanity might at first glance seem misdrawn. Are car and airplane not usually presented as *the* symbols of contemporary man? Has not the human of today even been called *homo viator*, the being who travels (Gabriel Marcel)?[30] But one must ask how justified this is. Nothing our contemporary values about travel hints at any interest in the regions he is traveling through. The same is true for the destinations he has himself mailed to as an express delivery or to which he delivers himself. He doesn't travel to gain experience; he travels because he hungers after rapid change and omnipresence *per se*. The very speed at which he moves is also the means by which he robs himself of the opportunity to gain experience (so much so that "speed" becomes his single experience and his last).— Not to mention that his travels effect a homogenization of the world that really does reduce the number of things worth experiencing and capable of generating experience. Wherever in the world our contemporary touches down, he finds himself back home—no experience required. "With us," a poster of a famous airline most confusingly declares by mixing globalism and provincialism, "you travel everywhere as if you were at home." "As if you were at home"—it is thus clearly not unreasonable to assume that our contemporary essentially regards traveling as an archaic method by which to obtain omnipresence: By leading him out into the world, travel always takes him the wrong way. Even when his journey consists of snoozing toward the destination inside a cocoon of electric heat high above the North Pole, he considers it an antiquated and uncomfortable nuisance. And the only reason he still puts up with it is because it is not yet possible to have everything home-delivered, even though he considers this to be his right, and even though every effort is made to comply with his demand.—

Yes, *homo viator* may typify us, but there's just no way that the radio and television consumer, ruling over a world in effigy at home, is any less typical. Poured into his armchair mold like other millions of his type, this lord of vast image hoards is as exemplary of us as motorists and aviators are. And instead of venturing out, he simply summons the world into his living room until he is tired of it and switches it off again. The simple fact that even the motorist makes the radio play for him while driving through the countryside makes the radio and television consumer all the more typical. For even the motorist seeks comfort in the knowledge that the world has to come to him when he has to go out into it. (And by means of punishment, he now makes it drive alongside him on the other side of the window, or has it trail him in the rearview mirror.) In a sense, therefore, the world now only happens [*aufspielen*] so it can play back for him [*abspielen*].[31]

"To play back for him." "To happen for him." "As if at home."

These expressions point to a mode of being, a relation to the world so abysmally perverse that not even Descartes's *mauvais génie trompeur* could render it more twisted. When we now use the word *idealistic* in the sense outlined above, it points to a mode of existence that has become "idealistic" in a most emphatic manner, and this in two ways at once:

1. Although we actually live in an alienated world [*einer entfremdeten Welt*],[32] the world is presented to us as if it were there for us, as if it were ours and just like us.
2. *We "take"* (i.e., perceive and accept) *this to be true,* even though we are sitting in our armchair at home, even though we cannot actually *take the world* in any way—we cannot make it ours like the "feeding animal" or like a conqueror; or at least we, the average radio and television consumer, cannot. We simply "take" the world to be so because that's how it is served up to us in images. This turns us into voyeuristic rulers of world-phantoms.

We will start with the first point; the whole of part II is dedicated to the second.

7. The Home-Delivered World Is Made "Cozy, Comfortable, and Perfectly Nice"

It is not possible to address the origins, etiology, and symptoms of alienation [*Entfremdung*] here. The literature is vast, and alienation exists; this much needs to be assumed.[33] What we wish to address here is a deception of a very particular kind. For we may live in an alienated world, but the moment we consume film, radio, or television (and this applies to many other situations as well), we nevertheless find ourselves on a most intimate footing with everything and anything: This intimacy extends to unknown people, unfamiliar parts of the world, unfamiliar situations and events, and even most outlandish occurrences (especially these). The explosion of the thermonuclear weapon on March 7, 1955, for instance, was conducted under the comfy name of "Grandpa"—that is, "Gramps."[34] A process of pseudofamiliarization has now come into sight, which, for reasons that will be explained in the next section, has no name. We will call this procedure "*die Verbiederung der Welt,*" "the cozification of the world." Yes, "cozification," not a "cozying up to." It is not us who are made to snuggle up to strange or unknown things; instead, unfamiliar people, things, events, and situations are presented to us as if we were already acquainted and as if they were already familiar to us; they are presented in a way that is "cozy, comfy, and perfectly nice."[35]

Illustrations: While the spheres of use and making have been fully torn apart (we only ever use prefabricated goods yet are fully oblivious to or only vaguely aware of the effects we coproduce); . . . while our next-door neighbor, whose flat we have been passing for years, doesn't know us and makes no effort to bridge the distance between us, movie stars present themselves to us as if we were old friends, as if we were "chums."[36] Although we've never met these strange girls[37] in person, we've seen them countless times, and we know their physique and internal makeup better than those of our colleagues. And so, it automatically seems as if we are acquainted; they informally say "you" as if we were on first-name terms, while we call them "Rita" or "Myrna" when speaking about them. The things transmitted to us *no longer maintain their distance,* and we lose our distance from them; the gap separating us is abolished. The value placed on extinguishing this gap is further demonstrated by 3D film. Its invention and introduction weren't just motivated by an interest in advancing technology or competition (against television) but by a desire to create a state of distancelessness with the highest possible level of sensory and spatial realism. If it were technologically feasible—and as fast as things are developing, who could predict what lies in store?—we would probably also be blessed with "telehaptic effects" that would allow us to enjoy the uppercuts in movies as actual blows to our chin; only this would be true closeness. But even today, 3D film promises, "You are with them, they are with you."[38]

If this mutually familiar "you" is to prevail, if I am to have someone to be on first-name terms with, images must initiate this. And indeed, no program lacks this "you"; not one of the beings delivered into our home fails to adopt it in its attempt to project intimacy. As I switch on the president, he is suddenly sitting right there beside me for a nice little chat by the fireplace, even though he is thousands of miles away. (By the way, this coziness is broadcast in millions of copies.) When the announcer appears on my screen, her deliberately unintentional affection invites us to look deep into her eyes, as if there were something going on between us. (She also has something going on with everyone else.) When the radio family lays out its worries before me, I am confided in as if I were their neighbor, their physician, or their priest. (This family confides in everyone: It is *the* family next door; *it exists only* to confide in us.) All of them arrive as intimate or indiscrete guests, all reach me in an already "precozified" state, and not one of them shows even the slightest speck of otherness. This isn't just true for people but for everything else as well, for the world as a whole. The magical force of cozification is so irresistible, its power to metamorphose has such a vast scope, that nothing escapes. Things, places, events, situations—all are transformed; everything arrives with a chummy smile and vulgar *tat tvam asi* on its lips, until we are as intimate with the stars in the night sky as with the stars in the firmament of film, until we feel just as entitled to speak of "good old Cassiopeia" as of Rita or Myrna. This isn't meant to be a joke.— In public and even academic discussions of flying saucers, it is considered possible (even probable) that, like us, the would-be inhabitants of distant planets have no other concern than that of interplanetary space travel. Such is the extent to which everyone and everything is treated as if it were "one of us" that this concern is deemed as topical there as it is for us here on Earth. The anthropomorphism that has now come into sight makes the anthropomorphisms of so-called primitive cultures look positively timid. The purveyors of the cozified cosmos have exchanged the Plotinian-Goethean formula of identity "if the eye were not sunlike . . ."[39] with the commercial motto "if you don't make the sun eyelike . . ." Anyone who fails to transform it into a pleasing product misses out on selling nature and lets a virtual commodity go to waste.— With this move, we, the consumers, are systematically transformed into *buddies of the globe* and *chums of the universe*. But we are never more than buddies, of course; a humanity conditioned in this way cannot possibly feel true union, pantheism, or love for what is most distant, let alone be "at one with the universe."—

What is true for the spatially most remote and socially most unlike us also applies to the most distant in time: *The past is also made to be our buddy.* I won't even mention historical movies where this applies as a general rule. Recently, however, I also came across this move in an otherwise serious American academic book written in a very vivid style. The book introduced Socrates as *"quite a guy,"*[40] a classification that is meant to make the great man of the distant past seem more relatable to its reader—for he, the reader, is, of course, "quite a guy" as well. This classification thus provides the reader with the unconscious satisfaction of knowing that Socrates, had he not coincidentally lived in ancient times, would essentially be no different from us and would have nothing fundamentally different to say—certainly nothing that might give him a claim to authority that we may lack.— There are also those who will assume that there must be some good reason why Socrates ended up living in the past, in a time that cannot quite be taken seriously. Such readers look down on Socrates and certainly not up to him—their belief in progress and their suspicion of any kind of privilege prohibit them from this.— Others (as they prove by their reactions to historical films and the like) deem historical figures positively comical. They treat them as *provincials of time*—people who didn't grow up in the metropolitan "present" and thus act like the village idiots of history or superstitious backwoodsmen. Every electronic device that had not yet been invented back then is treated as one of their touching defects. Some of our contemporaries even regard historical figures as nonconformists—suspicious eccentrics who obviously seem to expect some sort of special treatment. Instead of living in the present, as all honest people do today, they prefer inhabiting the cave of the has-been.— But regardless, really, of whether they are introduced as "quite a guy" or as provincials, both classifications convey neighborly proximity and are thus examples of cozification.

As for "Socrates, the guy,"[41] this phrase was evidently modeled on the

"Declaration of Human Rights," the grand political principle that "all men are born equal."[42] This principle has now been adventurously expanded into a proclamation of the "equality of all citizens of the commonwealth of times past and present."[43] This expansion not only creates an obviously false impression of historical proximity but also conveys an erroneous idea of what all human beings have in common: The important thing about Socrates is not what we have in common but what our sort lacks. The method that allegedly brings the object closer ends up concealing it instead; it alienates, if not even abolishes it. Yes, abolishes it—a past projected onto a single plane of *buddyness* ceases to exist *as* history. This may even be more plausible than our general thesis that the world is made to disappear *as* world when all its diverse and far-flung regions are brought equally close to us.

8. The Sources of Cozification: The Democratized Universe; Cozification and Commodity; Cozification and Science

So, what is behind this process of "cozification"?

Like every other historical phenomenon of such broad reach, cozification is overdetermined. It owes its existence to very different sources that all had to converge and connect to turn it into a historical reality.

Before we trace its primary root, we will bring three lateral roots to attention. We have mentioned the first just now as we spoke of Socrates. We will call this root the "*democratization of the universe,*" which means the following:

If everything and anything is close to me, regardless of whether it is close or distant—if everything and anything can claim an equal right to speak up and be immediately acknowledged by me as if it were already familiar; if any kind of preference already carries the odious stain of privilege—the idea of a structurally democratic universe has been internalized. By means of this no doubt unconscious bias, the (moral and political) principles of equality and tolerance for all are applied to the universe as a whole. There is nothing unusual about this kind of projection. Humans have a long history of fashioning the universe in the image of their own societies. If anything, it was the worldview that dominated Europe over the last few centuries that was unusual, one that divided the universe into a practical and a theoretical realm, the latter being fully alienated from the former. It's not surprising, therefore, that the United States, with its strong democratic tradition, also exhibits a tendency to expand its own principles onto everything else. There was even a branch of American academic philosophy that, on account of its *pluralism* (which mirrored democracy), would have represented a full rejection of the monistic and dualistic principles of classical philosophy had it been thought through to its end: the philosophy of William James.—

Second, since cozification makes everything equally close or seemingly close, it is a phenomenon that evidently also *neutralizes things politically*. To trace its root, it is thus necessary to search among worldly forces that have a fundamentally neutralizing effect. One of these forces is democracy (or at least its hollow projection onto nonpolitical horizons).—

Yet today's main neutralizing factor is not political but economic: It lies in the fact that all things have become commodities. Is commodification a root of cozification too?

At first glance this seems impossible: Commodification is known to be a vehicle of alienation, whereas cozification—which makes everything seem familiar and nice—seems like its very antithesis.

But things are not as simple as this first reaction might suggest. True, everything that becomes a commodity is alienated, but because every commodity also wants to be purchased and become part of our life, it also needs to make itself cozy and nice. Let's have a closer look at this process:

Every commodity tends to assume a form that makes it easy on the eye, tasty when consumed, handy when in use, and perfectly tailored to a specific need, lifestyle, or standard of living. The degree to which this fit is achieved is the main factor determining the quality of a commodity; or, the other way around, the lower the commodity's resistance to being used, and the fewer undigestible, alien scraps its consumption leaves behind, the better it is considered to be. Since radio and television programs are commodities too, they must likewise be served in a comfortably familiar, easily absorbable, and deseeded state optimal for consumption. They must arrive in "bite-sized"—or, rather, "eye-" and "ear-sized"—formats and present themselves as our *likeness:* They must seem custom-made for us; they must address us as if they were *one of us*.

Viewed this way, cozification now actually seems to lose much of its odious taint: It appears to be but an expression of the fundamental fact that we, as *hominis fabri*, "make something out of something," refashioning the world to our needs. In this, its widest possible sense, cozification simply seems to be another word for "culture." Yet, although every kind of making is clearly in a sense an act of cozification, this broader meaning would be utterly misplaced here. We evidently deem cozification to be something quite despicable, whereas we cannot accuse the act of "making" itself of some capital flaw for being what it is. We would, for instance, never think of accusing carpenters of fraud because they don't provide us with unworked wood but with wooden tables, which are a much more comfortable fit. *Yet such customization and fitting-to-measure becomes fraudulent the moment it presents something that is made as the thing it is made of.* Which is the case for the cozified world: It is a commodity that, in an attempt to secure a sale, presents itself to its customers in a way that seems to suit their requirements, as if it were a comfortable fit. It thus deceives by projecting precisely those qualities that the world emphatically lacks—for the world is what is uncomfortable—and yet this cozified-product version has the audacity or naivete to claim to be the world itself.—

The third and final root of cozification—for which everything is equally close—can be found in *the scientific attitude,* which is rooted in the warranted pride that science can turn even the most distant things into objects of close study, while its research can also defamiliarize the things closest to the scientist's own life. The scientific attitude is hence celebrated for its very ability to neutralize the difference between the close and the distant: It can fully dedicate itself to matters the scientist is personally unaffected by, while also remaining dispassionate toward everything else.— But if it is to be successfully adopted and maintained, this all-neutralizing stance—this "objectivity"—requires some major moral artifice, if not to say a "violation of self": It requires that one ascetically abstain from seeing the world the way we are normally used to seeing it. The belief that this neutrality can now simply be detached from its moral root and then passed on to everyone (and thus also to people who live an entirely nonascetic life, without any dispassionate interest in knowledge) doesn't just misunderstand science but also fails to grasp the moral tasks its popularization is meant to perform. Yet this misunderstanding has become the very principle by which popularization is practiced: There is no reader, nobody listening to radio or consuming television, no viewer of a cultural film who isn't transformed into some vulgar doppelgänger of a scientist. The consumer is now also expected to believe everything to be as close as it is distant. Of course, this doesn't usually mean that consumers are also asked to allow for everything to have an equal right to be noticed and acknowledged, only that everything has an equal right to be consumed.

And since "knowledge" today presents itself as a "pleasure," and "learning" as "fun,"[44] these boundaries are totally blurred.

9. *Cozification Is Itself a Cleverly Camouflaged Version of Alienation*

With these three observations, however, the main root of cozification hasn't yet been revealed. And neither has there been a plausible explanation for the strange circumstance of how a process that becomes manifest in so many different ways doesn't even have a name of its own. And because cozification appears to be the counterforce of alienation, this lack of a name is all the more conspicuous. How is it possible that a phenomenon that is no less powerful, fatal, and symptomatic of our age than alienation was able to remain so hidden, when alienation itself was not—even if the term "alienation" has itself been cozified and thus rendered harmless?

But is cozification really the antagonist of alienation?

That's the very last thing it is. With this, we have arrived at the main root of cozification, the one that also allows us to understand why it hasn't had a name: *The main root of cozification*—paradoxical as this may sound—*is alienation itself.*

Anyone who cozily believes in cozification, anyone who views it as an actual counterforce to alienation, falls victim to the misdirection it is meant to create. Any notion of enmity between the two can be dispelled by the simple question of whether cozification harms alienation or benefits it, for the unanimous answer must be that it comes to its aid. Its main feat is to dim the causes and symptoms of alienation and the very misery it creates. It robs humans who are forced to live in an alienated world (and who are also alienated from the world) of their ability to notice that this is the case. In short, *it drapes alienation in a cloak of invisibility and denies its reality so as to allow it to run its course without restraint.* To accomplish this, it endlessly populates the world with images that make all things seem familiar; it presents the world—even its most spatially and temporally remote regions—as one big home, as a *universe of coziness and comfort.* Cozification exists to perform this function, and its instigator is alienation. It would be unreasonable to believe—as naive as undialectical—that these two forces are separate or even hostile siblings. They rather cooperate as harmoniously as a pair of hands. *The wound of alienation inflicted by the one is soothed by the balm of familiarity the other applies*—if it is not even just *one* hand that is at work here, as the two processes may ultimately be one and the same: Cozification can be seen as an act of camouflage undertaken by alienation itself. To distract from its dictatorial rule, alienation steps forth in the guise of its harmless adversary, taking the stand against itself and declaring that the two forces are in perfect balance . . . just as Metternich did when he established a liberal opposition newspaper that seemed to take aim at his own politics.

In a Molussian fairy tale, there is an evil fairy godmother who heals a blind man without restoring his sight. Instead of piercing his cataracts, she curses him with a second blindness—one that blinds him to the fact that he is blind—by sending him a constant stream of dreams that make him forget what reality had looked like. There is a striking resemblance between this fairy tale and the way alienation acts in the guise of cozification, for it also uses images to create for people who were robbed of their world the illusion that they still have a world of their own—a universe, even, in which everything is theirs, familiar, and just like them. It makes us forget what an unalienated existence and unalienated world actually look like.— The situation we find ourselves in is, of course, a bit more jinxed than the fate of the blind man in the fairy tale, for in our case, the fairy blinding us to our own blindness is that same fairy who blinded us in the first place.—

Considering all this, is it really a surprise that alienation pursues its mission of self-denial in secret? That this self-denial doesn't even remind us of its existence by having a name of its own? Why would the forces responsible for alienating our world want to attract our attention? Why would they ever highlight—even if merely by a specific term—their need to conceal their effects by supplying surrogate images, something they are doing with great success? What *is* surprising, however, is how successfully they are keeping such a widespread, openly visible, everyday phenomenon under wraps by simply leaving it unnamed. Yet this is undeniably the case—that's why they supply their images but fail to disclose why they are being supplied. Which they can do with all the more ease the more we, their recipients, really do let ourselves be deceived and feel perfectly healthy as the deceived. It's as if the injury of alienation has made us incapable of realizing that we're under the influence of the drug of cozification; yet we also seem too numbed by the drug to be able to feel that we are injured—it's as if the two states were canceling each other out.

Yet, even if it weren't being generated by alienation's attempts at subterfuge, cozification is undeniably also a source of alienation itself. The same *neutralizing* effect occurs regardless of whether *what is close is made to seem distant* (as alienation usually does) or *what is distant is made to seem intimate and close* (cozification). To be-in-the-world means to find oneself at the center of a set of concentric circles of varying degrees of closeness and distance that continually differ, overlap, and accumulate. It is this fundamental relation between the world and the human that is distorted by the neutralization cozification performs: If everything is as close as it is distant, if everything always affects one to the same degree, one is either an indifferent god or a totally denatured human (and we are obviously not talking about stoic gods).

Indeed, there's nothing that could alienate us from ourselves and the world in a more fatal manner than the fact that we now almost constantly surround ourselves with an entourage of false confidants. I am speaking of those phantom-slaves we summon into our room for a morning audience, even as we fumble for the buttons of the radio while still half asleep. We bid them in as the day's first bit of world, so they can speak to us, look at us, sing to us, console us, cheer us up, calm us down, or excite us enough to make us get up and start the day—a day that is not ours. The alternative between being asleep or awake has been supplanted by either sleeping or listening to the radio. And the fact that we then continue our day under the aegis of these make-believe friends fully consolidates our alienation from ourselves. Even when real company is available, we now prefer our portable chums,[45] for we no longer see them as substitutes for actual human company; they feel like our real friends.

I remember how I once entered a Pullman compartment to take the seat across from a woman who was fully absorbed by the voice emanating from her tiny portable (it was clearly one of her favorites). As I uttered a greeting, she was so startled that it was as if *I* were the phantom and not the man in the box. It was as if I had made myself guilty of a most dreadful trespass into her world—her love life even. I'm convinced that countless humans would feel much more cruelly punished today if one were to confiscate their radios than would prisoners who may have lost their freedom but are still allowed to keep their apparatuses. Such prisoners could still enjoy a most extroverted existence; their world and their friends would still be available, still on call to be granted an audience—so what exactly would have changed? But if some poor fellow were to be deprived of his device, he would instantly be gripped by panicked anxiety and left feeling numb amid nothingness; his loneliness and worldlessness would instantly suffocate him. From my New York days, I remember how an eight-year-old boy from Puerto Rico burst into my landlady's flat, completely distraught. His radio had for some unknown reason suddenly stopped working (perhaps the world had come to an end?), and so, he now crashed through the door to try to catch the voice of one of his beloved phantom-friends from Los Angeles on my landlady's machine—the voice was not to be missed at any cost. As he found it with the first twist of the tuning knob (he didn't just know the frequency it lived on; his hand, of course, found the address blindly), a soft sigh of relief could be heard, and the boy began quietly whimpering to himself. He was like a shipwrecked sailor who, having been washed up on a beach and feeling solid ground beneath his feet, dissolves into tears. Throughout the entire episode, the boy, of course, didn't once think to acknowledge the presence of my landlady, let alone me. Next to the unmet friend the radio had reunited him with, we were both quite unreal.—

10. *Is Alienation Even Still a Thing?*

Maybe the premise that our dependence on "cozified friends" and a "cozified world" is alienating us from ourselves isn't entirely accurate after all—not because it goes too far, but because it doesn't go far enough. Isn't it perhaps unwarranted optimism to assume that we, who exclusively feed on surrogates, templates, and phantoms, still *have a self* we could lose—and that we could be prevented from "being ourselves"? Has the moment at which "alienation" could take hold of us and be inflicted not already passed? At least in certain countries? Is ours not already a state in which we are no longer actually "ourselves" anymore, but only "the self" that forms from the surrogates we are daily spoon-fed? Can one rob someone who has already been robbed? Undress someone who is already undressed? Can mass-humans be alienated from being themselves? *Is alienation really still a thing? Or is it not already a fait accompli?*

For quite a while now, we've been ridiculing approaches to psychology that treat categories such as the "ego" or "self" as laughable metaphysical remnants because their apparent falsification of the human made us mock them as "psychologies without soul." But were we right? Was this derision not pure sentimentalism? Were these psychologists actually falsifying the human, or was their praxis not already a psychology of the falsified human? Were they as robots not entitled to conduct robotics instead of psychology? And since they were already studying and treating the human in a state of untruth, were they not entitled to their untruths?

II. The Phantom

The world is delivered home to us; events are paraded in front of us.

But *what* are they paraded as? *As* events? Or just as representations of events? Or as news reports about them?

To answer this question, the guiding question of this section, we will first translate it into a slightly different language and ask: In what way are broadcast events *present with* their recipients? In what way is their recipient *present at* them? Is the recipient actually present? Seemingly present? Absent? Present or absent in what way?

11. *The Human–World Relation Becomes Unilateral; the World, Neither Present nor Absent, Turns into a Phantom*

On the one hand, events really do seem to be "present": As we tune in to a battle scene or parliamentary debate, we aren't simply listening to reports *about* explosions or *about* the speakers but hearing them directly.— Doesn't this mean that events, which we were once neither able nor allowed to participate in (nor had to be part of), are now actually here with us? And are we not there with them?

Yet, at the same time, we are not. When the world's voices have free access to us—when they have the right to be with us, although we remain entirely without rights and without a say in any events supplied—can such a situation really be called "live"? Is it really "live presence" if we remain unable to respond to anyone, regardless of who is speaking—not even to people who seem to be addressing us directly? Finally, are we really present if we find ourselves immersed in the noise of events but aren't allowed to intervene in any of them? Doesn't true presence need the human–world relation to be *reciprocal*? Hasn't this relation been severed? Hasn't it become *unilateral,* in other words—transformed into a situation in which *one listens to the world while remaining imperceptible to it*? "Don't talk back"[46]—isn't this what the listener has been categorically condemned to? Does such inability to speak not equate to powerlessness? Hasn't the omnipresence we have been gifted with installed a state of unfreedom? The unfree—those without a voice, those treated as nothing but air—aren't they absent?

They clearly also seem absent.— It would, however, again be possible to interpret this one-sidedness the other way around, as a guarantee of freedom and presence. Is it not freedom to be able to participate in any event from a *safe distance*, invulnerable and without being exposed to danger? Is this not even a privileged form of freedom, since it also entails the added license to transform any such event into a means of entertainment and a consumer good? If a person cannot be made to want to run away from anything they are witnessing, if they cannot be chased into absence, is *this* person not truly present?

This also sounds plausible. And if some heckler were now to interrupt these questions and declare our back-and-forth about presence and absence baseless and entirely flawed, this would be easy to understand. "Radio and television supply us with *images*," or so I already hear him say, "with representations, not presence! It's blindingly obvious that we are but air to them and cannot intervene; you're talking about a phenomenon called 'aesthetic appearance'—something that has been known for ages."

As persuasive as his argument might seem, it is incorrect. First—and this is a basic phenomenological fact—there is no such thing as an "acoustic image": The gramophone provides us not with an image of the symphony, but with the symphony itself. When a mass-rally comes into our house through the radio, we believe we hear the actual crowd—the actual noise it makes—and not just its "image" (even though the crowd itself does not physically enter). Unless the broadcast consists of a work of art that explicitly conveys its staged and artistic nature (e.g., a radio play), an aesthetic attitude is the very last thing listeners will adopt. People listening to a game of football on the radio listen as excited supporters for whom the game is really happening—the "as if" character of art is the furthest thing from their mind.

No, the heckler isn't right. What we receive via the radio aren't just images, but we aren't really present at actual events either. The question "Are we present or absent?" is thus baseless indeed. But not because the answer is obvious—that is, we receive "images" (and are clearly "absent")—but because the broadcast creates a situation of *ontological ambiguity,* an ambiguity that is the defining characteristic of the situation itself. Broadcast events are at once present *and* absent, actual *and* virtual, at once here *and* not here at all. In short, they are *phantoms.*—

12. *In Television, Reality and Representations of Reality Are Synchronous: Synchrony Is an Atrophied Version of the Present*

"But," the heckler will no doubt continue, "just because something is true for sound and for radio broadcasts doesn't mean it automatically also applies to television. That television supplies us with images surely cannot be denied."

Hardly. And yet, this doesn't mean that televised images are "images" in the conventional sense. Throughout the entire history of human image-making, there has been a fundamental constant that hasn't changed until today: It has been an essential trait of images that they are the product of some sort of asynchrony—a "*temporal differential,*" a "*time gradient,*" that is, a "*time gap*" [*Zeitgefälle*]. And although it was only implicit, this temporal difference also separated the object from its representation and image. The German language conveniently makes this temporal gap explicit in the way it uses the preposition "*nach.*" When painting, one either *follows*—that is, *paints from*—a model [*nachmalen*], or else something real is created *on the basis of* a blueprint or a model [*nachbilden*]. Until television was introduced, the image was thus either a *Nachbild,* a *representation and likeness* that sought to "emulate" and thus "follow" an existing object in some way; or else it served as a *Vorbild,* an idea or blueprint[47] that *prefigured* the actual object it brought forth. It was thus either used as a memorial that attempted to bring something back from the past and reverse its transient nature (in other words, to cling to its presence and preserve it); or else, it was used as a device that could magically provoke the existence of things. In the latter use, it preceded its subject only to vanish the moment the event or the thing it depicted was realized, for as soon as this happened, it would also cease to be a mere image. There was, of course, also a third function: The image also acted as a vehicle with which we could transport ourselves to a dimension beyond the present and even beyond time, or at least we believed it to be doing so; for even in this neutralizing mode, it was a product of a relationship to time. It would, in fact, be very difficult to name images in which none of these temporal relations between the human and world are active. Whether something lacking this temporal gap can be called "an image" is likewise doubtful. Yet the images transmitted via television are precisely objects [*Gebilde*] of this kind.

Even though television images unfold cinematically *in* time, *one can no longer speak of a temporal relation between the image and what it reproduces.* What was introduced as the "time gap" above is here reduced to zero; television images reach us in "real time": They appear *synchronously* and *simultaneously* with the things they depict. Just like a telescope, they show something that is *currently present.*[48] Doesn't this now also mean that we are "in the presence of" whatever is depicted? Can objects [*Gebilde*] that show us something that is actually present really be called images?[49]

This problematic has not gone unnoticed, but it has been inadequately addressed. Typically, one simply fell back on the existing term "snapshot" [*Momentaufnahme*] and (with this no doubt obvious move) the matter was considered dealt with. But rather than solving anything, this term, in fact, emphatically obscures the problematic. Snapshots aim to capture and arrest the passing moment; this makes them images in the most legitimate sense. As such, they are more closely related to thought-images, memorials, and even ancient mummies than to TV-phantoms. It would be wrong to say that TV-phantoms seek to arrest the passing moment: They don't just appear in real time, simultaneous with the events they reproduce; they disappear alongside them as well. Unless they are themselves captured somehow, these phantoms are as fleeting as the instant they reproduce. If they were snapshots, they would at best snap the moment *for* the moment. What they most closely resemble, therefore, are the images created by mirrors: These are also synchronous and simultaneous, and they perish the moment one looks away; they are in every respect pure *presence* [*Gegenwart*].

But is this not mere wordplay? Are we not just exploiting the fact that the German word "*Gegenwart*" ["presence"] coincidentally hovers between two meanings to conjure up imaginary problems? As we are evidently using the word *present* in two distinct ways at once, these questions will inevitably arise. On the one hand, we are using it in the sense of *concrete presence*—that is, to describe a situation in which a human is present, in actual physical contact with another human being or the world; a situation in which they are encountering, concerning, and affecting each other in a way that makes them form (= *concrescunt*) a "concrete situation." But then we are also using the word to refer to the *abstract, formal simultaneity* of "the present moment in time." Here, "presence" simply implies that human beings are always already balanced on the same temporal needle-tip as everything else that is happening "right now"; and as such, all these discrete and separate entities are formally part of the same world-instant. They all happen in the same blink of an eye [*Weltaugenblick*].— Yet it is no mere coincidence that the word "presence" holds this double meaning (and not just in the German language). This is the case, rather, because it is genuinely impossible to clearly identify the boundary beyond which a given event or thing in the world becomes of such slight concern that it morphs from something that we are "attentive and present to," into something that is merely "part of the present" (in the sense of happening "right now" but out of sight and out of mind). When something we are attentive to is in the process of transitioning into something that is merely part of the present, we encounter the borderline case. And it is always what is currently of the least concern to me, the thing most distant to me, that is in the process of transitioning in this way. Yet, as long as it hasn't yet fully withdrawn into "what is not given," it is evidently still something that is of concern to me.[50]

But even if it were possible to clearly demarcate the two meanings, we are not the ones playing with them—television is. Such play is in fact the very principle of broadcasts. Broadcasts supply us with *something that is currently happening elsewhere, something we are unaware or barely aware of, and they make this something seem like true presence*; they have the capacity to make things that are only formally "part of the present" seem concrete and actually present. They can fully erase the already indistinct boundary between these two forms of "presence." In doing so, they also erase the line separating the relevant from the irrelevant. Every transmitted image says, "*Jetzt bin ich*"—"I am happening now / now I matter—and not just 'I,' the transmission, but 'I,' the transmitted event." And it has every right to say so. It is with this proclamation, "I am now / now I matter," and this topicality, that the broadcast image turns itself into a phenomenon that far exceeds anything image-like. But because it is not actually present either, it turns into something intermediate between being and semblance—into something we have just called a "phantom" in our discussion of radio broadcasts.—

The blurring of the boundary between these two forms of presence would in itself be nothing to object to—done correctly, it would even be welcome. There are far too many things today that we falsely treat as *adiaphoron*, as matters we are morally indifferent to. We push such things aside as if they were "merely happening in the present," even though they actually affect us and impact us (and can be affected by us); that is to say, we ignore them even though they are *nostra res* and present in a most concrete and threatening manner. Provincialization is no less dangerous than false globalization. So there is certainly a need for technologies that can extend the horizon of what we feel morally responsible for [*moralischer Gegenwartshorizont*], a horizon that should reach far beyond the range of what we can sense and feel. But television does not achieve this kind of expansion. On the contrary, it dissolves our horizon so effectively that we no longer recognize true presence. Even events that should genuinely concern us are now only met with the imaginary interest we've been taught to muster by the pseudopresences delivered to us in our homes.

I hardly need to add that there are an unlimited number of such phantom-presents. And because the common ground between consumers and events is merely created by the abstract and isolated principle of the shared "now," this principle is also universally applicable. This global now is all-encompassing—there are no events that cannot be included in it, and so there is also nothing that cannot be transformed into something that is supposedly present. But *the more things are made present, the less present each thing becomes.* Among the radio and television fans I've met, I don't know a single person who was taught how to be a friend of the world by their daily dose of simultaneity—or even a mere *Zeitgenosse*, a *contemporary*. I've met many, however, who've become worldless, scatterbrained, unconnected, and lonely because of this daily feed—that is to say, this simultaneity has turned them into mere *Jetztgenossen*, mere *simultemporaries*, mere *instant-mates.*

13. *Excursus: A Flashback to a Long Burned-Out Passion; the Distracted Live Exclusively in the "Now"; Our Devices Are Producing Artificial Schizophrenia—the Individual Is Turning into a "Dividual"*

Many decades ago, there were a number of poets (such as Apollinaire and young Werfel) who, to vary an old saying, always "liked to abscond to several weddings at once." Expressed more formally, their attention was, in the metaphysical sense of "*ubique simul*," always scattered and "fleeting," always elsewhere and everywhere but here. In their poems, which often opened with the small vocable "now," they listed everything that was happening right now in Paris, Prague, Cape Town, Shanghai, or anywhere else. It can hardly be denied that it was some real metaphysical agitation that drove them to produce their peculiar, hymnlike catalogs of world-parts. Perhaps these poets misunderstood "*non percipi*" to mean "*non esse*" and thus feared that anything they didn't take note of was inexistent, lost. This much is clear: Condemned to living "right here"—that is to say, in the single, contingent place they happened to be in—they were deeply troubled by the fact that they had to miss out on everything that was happening elsewhere. And so, they hoped to make the dispersed absent "theres" present with a kind of incantation. In their poems, they desperately tried to *gather and collect* all these disparate places and everything happening in them by forcing them *into focus together with the help of the one omnipresent moment "now."* One can almost speak of an attempt at metaphysical magic: By invoking the "now" and its omnipresence, they tried to undo what was to them the unbearable fact that the world consists of spatially discrete, separate (and thus absent) events; they attempted to utilize *the instant as a magic cure for the "principium individuationis" of spatiality.* This desire may have been misguided, but it can still be viewed as a final version of Eleatic passion[51]—the desire to *metaphysically discredit multiplicity.* It's almost tragic how they viewed something most unreal and illusive (the isolated instant "now") as the "true realm of being" in which multiplicity would reveal itself to be nothing but an illusion. And this indicates that they no longer had recourse to true metaphysical principles, not even cheap pantheistic ones, nor to the late auxiliary concept of the "system" that turns "the whole into the true." These poets clearly already belonged to a dying species. But still, how incomparably more alive they feel than today's now-fans,[52] in whom it would be difficult to find even the smallest flicker of our poets' passion for the now.—

It is, of course, no coincidence that these poets appeared at the exact historical moment in which entertainment technologies—that is, technologies of diversion and dispersion [*Zerstreuungstechnik*]—began to attain a mass-dimension (by means of illustrated newspapers and the like). While the poets were desperately trying to force together what was dispersed, these technologies and forms of entertainment had the exact opposite aim of actually creating and facilitating diversion and dispersal of this kind. "*Zerstreuung*," "enjoyment of entertainment" (and one is usually far too "distracted" and "diverted" by this term to understand its reference to "dispersal" and "scattering" as anything other than a metaphor) had but one aim: It sought to strip humans of their individuation. More precisely, it sought to expunge all awareness of this loss by negating their *principium individuationis*—their singular position in space. It made humans assume a position in which they were *ubique simul*, always also somewhere else; a position in which they were no longer occupying a specific point in space, in which they were never with themselves and never concentrating on one thing. In short, a situation in which they were *nowhere*.— One will object that the victims of this entertainment technology [*Zerstreuungstechnik*] weren't victims at all—that by supplying its distracting and entertaining products, this industry was simply meeting a demand. This may not be entirely untrue, but because this demand is itself fabricated it isn't the full truth either.

If humans are pressed daily by their work ever further into the confines of some specialized activity that has little to do with them, thereby exposing them to boredom, one cannot expect that they simply find their way back into their *proportio humana* at the end of the workday. When this boredom and pressure are suddenly released, such creatures cannot be expected to find, or even *want* to find, a way back to being themselves (should such a thing still exist). No. Because the end of this confining pressure feels much more like an explosion, and because those who are suddenly set free from their work no longer know anything but a sense of alienation, they now pounce onto any one of a thousand alien things—if they aren't too exhausted even for that. After the airless lull of boredom, they seize on anything that can set time in motion again and give it a different speed: They pounce on rapidly changing scenes.

There is nothing that can satiate this understandable hunger for omnipresence and rapid change as effectively as radio and television. Their programs can simultaneously accommodate craving and exhaustion, tension and relaxation, speed and inaction, leisure and being patronized—they serve everything up at once. Yes, they even save us the effort of having to throw ourselves at these things, for diversion throws itself at us. In short, such multiplicative temptation cannot possibly be resisted. Small wonder, then, that the curse our poets suffered—the curse of having to dance at two or even a hundred weddings at once—has now morphed into the (seemingly) most innocuous and ordinary state of leisure. This is the state of everyone who is sitting here but present there, of all those who have become so used to being everywhere at once that they no longer really live anywhere, at least not in an identifiable place, let alone a home. They can at best be said to be living in *the now*—in their own, uninhabitable moment in time which changes every instant.

But this doesn't yet fully capture our contemporary's specific mode of "distraction" and "dispersion." The condition culminates in what can only be called *"artificially generated schizophrenia"*—a "schizophrenia" that is no mere side effect of our entertainment devices [*Zerstreuungsgeräte*], but which is in fact produced by design and upon customer request (although it is of course not requested under this name).

What does the term "schizophrenia" signify here?

It refers to a condition in which the self is cut up into two or more partial beings—or at least two or more partial functions. This schism creates entities or functions that not only lack coordination, but which cannot possibly be coordinated. Such coordination is not just impossible; the self also has no wish to be coordinated; in fact, it emphatically resists it.

In his Second Meditation, Descartes declared it impossible "*à concevoir la moitié d'aucune âme.*"[53] Today, such split souls are an everyday phenomenon. There is, in fact, no trait more characteristic of contemporary man, at least not of the idle one, than *his tendency to indulge in two or even multiple disparate activities at once.*

Take, for instance, the sunbather, who lets his back tan while his eyes swim through a glossy magazine, his ears participate in some sporting event, and his jaw chews gum[54]—this *passive actor playing many parts at once* [*Simultanspieler*], this *busy do-nothing,* is a common sight around the globe.

The fact that one takes the sight he represents for granted and accepts it as normal doesn't make it uninteresting; on the contrary, it makes it all the more in need of explanation.

If one were to ask this sunbather what he was "actually" [*eigentlich*] occupied with, where his heart now "actually" was, he could not provide an answer. In fact, the very line of questioning—asking about "what is actual" [*Eigentlichem*]—is already built on a false premise: For it assumes that *he,* the sunbather, is a subject who is occupied with something and who is present somewhere. If one can still speak of a "subject" or "subjects" at all, they would merely be his organs: his eyes (absorbed by their glossy pictures), his ears (at their sports event), his jaws (chewing gum)—in short, his identity is so utterly disorganized that one would be looking for something inexistent if one went looking for "his actual self." He is thus not only *dispersed across many places in the world* (as noted above), but *he is also split into many individual functions.*[55]

The answer to the question of what actually drives this man to his disorganized busyness and what makes his individual functions work so independently from each other (or seemingly autonomously) has in a sense already been provided. But let's repeat it anyway: It is *horror vacui*—the fear of freedom, the fear of having to occupy himself and having to rely on himself. Or to be more precise, because freedom and being at leisure provide him with space, he now fears that he himself must structure the free time and space of freedom that the vacuum of leisure exposes him to.

His work has so firmly in him implanted the habit of *being* employed—of being occupied and not needing to rely on himself—that he is no match for this task the moment work is over, for he no longer has a "self" to draw on. *Today, all forms of leisure have a secret family resemblance to being unemployed.*

If he is left to rely on himself in this situation, he falls apart into his separate functions because he himself has ceased to exist as their organizing principle. But these individual functions are of course as used to *being employed* as he himself is. As a consequence, each one of them snatches at the first available content [*Inhalt*] the moment they find themselves "unemployed"; and the first available content is also the best because it provides a support [*Halt*] for these individual functions to cling to.[56] *One* content, *one* form of support, is never enough—every organ needs its own. If even one organ were to remain unemployed, this would constitute a breach through which nothingness could flood in. It would thus be entirely inadequate for him to *only* listen, to *only* watch.—At any rate, such an "*only*" would demand exclusivity; it would require him to have the capacity to concentrate and abstract. And in the absence of a self, of an organizing center, this power is beyond his reach. This is also why we were always in need of music in silent cinema, and why we would gasp for air whenever the music stopped and the experience continued in a "visual-only" form. In short, to be sealed off against nothingness, *every* organ must be "*occupied*." And to describe this state as one of "being under occupation" is much more accurate than to call it "being occupied with" something.

But because such occupation should not consist of work—for it happens in a time of leisure—only pleasurable consumer goods [*Genußobjekte*] are allowed to occupy the organs. Each organ, each function, thus pursues its own type of consumption and its own desire to consume.

Consumption here, of course, doesn't necessarily need to generate any actual pleasure. It only—and language sadly has no word for this—needs to keep at bay the anxiety or hunger that is felt the moment the *object of consumption is missing*—just as breathing doesn't need to generate any distinct sense of pleasure either (and indeed rarely does), whereas a lack of air immediately leads to a hunger for air and gasping panic.

The key word here is "*hunger.*" The moment an organ is *free,* the moment it is exposed to a vacuum instead of being engaged, it believes itself to be suffering from hunger. *Every momentary nonconsumption is already experienced as distressing;* the chain-smoker is the prime example. In this way, *freedom—horribile dictu—becomes identical to distress* (freedom = free time = idleness = nonconsumption). It is this equation that also generates the demand for the kind of consumer goods that can be consumed *nonstop*—goods that don't pose any danger of ever satiating hunger. I say "danger" because satiation would limit the duration of enjoyment, which would dialectically revert into nonconsumption and thus cause distress: This explains the purpose of never-ending chewing gum and the radio that plays continuously on and on.[57]

This perverse equation of freedom with distress—or, rather, of the loss of freedom with bliss—was of course not just invented today:

The "*Gesamtkunstwerk,*" the "total work of art" of the nineteenth century, already had its hopes set on *horror vacui* and delivered works that sought to take *total* possession of the human, works that assaulted all senses at once. The profound state of intoxication this onslaught resulted in is a historical fact of common knowledge, as is the great joy that the overwhelmed would derive from this total deprivation of freedom. At the time, it was also considered the done thing to pay the highest entry fees for "overwhelming" performances. To understand what I mean, one need only listen carefully to the original meaning of this common expression (a meaning barely registered today). Nietzsche was the first (and to this day almost only one) to sense how dubious the state of being "overwhelmed" is, and he sensed this with his own body and then also put what he experienced into words. But compared to today, of course, the "overwhelming" art consecrated in Bayreuth was still quite humane. The "total work of art" was at least still premised on an archaic idea of the human that was deemed worthy of respect. That is to say, even when being overwhelmed and having their various senses assaulted all at once, *humans remained one single being.* They could still claim the right to receive one unified work and could thus still be *one unified being. Even in the midst of the onslaught, they were still deemed worthy of suffering one single homogeneous defeat.*

Even this remnant of the human has now been surrendered. Today, the modest principle of pure addition is deemed perfectly sufficient: *The simultaneous delivery of divergent elements has become the norm.* These don't simply diverge in their content but also in their respective sentiments and moods, and not just in their moods, but also in terms of their intellectual levels. *Nobody bats an eyelid if the jungle girl in the cartoon they "experience" while eating breakfast has a knife stuck into her oversexualized ribs while, at the same time, the Moonlight Sonata's gentle tercets gently trickle into their ear.* And nobody has any difficulty taking both in at once.— Until very recently, scholarly psychology has denied that the simultaneous consumption of two completely divergent contents and moods was even possible. The fact that this very situation can now be witnessed a million times over at any given moment makes it seem at least likely that it is possible after all.

To this day cultural criticism treats the destruction of the human exclusively as a consequence of standardization; that is to say, it equates it to the fact that individuals—through their transformation into serialized beings—are only left with individuality in a numerical sense. But even this numerical individuality has now been gambled away because the numerical remnant is being "*divided*" further. *The individual has been transformed into a "dividual"* and cut up into multiple functions. It seems that the destruction of the human cannot go further, and that the human cannot become more inhuman. All this makes the "Renaissance of holistic perspectives," which today's psychology celebrates with such pathos and poise, even more fanciful and hypocritical. For it is indeed nothing but a maneuver to hide the shards of the shattered human beneath the academic toga of theory.

14. Everything Real Is Becoming Phantomlike, Everything Fictitious Is Becoming Real. Deceived Grandmothers Crochet for Phantoms while Being Educated in Idolatry

After this long (but not unnecessary) excursus on the "dividedness" of humans who are being distracted and entertained, we will return to our narrower theme: the danger radio and television present to the human.

We had already established that radio and television "send" something home to us that is ontologically so ambiguous that we were unable to decide whether to address it as present or absent, as something real or as an image. Because of this undecidability, we decided to give this ambiguous entity its own name and called it a *"phantom."*

At this point, however, a heckler intervened to challenge the notion of ontological ambiguity that our thesis hinges on. He claimed that it isn't necessary to examine the meaning of the "presence or absence" of broadcasts, as these are but *"aesthetic appearances"* and are hence met with an aesthetic attitude; he then added that this topic has already been exhaustively studied by aesthetic theory.

Yet this line of argument only puts new wine into old bottles. These old categories are no longer up to scratch. No viewer who impartially observes their own attitude while watching television would ever arrive at the conclusion that they are consuming an "aesthetic appearance"—this idea *couldn't possibly arise.* What makes broadcasts so unique and so uniquely unsettling is precisely the fact that they couldn't care less about the distinction between "being and semblance." *It is as true to say that broadcast events become phantomlike, as it is false to assume that this quality also makes them take on the "as-if" character of art.* The frame of mind with which we bear witness to a political tribunal via a broadcast is categorically different from the one we assume in the theater when watching the tribunal scene of Büchner's *Danton's Death.* What makes this attitude so difficult to describe unambiguously isn't just that our theoretical concepts are lagging behind—that they haven't yet caught up with this new reality (and they really haven't)—but also that our broadcasts actually aim to produce ambiguous attitudes of this kind. They seek to create *serious unseriousness* and *unserious seriousness.* They are meant to induce a state of suspense and oscillation in which the distinction between the consequential and inconsequential has become irrelevant; a state in which listeners can no longer say—or even ask—how a broadcast program ought to affect them (as reality or as an appearance, as information or as "fun");[58] a state in which they no longer know or ask "whom" they should receive the delivery as (as a moral and political being, or as a consumer who is being entertained).

This ambiguity of seriousness and fun is especially pronounced in radio and television dramas—in formats, therefore, that still try to hold on to notions of "appearance" stemming from the theater. Because these dramas are aired via the very technology that also makes real events seem phantomlike, these allegedly "fictional" events also dialectically switch into something that appears to be real. When life is taken to be but a dream, dreams also take on the quality of lived experience, and these broadcasts now mirror this very logic. Because every reality now appears in the form of a phantom, every phantom now also seems real. If, when broadcast, every real event also takes on the semblance of an appearance, any broadcast that *seems like an event* (e.g., a fictional dramatic scene) must now necessarily also *lose the traits that reveal it to be an aesthetic appearance.* And it really has become impossible to feel that broadcast events only *seem* real; or, at least, this can only be felt so faintly that we now believe the fictional events we witness to be real—a reality we actually believe ourselves to be part of. We might even believe that we are its actual victims. I am here especially thinking of *The War of the Worlds,* Orson Welles's 1938 radio play about the invasion of planet Earth. By means of a crude adaptation of the *Hamlet*-principle of conducting "a play within a play," radio on this occasion simply aired the play in the style of a news report. This resulted in a broadcast that was in effect indistinguishable from actual news coverage (apparently, its artistic achievement had consisted in the creation of this veneer of realism). It's anyone's guess whether one was supposed to be able to tell it apart from real coverage, or what was more pronounced: the stupidity of its creators or their lack of scruples. At any rate, interrupting the action with explanatory announcements alerting listeners to the fictitious nature of events wouldn't have helped. Upon hearing the news "the Martians are here!" no listener who already believed such an invasion to be possible would have continued to calmly sit in front of the radio until the next announcement was aired. One thing is clear: This illusory illusion reached us partially as a real event and partially as a real news report about it, causing real panic. This was the first "soloistic mass-panic" ever to erupt: Those who felt it were gripped by panic within the space of their own four walls, without having any contact with their neighbors. This panic had about as much to do with an "aesthetic attitude" as the screams of terror caused by an inferno have to do with the jubilations brought on by a firework display.—

Yet this already "classic" event in the history of broadcasting is not an isolated case. What is true for it also applies to all other forms of radio drama, or at least to every nonstylized play set in the present—even to those with seemingly harmless content. Such plays likewise muddy the boundaries between being and semblance, participation and pseudoparticipation, so profoundly that they cheat listeners out of the option of taking things seriously. To be clear, in this case, *the frivolous element is not rooted in the fact* that serious issues are served up as light entertainment and consumed as such, *but in the fact that such content is being offered and received in a format that is far too serious.* The creation of this kind of credibility is apparently the whole point. I'm speaking about those categorically nongory, often tearfully sentimental serials in which the staged daily lives of fictional families unfold over many years. Such shows are anything but inconsequential light entertainment. I know a number of lonely old ladies in the United States whose social circle—that is to say, whose "world"—exclusively consists of such inexistent beings. They care for the well-being of their phantom-family so much that they suffer sleepless nights whenever one of these family members dies, let alone gets engaged. Their social interactions are with phantoms, and this phantom-family gives their lives meaning. Without them, they wouldn't have anyone, and their life wouldn't be worth living. When it starts to get cold, they knit gloves for their phantoms, and when a phantom-baby is expected, parcels filled with baby clothes—little crochet jackets and bonnets—begin to pile up in the broadcasting houses. These items are then passed on to children's homes behind the backs of the deceived donors—to babies who remain totally unknown to them but are at least real.

"How is Walt?"[59] one of these poor ladies was asked in 1943.

"He's a prisoner of war in Germany," she replied, without hesitation.

This was puzzling to the person who had asked the question. "In Germany? I thought he was in the Pacific."

"Oh, *that* Walt! Why didn't you say so! I thought you were asking about Walt."

"Walt" was a character in the soap opera *Portia Faces Life,* a radio series known across the nation, which practically turned Walt into a relative of every radio listener.

Many people may just believe these old ladies to be odd or pathetic—yet to me, they seem spooky and ghostlike; they seem like the three Fates from ancient myth who weave the thread of life, only that they are crocheting our phantom-world. Earlier, we used the term "unilateral" to describe the paradoxical situation in which humans can apparently experience the world without being actually able to address it, whereas, in turn, the broadcast world doesn't actually take note of humans but addresses them continually anyway. These ladies, these Fates, embody the absurdity of this situation in a most shocking way: Although they in some sense clearly seem unable to comprehend the one-sided nature of their relationship (or else they would not actually crochet), this one-sidedness also seems to have been accepted as self-explanatory and obvious. I have not once heard any of these ladies complain that their phantom-family takes no notice of them or treats them as if they didn't exist. Not once have I heard them bemoan the fact that no actual contact is possible, or that they need to be definitively satisfied with playing the part of an eavesdropper with their ears to the wall. What makes this situation deplorable, scandalous even, is the fact that this make-believe family [*Attrappenfamilie*] can actually replace the real thing: for it is able to excite, occupy, and satisfy these ladies' starved motherly and grandmotherly feelings, feelings that would be a regular part of family life. And because this fake family is nothing but an "image," it pays no heed to those who give it their love and thus mocks the real feelings it receives (feelings it mass-produces for solitary consumption).

"What's wrong with that?" I hear the objection. "Why should one begrudge these old ladies such pleasant feelings? Isn't feeling something good? Aren't their emotions positive? So are their feelings really just fake mock-ups or phantoms too?"— Old-fashioned love of the truth, a love that cannot be justified, leaves only one possible response: People who inhabit *feelings* that pour out into a void and connect to nothing real have been even more categorically and shamefully cheated than people who only inhabit *false beliefs*. They are being deceived regardless of how real, pleasant, or good these feelings are. Lies do not become more defendable when they are genuinely taken for truth by those who are being deceived. Quite the opposite. Their sole aim is to evoke such false sincerity, and they triumph whenever they succeed.— *Because their subjectivity and their world have been definitively torn apart, these ladies, these phantom-devotees, have in fact been cheated out of their own humanity.* It's difficult to decide what makes this situation more scandalous: the fact that feelings mass-produced by machines are being utilized to instill one identical love for one identical grandchild in millions of women, or that in the end, rather than loving "their own" grandchildren—children who do not exist—these women are all coerced into becoming sensitive and sentimental by being made to love their own love for grandchildren.

The evil game being played with the human dignity of feeling here is depressing; the transformation of humans of all age groups into solo-istic empaths or eavesdroppers and voyeurs is disgusting; what makes this all completely disheartening, however, is that critiquing such phenomena is considered a sign of resentment.

For many thousands of years, false idols were able to create, demand, and thus abuse real feelings—feelings of respect and humility. But this seemed to have become a thing of the past until the graven images of gods were replaced with mock-up humans. Today, the little crocheted jackets for nonexistent children accumulating in the broadcasting houses don't differ much from the idolatrous gifts that once accumulated on the steps of false altars. The emotional abuse practiced today is no less acute than it was in the past. And I don't know why our indignation about this should today be any less intense and less justified than that of past eras.[60]

15. *Today's Ghost Stories: Phantom-World and World Collide—a Phantom Is Being Extorted*

But the cheated grandmothers, who aren't really part of this world anymore or only cling to it because they savor the prospect of the phantom-feelings it offers, are a rare and far too clear-cut case. Only in exceptional cases can phantoms supplant their competitors and fully outmaneuver reality, thereby securing a monopolistic hold on the consumer's emotions. Under normal conditions, something else happens—a hybrid case: The creatures of the competing worlds clash, collide, begin to compete, and blend into one. To be clear, these creatures originate from two worlds that are ontologically different and not from two different home planets, as is the case in Science Fiction (when compared to the fantastical nature of today's reality, Science Fiction is, in fact, quite unimaginative). What we are usually dealing with, then, are *ghost stories*. The word "ghost" shouldn't be understood in a figurative sense; it belongs to the essence—no, the unsettling nonessence [*Unwesen*]—of ghosts, that they float across the threshold of their own world into ours and, as they leave the company of their own kind behind, get into conflict with what is real. This is what is happening today; ghost fights of this kind are happening at every given moment in everyone's world. They usually only go unnoticed because these fights already belong to the fabric of our everyday lives (like the struggle between spirit and flesh). Many of the creatures who now make up the real world have, in fact, already completely lost the fight against the phantoms. They now look exactly like them and, as such, the difference between the two opposing parties has already become unrecognizable. There are, for instance, countless girls[61] who have made themselves look like film images and now walk around as *reproductions of reproductions.*— They do this because they wouldn't be able to compete with the sex appeal of the phantoms if they were content to just look like themselves, and it hardly needs to be demonstrated that this would put them at a disadvantage (a very real and most unphantomlike disadvantage) because they would have to contend with it in their real, hard lives.—

An especially blatant example of a clash between a phantom and reality—that is to say, between a TV-phantom and a citizen of London—recently made headlines in the press:

There was—or, rather, there is—a woman in London, an ordinary middle-class housewife, who was so fascinated by an Apollonian TV-star that she never missed an opportunity to see him in her home. No sale could tempt her out; none of her husband's threats could intimidate her. Every morning at the designated hour, her squalid kitchen would transform into a place where, for a heavenly quarter of an hour, she could freely receive her visitor; and although this was for a lover who was only an effigy, she would use her best soap and dress up in her Sunday best—for her, the affair was most real.

Had she been asked, she probably wouldn't have denied that she was sharing her lover with a hundred thousand other women. But as she only ever saw him in private—that is, by means of "soloistic mass-consumption"—she had only a very rudimentary sense of this shared ownership (which would have been impossible to miss in the theater or cinema). In short, she "had" something with him. The situation was all the more flattering for her because *he* had started things: It was *he* who had approached her; it was *he* who was visiting *her* and talking to her every day. Yet she also couldn't have denied that her affair had remained entirely voyeuristic and none of her proclamations of love had ever been acknowledged. So, one can see, things have already become quite complicated and spooky. But then there was also her real husband—a small, beleaguered employee in the local gasworks, a man with whom she had not exactly lived enthusiastically, but not particularly badly either. This husband, of course, had no chance of competing with a lover of such charm and gentlemanly wit—a lover with a permanently sunny disposition who endlessly found new, ingenious ways to flirt. And before she realized what was happening, her real husband had begun to annoy her, and annoyance soon turned into steady hate.— Not least because her husband had a tendency—and this was clearly a sign of malice—to come home hungry from work and call for food just as her lover was dropping in for their afternoon rendezvous. (Her lover, on account of being a phantom, had the unrivaled virtue of never demanding food and never shouting at her.) The real husband and the phantom thus came face to face. The *collision* had happened, even though this collision was itself phantomesque, or at least half-phantomesque: For while the real man grated his teeth, the phantom continued to whistle without the slightest care in the world and without taking any notice of him. While the real man had to watch his wife cling to each and every one of his rival's words, the phantom didn't have to do anything at all. While the man was rendered defenseless because the other man was a phantom, the phantom remained composed and in control for precisely the same reason. And thus the stage was set for a *clownesque* scene: The husband would switch the hated rival off, while the wife would immediately switch him back on, and not just once, but time and time again.— This back-and-forth, however, was only the oft-repeated, playful overture that would soon build into a furioso. The husband, as one would expect, of course felt compelled to "teach his rival a lesson," yet he was also unable to give in to this urge. The television set was his, after all, and he didn't just own it; it was his best possession—his pride and joy, his prestige piece—and not even half paid off. The TV also provided him with his only occupation, his sole source of solace after work. Had he given in to his rage, he would have harmed himself. And since nothing makes people more vicious than the quietly raging struggle between the desire to destroy and the desire to possess—because nothing ferments wild fury as much as a rage that is suppressed—or, in short, because he needed to hit something, preferably something less valuable and more durable than his device—he began to beat his wife. But since his wife, gazing toward her lover with a martyr's look, suffered these blows without saying a word (the lover, all the while, softly whistled on and took no notice of anything), his blows remained futile too. The testimonies in court later revealed that she was able to suffer all this because the wifebeater had apparently never fully forgotten that her durability wasn't unlimited either; that she, too, had some value that ought not to be underestimated. And this, apparently, had made him limit the force of his blows. Needless to say, he failed to put an end to his wife's visits from the phantom and failed all the more to beat her old love for him back into her.—

It seems likely that in his futile rage, her husband would have much preferred a rival made of flesh and blood—a competitor from the real-and-decent world, even one who had actually seduced his wife, a suitor he could have then really thrown down the stairs. But he had to contend with this incorporeal one instead—a rival not prohibited from entering the home, but who had spoiled his home for him. A rival who didn't eat but still came between him and his dinner; a lover who didn't love, but who had still destroyed his marriage; a rival who couldn't see but had transformed his wife into a creature whose eyes were hungry with desire. And so, the desperate husband finally saw no other option than to send a blackmail letter to the cursed phantom with an ultimatum: "Get out or . . ."[62] Because the alternative listed after the "or" culminated in a death threat—and because the postman, being unfamiliar with the subtle distinction between a phantom and a real human, had delivered the letter to the actor X—the matter ended up in court. And although the actor had never heard about the existence of the woman with whom he was having an affair, he genuinely feared for his real, unphantom-like life. The court case that followed was widely covered in the English press. The verdict is pending.[63]

16. *On Account of Its Small Format, the TV-Screen Transforms Every Event into a Synchronous Tableau of Trinkets*

Television productions, or so we just noted, aim to create a sense of "unserious seriousness" or "serious unseriousness." Only by secretly getting consumers accustomed to this ambiguous and oscillating state can these consumers also be counted on as mass-humans—that is, as beings who have become unable to reach decisions. The ambiguous nature of being and semblance, which in itself perhaps might only be a contingent phenomenological quirk of broadcasts, is thus put to use as a morally opportune effect.

Orson Welles's radio play and the little jackets the Fates crochet for phantoms showed us how fictions are transformed into something appalling or half serious; the extorted phantom illustrated how something half serious comes into conflict with reality, with potentially real and serious consequences. We'll now show how real things are transformed into something inconsequential and harmless—how they are *"cozified."* We are thus briefly returning to a phenomenon we are already familiar with. Yet, rather than offering general observations, we now seek to expose a technological cozification trick that has to do with *the small size* of the images that appear on the screen.

One will now, of course, say that this small size is not some technological trick but rather a defect—a very temporary one at that—that will certainly soon be fixed. That it could be fixed is certain; yet whether one wants to fix it and will do so is an entirely different question.[64] For although it was not actually created by design, the tininess of the screen has turned out to be a *highly welcome defect,* which has assumed a very special role: *It presents the macrocosmos as a microcosm, thereby transforming every world event into a miniature diorama—a cute little Nippes scene.*[65] I say "Nippes" because the small size of the screen now fulfills the same function as those small little Nippes figurines—the little porcelain Napoleon heads, for instance—that used to be displayed on the mantelpieces in our great-grandfathers' houses; figurines that blotted out and distorted the catastrophic fate of the *Grande Armée* much more successfully than even the thickest volumes of history. But things are much more expedient today: If someone is to be persuaded that they are living a harmless existence in a harmless world, there is no need to wait until events have taken their course for a harmless version of events is now supplied even as these events unfold. One is supplied with *"real-time Nippes,"* so to speak (at times, this supply is especially "forthcoming," and a reassuringly miniature version of events is provided prophylactically even before anything has occurred). As we sit in front of the tiny screen, new eyes are implanted into our head—eyes that are like opera glasses held the wrong way round, enabling us to see any worldly scene as harmless and humane. Or to be more precise (since most gifts today are concealed impediments), these eyes we are given prevent us from seeing any other way. They stop us from recognizing that the world, the events, the decisions, and the malice we are made to witness and made victims of are actually unsurveyable and unforeseeable. What we're supplied with is thus a false *overview*—an overview that isn't false because its perspective makes us "overlook"—that is, "miss"—this or that detail, but because it convinces us that we have "oversight" over the world's unforeseeable and incalculable nature (in the sense of "mentally mastering" it). Even if the screen were able to achieve by optical means what philosophical systems have tried to accomplish—to present us with the world "as a whole"—the "whole" that is displayed would not present us with "the true" in Hegel's sense.[66] In fact, it is precisely *by* presenting us with the whole—precisely by providing us with a model that can be grasped—that such an image would suppress the size of the world and the unforeseeable nature of our acts. Television screens, of course, are not the only devices that manipulate proportions in this way: Maps seem guilty of the same thing. But maps present themselves truthfully, clearly signaling that they are scaled-down overview models, while the scenes appearing on television alongside unfolding events present themselves as these events themselves.—

Although sensationalism certainly characterizes our age, a closely allied and no less dangerous force has now come into sight: antisensationalism. Whereas sensationalism falsely makes something out of nothing and turns flies into elephants, antisensationalism deceitfully turns elephants into flies. This process is something today's cultural criticism is paying far too little attention to. When sitting in front of the screen, it really is difficult to resist the utter phantomization of the world that the trick of scaling things down achieves. And even for those who are fully aware of what is happening, it takes great effort to escape its effect. Anyone who has had the dubious pleasure of experiencing one of those toy-car-like races that happen on TV will also be familiar with the incredulous realization of how even fatal accidents don't seem all that bad: In such moments, we may very well know that we have just coexperienced something that actually happened the very moment we saw it on screen, yet we at the same time *only* know this. Although this is something we know, we only know it—the knowledge remains utterly lifeless. Any attempt to make the tiny image congruent with what is happening someplace back there is destined to fail—and the same is true for comprehending that what is happening *here is part of the very moment that is also happening there.* And so, the shock we feel also stays small and imaginary—much smaller even than the kind of shock that is induced by the fictional catastrophes that happen only on the theater stage.

But such congruence is also *not meant* to be established. The projected image is actually meant to rob us of our ability to ever conceive of such a reality, not to mention our ability to contemplate that there is a reality "besides" the things we are supplied with. And this is precisely what it achieves. Because we are constantly supplied with images—an entire worldview even—reality is intentionally covered up with the help of something that is presented as reality (as we stated in the opening paragraph). The world is made to disappear beneath its own image.

We are unable to imagine a nuclear explosion; this much is certain. But it's just as certain that this failing imagination and our despair at this failure bring us much closer to the terrible limitlessness of this event and that they are much more appropriate than the eyewitness view the TV image seems to provide. Because this image is easy to take in and behold in its full entirety, it falsifies the event's unforeseeable and incalculable scale: It betrays us by even putting us into the picture.

III. The Notification

"As what?" we asked at the beginning of the last section—"as what kind of entity do the events that are delivered into our home reach us?" The ambiguous response was: "as phantoms."— This term was meant to indicate that events arrive neither as themselves nor merely in the form of images, but as entities of a third kind.

But is this really astonishing? Is this not an entirely mundane occurrence that was made to appear strange with the help of a strange expression? Does something similar not happen all the time? With every *notification,* every *message* we receive?[67]

What does this mean?

Let's assume our coal cellar is empty and we are notified about this. *What* is being communicated here? What does the message "bring home to us"? The object itself? The empty coal cellar?

An image of the depleted cellar?

Neither one nor the other. What we receive is an "object" *sui generis,* an object "of a third kind" that lies peculiarly outside of this alternative, for it consists of the *fact* that the coal box is empty. It is a phenomenological given that this fact is not identical with the coal cellar itself, for unlike the cellar, the fact is not empty. But it's just as evident that the fact conveyed by the message isn't exclusively an image either.

So the message also supplies us with something that is neither the thing itself nor its image. Does this structural similarity not suggest that broadcast programs are but notifications, messages?

To answer this question, we need to examine the *nature of messages* as such. This digression is all the more necessary because our argument might have created the impression that we're calling for immediacy and an absence of mediation [*Unmittelbarkeit*] to the exclusion of everything else (which wouldn't be permitted).

17. A Pragmatic Theory of Judgment: The Recipient of a Message Is Free Because the Message Places Something Absent at Their Disposal; They Are Unfree Because They Don't Receive the Thing Itself, but Only Its Predicate

What, then, is a message? What does it perform?

It informs its recipient about *something that is absent*. It enables perception by proxy and ensures that its recipient knows about this absent thing indirectly, without having any firsthand experience of it.— The phrase "*something absent*" already signals that we haven't left our field of inquiry, which also concerns the ambiguity of presence and absence.— And our definition of the message calls for further clarification.

To speak means to speak of something absent. It means to introduce something not present here to someone not present there.

Even the most direct, noninformative mode of speech—that is, the imperative—maintains a relation to presence and absence. The imperative addresses someone who is averted, inattentive—that is, someone absent—and asks them to listen and participate—that is, to be present. While it summons its addressee forth out of absence, the communicated message brings *what it speaks of* from absence into presence. Speech would, in fact, be no more than gossip and not exist at all if the thing being spoken about were not absent. Every speech act [*Rede*] in a sense happens behind the back of the "third," the categorically absent "person" or thing that is spoken about; every speech act pursues the aim of making what is absent present.— Language has, of course, inherited this relation to absence from pointing: *dico—δείκνυμι.*[68] A person who is pointing only ever points at something present because it is absent (absent from the attention or field of vision of the person being alerted to it). One only points to bring the other person into the presence of the thing being pointed out—to give them an opportunity to experience it directly or actually grasp it.

The message recipient doesn't seem to be granted this opportunity: A message neither brings the recipient to the object nor the object to the recipient. But is this really the case?

True, the message doesn't deliver the actual thing itself, but only something *belonging to* it—only something *about* it, a new and peculiar kind of object. And because this new object is "made out of" the old object, it is no coincidence that this thing is called "*factum*"—that is, a fact.[69] What makes this object "peculiar," therefore, is that it is categorically mobile and communicable, whereas the object it was derived from is not. Yet, despite this difference, its recipient, the addressee of this new object—this "*factum*"—now receives the old object as well. To be precise, the new object ensures that the addressee *has* something of the original object—in fact, the addressee receives an extraordinary amount.

By communicating the "factum," the message enables the addressee to behave as if the object were present—it can now be taken into account and be made part of one's plans and calculations. *The message exists to create the opportunity to adjust oneself to it.*

Pragmatically speaking, it really does make the object "present"—or, rather, it brings it into the presence of the addressee. The addressee now knows "about" [*über*] the matter it conveys. And this small word "*über*" is no mere caprice of language. It indicates that messages really do put their recipients "on top of" [*über*] things; it points to the power the addressee now has *over* [*über*] the situation and matter conveyed—a situation that has changed because of what the message communicates. Having received the message "the coal cellar is empty," I can now preside over the purchase of new coal.— In other words, when the message supplies me with something that is detached from its subject (the coal cellar)—something "about" the subject instead of the thing itself—I am not just supplied with some defective substitute, but I learn what is "going on" with it. I thus receive that aspect of the object that actually or allegedly concerns me—the thing I have a reason to follow up or adjust myself to.[70] The thing that concerns me has thus already been made explicit; it has already been processed and prearranged; and it is sent to me in this processed form. In the language of logic—which has discussed this amazing feat countless times but admired it on far too few occasions—the thing detached from the object that communicates "the state of affairs" is called "*the predicate*."[71] The *predicate* is thus a *processed, ready-made product* [*Fertigware*] created for its recipient. Because the message delivers a ready-made product—a "fact" split off from the original object—it necessarily requires a *separation* [*Teilung*]: The act of executing this split is called "*making a judgment*" [*Urteilen*]. That's why the message breaks down into two parts: *S* and *p*. Instead of the one object "coal cellar," the addressee learns a fact composed of two parts: "*The coal cellar is empty.*"— And the message doesn't consist of two parts on account of being a judgment; the judgment consists of two parts because it is a message.—

In other words, the predicate, which is usually discussed only in the context of formal logic, is of much wider significance. As our discussion of the "*über*" has already sought to highlight, the predicate indicates *freedom* (of use): Anyone who has power over what is absent via a predicate, anyone who can take absent things into account and adjust themselves accordingly, has increased their own sphere of presence and power. This person has been liberated from their own contingent position in space and is now here *and* there. When one learns what is relevant ("the state of affairs") via an already separate, isolated, preprocessed, predetermined item—through a ready-made outcome of *λέγειν*[72]—one is freed up: One doesn't need to deal with the weight of the irrelevant that every object of perception drags along with itself; one is *freed* from having to invest one's own labor and cognitive effort.

Yet at the same time—and this second point is decisive for us—every message also represents a *deprivation of freedom*. And surprisingly, it curtails our freedom on account of the very thing that turns it into a vehicle of freedom: once again, therefore, because it doesn't supply what is absent, but only something *about* it, something *belonging to* it. This time, however, we place a different emphasis: We wish to foreground that the message *only* provides us with *a part* of the absent object. It *only* supplies us with the "predicate"—the processed part, the part that was originally split off [*Ur-Teil*], the part on account of which the judgment [*Urteil*] is called a judgment. This is the only thing the message supplies. It aligns us with its preselection even before we can form any judgment of our own. This also means we've been worked on and subjected to preparation, for the message commits us to its selection *a limine*—that is, without a case ever being heard. When we receive a message, the predicate is not subsumed by its subject; instead, the subject consists exclusively of the predicate—that is, it consists of the part that was detached from it. Because every message "only delivers a part," it is also a *prejudgment*, a *preconception* [*Vor-Urteil*] that can be true but also untrue. Every predicate is in itself a *prejudice;* and the content of the message, regardless of what it is, inevitably also *withholds* the subject from its addressee, as this is left obscured by the predicate—the only thing that is delivered.

Because the recipient is forced into assuming a certain perspective (the perspective opened by the predicate) and because the thing the judgment apparently contains is also withheld from them, the recipient is made *dependent* and *is deprived of autonomy.*

"Take it or leave it,"[73] or so the message seems to say. "You either accept the part of the absent matter I deliver in its already detached, prejudged, and ready-made state, or you receive nothing at all." The messenger is thus the master of the master.

Usually, the difference between immediate and mediated experience is absolutely clear, and the two are very easy to tell apart. In fact, it's scarcely possible to mistake one for the other or to doubt to which type one's current experience belongs. Our immediate experience—that is, our perception—takes in prepredicated images, whereas the type of mediated experience facilitated by the message breaks down into the formula "*S* is *p*." And this is true even for the most dedicated bookworm or newspaper addict who lives within the horizon of mediated experiences and sustains themself with them. This, of course, doesn't mean that we may at times not retroactively struggle to identify whether this or that experience that has sunk in and settled into knowledge was the product of direct or indirect experience; but when reading a book or a newspaper, we would only very seldom arrive at the idea that we are experiencing the content directly in an unmediated form (or, vice versa, mistake ourselves for reading when we are in fact directly experiencing something).

And with this, we have arrived at the point we have been getting at.

18. Broadcasts Erase the Distinction between the Thing and the Message. They Are Judgments in Disguise

The ontological ambiguity that is a defining trait of radio and television broadcasts stems from the way they affect their recipients: They induce a state in which the difference between experiencing something and being notified, between immediacy and mediation, has been erased—a state in which it is unclear whether they are faced with the thing itself or with a fact (i.e., *factum*). What does this mean?

As we have seen, facts, in contrast to most other objects, have the quality of being mobile: Although a messenger cannot transport a burning house, they can speedily convey the fact that it is burning and pass this on to a recipient. In broadcast programs, however, the things themselves (or at least their phantoms) are transported—what reaches me is the symphony and not the fact that it is being played; the speaker, and not the fact that they are delivering a speech. *The trait of transportability once specific to facts now seems to have infected the things themselves.* So has this mobility not also turned these things into facts?

This sounds like a strange question—for facts (or at least the messages transporting them) are judgments, and as such they are split into the components *S* and *p*. Yet broadcast objects [*gesendete Gebilde*] don't seem to follow this pattern: Am I not hearing the speaker "himself" rather than hearing something "about" him? Or is this not correct?

Let's assume candidate Smith appears on television to introduce himself to the electorate. It's self-evident that this Smith will seek to show what a pleasing personality[74] he has—that he has a duty to smile at us as charmingly as possible. But with this simple observation, his on-screen demeanor hasn't yet been adequately described. He will, in fact, make his charm leap out at us as his only attribute; he will seek to make us forget that he also consists of other things than just his smile. Although it may seem as if candidate Smith (let's call him *S*) were presenting his whole self, what actually appears, is exclusively the fact or pretense that he is a "pleasing personality"[75] (let's call this *p*). What appears on screen is thus only the equation "*S* equals *p*"; or, rather, *p* appears in place of *S*.— On television (and here we can insert the formula we arrived at while establishing that messages are a type of judgment), we thus get to see "a subject that exclusively consists of a predicate."— And we may even be right to only see the *p*, for the *quid pro quo* between subject and predicate often becomes a reality. In other words, it often happens that *S* ends up turning into his own predicate and becomes unable to be anything else. Condemned to this existence as *p*, he then walks around as a professional smiler. The fabricated story often ends up in a lie that is lied true.

The candidate's performance thus achieves the same thing as the message; no, it actually accomplishes even more. It's a message that manages to *cover up* that it is actually a judgment. In terms of efficacy, this is a substantial bonus, for this now also conceals the effects that, as we have just shown, ensue from any judgment: its prejudicial character and the fact that it depletes freedom. *To persuade consumers that they aren't being persuaded of anything, the judgment gives up the form of a judgment as it transforms into an image*—it appears to turn into the *S* it is about, the *S* doing things right there on screen. Its liveliness doesn't disclose the splitting of *S* and *p*, or at least reveals it much less obtrusively than any regular judgment.

Although an everyday occurrence, this process is philosophically highly peculiar. It represents an inversion of the normal sequence. While the message normally takes its cue from the *factum* that it communicates and thus follows on from, the *factum* now takes its cue from the judgment. The desired proposition "Senator Smith is a pleasing personality" is given precedence, and *S* must follow suit—or rather, *S*'s image, which acts as if it were the man himself (and thus something that isn't yet a judgment). In reality, however, this man (S) is nothing but his *p*, but a *p* dressed up to the point that the characteristic structure of a judgment no longer shines through. As it transforms into an image, the judgment thus projects (i.e., "pretends") that it is not projecting (i.e., "prearranging," "prejudicing," and "predicating") anything. And because of this, the expression "to disguise" isn't entirely accurate here, for the judgment, in fact, performs *the negative form of "dressing up"*: It *drapes itself* in apparent nudity, adorning itself with *the sash of missing predicates.—*

19. *Commodities Are Disguised Judgments. Phantoms Are Commodities. Phantoms Are Judgments in Disguise*

Our example will now be deemed entirely unrepresentative. Not every phantom, one will object, is an exhibition of a *p;* not all phantoms are *adverts* (for this is the category our example belongs to) and, as a consequence, they aren't all judgments—that is, prejudices—either. I admit, not every phantom is as blatantly an advert as our candidate Smith (who was specifically made up to illustrate this trait). The fact remains, however, that on account of being home-delivered, all phantoms are *commodities*—and this is decisive, for *this makes them judgments.*

This again sounds strange. What do judgments—belonging to the realm of logic—have in common with commodities, which belong to the realm of economics?

The answer is: the *predicate.*

Every commodity—insofar as it is on display and offers itself for sale (and to be a commodity it must be for sale)—*is also its own specific form of self-evaluation: It praises itself.* Its very appearance is a form of self-recommendation: It's in the shop window as a visible preappraisal [*Vorurteil*] of its own quality. Of course, just as is the case for our candidate Smith, it doesn't necessarily break itself down into the axiom "*S* equals *p*." When it is on display, the commodity doesn't necessarily state its own quality explicitly (although accompanying advertising slogans often do). But this much is certain: *Its display is carefully arranged.* And arrangement means that, to function as a lure, its "*p*"—its state of affairs (the thing the commodity has "going for" it), its true or projected quality—must be detached and accentuated to such an extent that only this quality is visible, not the commodity as a whole. What viewers [*Betrachter*] are offered is thus primarily the position from which they should "consider buying" [*in Betracht ziehen*] the commodity—a position that is already supplied and predetermined before the commodity itself is ever supplied.

That commodities have the character traits of judgments thus cannot be denied. In the previous section, the negative effect of messages was equated to a curtailment of freedom. We noted that messages assign their recipient the perspective from which the absent thing is to be viewed—an absent thing that is qualified through a predicate—and we observed that messages thereby supply a ready-made position. Our discussion of messages also applies to commodities. In place of the recipient, we now meet the customer, yet the parallel isn't affected by this distinction. Still "absent" and still separated from the customer by the shop window, the displayed *p* is now meant to lure customers out of their absence and into the shop by transforming them into buyers.

At the outset of this study, we had already established that events become commodities when they are turned into phantoms and are home delivered. What is true of all commodities—they are judgments, albeit judgments in disguise—is also true of broadcast events.[76] Although they "drape themselves in nudity and wear the sash of missing predicates" to present themselves as actual events, they, too, are assertions [*Aussagen*] *about* events. Because no judgment is as unsuspicious, inconspicuous, and seductive as the one that purports to be the thing itself (and not a judgment at all), it gains its power to deceive from its renunciation of a tangible "*S* equals *p*" schematic. As we sit in front of the television or radio, we consume an arranged montage instead of the scene it presents; we consume the predicate (*p*) instead of the thing (*S*) we appear to see. In short, we consume a prejudice that appears in the guise of an image. And like every other prejudice, it hides the fact that it is a judgment. And because it secretly remains a judgment, it also keeps consumers from investing the effort to form their own judgments. And as is the case with other prefabricated goods—with preserved fruit, for instance, which saves customers from having to boil down fruit themselves—the idea of doing so now really does not occur.— What is true of the message (it makes us unfree because it either shows us absent things in a prejudiced and prearranged manner, akin to prefabricated goods, or else doesn't show them at all) is even more true of broadcast programs: They relieve us of reaching our own judgments, all the more so because we cannot defend ourselves against accepting the supplied judgment as reality itself.

IV. The Matrix

20. *The Whole Is Less True Than the Combined Truths of Its Parts—the Realistic Disguise of Templates Aims to Create Templates for Experience*

Ultimately, it is not the individual broadcast that is prepped for sale. Under certain circumstances, broadcasts are even left unedited and are factually true—and since lies love nothing more than the alibi of truth or partial truth, many broadcasts do indeed remain unmodified in this way. No self-respecting lie contains untruths. What is modified is ultimately *the worldview as a whole*—the world picture assembled from individual programs—and also *the whole human* (I'm speaking of humans who are exclusively nourished by mockups and phantoms). Even if every broadcast were to faithfully reproduce the truth, the simple act of only showing a portion of the world and *not* showing many other things that are real would transform the whole world into a modified world, which would also transform the consumer of this totality into a totally modified human. Once this state is achieved, *the whole becomes less true than the sum of its individually true parts*—or, to vary Hegel's famous phrase, "*the whole is the lie, only the whole.*"[77] The sole task of those who supply us with a worldview is thus to cook up one whole lie from as many individual truths as possible.

The whole that is meant to be achieved is, of course, *a pragmatic worldview,* not a theoretical one. This means that the world we are now allegedly presented with in place of the truth isn't merely a "subjective worldview"[78] but is delivered as a tool for practical use. It's a training device and exercise tool designed to shape our actions, our resilience, our behavior, our resistance to taking action, our taste—in short, our entire praxis. Yet to conceal its function, this tool also presents itself in disguise as "the world"—it's an *instrument in the form of a pseudomicrocosmic model, which, in turn, purports to be the world itself.*

This formula sounds quite obscure, so let's clarify it with an analogy to planetariums, where objects of a similar kind can be found. Planetariums contain projections of the night sky that are meant to hone our knowledge (of the stars) and our praxis (of detecting them). This makes them training devices. Simultaneously, however, they also present themselves as microcosmic models that (innocuously) seek to create the illusion of being the starry sky itself.— The comparison would be perfect if it involved a pseudoplanetarium—say, an astrological one—falsely pretending to be an accurate model of the night sky so as to train us to see the real world of stars in accordance with its image. It's this peculiar type of object that the "world" that is constructed and mediated by broadcasts belongs to: It's a "*stimulus model*" we're expected to attune ourselves to, a model with which we're meant to practice "behavior patterns"[79] and bring our "reflexes"[80] into line. These patterns are meant to become so deeply ingrained that our behavior in the real world begins to match our behavior in front of the model, until we become unable to behave any other way. At the same time, we're also not meant to let ourselves be taken or put to use by the world in ways the model does not envisage. *The real world is thus meant to become congruent with the stimulus model.* Yet this congruence is not just meant to be achieved by means of some theoretical declaration of identity (for such a declaration would inevitably have to concede a prior difference) but by means of a "*pragmatic equalization.*" It is meant to become an actual attitude within the world and a way of behaving toward it, so that any suspicion of the world not being identical to the "stimulus model" cannot even arise (or will not gain traction). A "pragmatic equalization" of this kind is known from Nazi Germany. For readers of *Der Stürmer*[81] who received their "conditioning"[82] from this publication and were molded by the model of a "Judaized world" [*verjudete Welt*] and the Jew models published therein, the difference between real Jews and the published "stimulus model" wasn't just negligible—it simply didn't exist. *Stürmer* readers were so oblivious to the fact that reality and image are separate entities that they could and actually did treat Jewish people as if they were nothing but these images. This process might even be called a form of "*inverted magic.*" Whereas a magic hex inflicts the harm intended for a person onto an image of them, in this case, it was the people themselves who were made to suffer the violence that was intended for their images (if such a distinction can still be made).[83]

In a sense, however, the pictures in *Der Stürmer* were still quite archaic; they didn't reach the heights of psychotechnology that National Socialism had already attained in other domains. So it is certainly not inconceivable that the contempt Julius Streicher[84] was made to feel—a contempt even held by those who were actually executing the liquidation he was calling for—was ultimately directed at his antiquated methods. When fabricating stimulus models and reaction patterns, nothing is more important than the successful suppression of the fact that these are fabricated. *Der Stürmer* attempted no such disguise. Motivated by a (sadly justified) disdain for the standards of its consumers, it didn't even deem it necessary to conceal the fact that it was lying—and it was this laziness that caused irritation even among mass-murderers.— To spell out the implication, it is in the utmost interest *of the industry that creates patterns and templates to invest them with the highest possible degree of realism.* If the patternlike stimulus model is to become effective as a whole, it must be presented as "reality." And National Socialism did indeed otherwise adhere to this principle, and the photographs it manipulated and edited in the pursuit of its own goals are classic examples in the inventory of stimulus models that lie realistically.

Today, obsolete models of the *Streicher* type are almost entirely out of circulation.[85] The principle that templates only reach their peak effect if they dress themselves with as much realism as possible now informs production across the board. There are scarcely any illustrated newspapers, any films, and certainly no news broadcasts that fail to adopt this principle. *We're not living in the age of surrealism but in the age of pseudorealism—an age of disguise that has disguised itself as the age of full disclosure.* When lies are resorted to—and when aren't they?—these lies no longer seem incredible, as if they were in tabloid print; they now seem like they were photographed. No, they don't just *seem like* photographs—they actually *are.* The medium of photography is so believable in itself, so "objective," that it has a much greater ability to absorb untruths and harbor lies than any medium predating it. Anyone who wishes reality to assume template-like traits uses photography as a device to disguise their templates as reality. To successfully dim reality with an image that purports to be an image of reality, a special type of image is required—a hyperreal [*überwirkliches*] or "surreal" image if one prefers. One thing is certain: This image must be dazzling and blinding; it must be *sensational.* Although this image may under certain circumstances even be genuine, it is nevertheless utterly untrue: It screens off reality, and (together with other sensational images) it contributes to an image of the world as a whole to which nothing in reality corresponds. *Wherever one seeks to produce templates, the sensational thus becomes the epitome of reality.* Because the word "template" is usually imagined to refer to something dull and boring, this may seem like a strange claim. But things are not that straightforward, for the sensational is in fact an essential aspect of any template. It not only serves to screen the template and shield it from view but also has a natural inclination to turn into a template itself. There is nothing more stereotypical than the supposedly newest thing, nothing more indistinguishable from yesterday's supermysterious murder than today's. And indeed, if some future historian were to utilize the crop of today's magazines to assemble a mosaic of "today's reality" in a hundred years' time, the resulting image wouldn't just seem generally absurd and far too hair-raising but also far too monotonous.—

Yet, although the template-producers only roll out their sensational pseudorealism to conceal that they wish to produce a template-world (they seek to prevent their customers from suspecting that they are being fobbed off with templates), *these same customers* now expect and even request very specific types of "surreality": They demand dazzlingly loud realities—that is to say, templates. This isn't really surprising, for the daily supply of die-cutting machines has already pressed consumer demand into shape. Customers now also demand sensationalism *and* the template—and they wish to receive both at once, ideally in the same object. When buying an illustrated magazine, they expect *the good old novelty of something that has never been seen before*; they expect the kind of outrage they received yesterday and the day before. They want the same limited "run-of-the-mill world" consisting of murderers, stars, flying "saucers," and all other kinds of planetary crockery—they demand that world that presents itself as "vast," "wide," "colorful," "renowned," and "great," even though its ingredients don't even represent an infinitesimally small fraction of the world. Those who attempt to break with this highly restricted range of topics and presentation styles—and luckily such attempts never cease—must be prepared for the fierce resistance of not only the template-producers (whose rules they fail to respect) but also the customers themselves. Their horizon of expectation has already become just as rigid, so anything that deviates from the accustomed forms of unusualness is either perceived as an imposition or deemed untrue. In fact, it may not even be registered at all: for anything that is untypical usually simply isn't "there." This raises the question of what method the truth should now adopt so as to compete with the lie and also be believed: Does this mean, for instance, that it is allowed to dress up as a lie (if it were able to) because the world of lies is pieced together out of truths? This question is not just unanswered, it hasn't even been adequately posed.—

So lies that once seemed as unbelievable as tabloid print now look like—no, they are—photographs, but even this formula is already outdated. The highest level of pseudorealism is, of course, reserved for TV-phantoms, which can convince their consumers that they aren't representations of reality but reality itself. How could reality itself, or so the consumer mindlessly believes, ever be unrealistic? How could it ever testify against itself?— And so the lie has never had a better tool at its disposal: *It no longer uses false images to lie against reality; it now relies on the help of reality itself.*[86]

While the "pragmatic equation" (the identification of "stimulus model" with reality) used to be subject to a certain amount of friction and doubt—for as long as an image is perceived to be an image, it will also invite some sort of skepticism—it has now achieved an almost ideal lack of friction. While watching the model, consumers think they are watching the world itself; while reacting to the model, they believe themselves to be reacting to the actual world. When model-phantoms make them feel embittered or enthusiastic, they believe themselves embittered or enthused by reality. And when the world now really does confront them—and the templates, as training devices, were created precisely for such moments—they now only see what the templates have taught them to see and only feel the feelings the templates have traced out for them. Templates are thus *a priori conditions*; they don't just give shape to perception, reason, and feeling but also serve as templates for *behavior and action*. The breadth of application and universal capacity of these matrixes is so vast that this was not even foreseen by the most speculative philosophers—and least of all foreseen for our own time, the alleged age of empiricism.

As long as Frazer's, Lévy-Bruhl's, and Cassirer's presumptions are correct, the only mentality that compares to the one produced by this conditioning is the one that once belonged to "primitive peoples."[87] These people apparently lived by a moral code and outlook so restricted and rigid that they were unable to "consider" anything (in theory or praxis) not included in it.

My use of the term "*a priori* conditions" shouldn't be taken literally, of course—that is, not in a Kantian sense. It's scarcely possible to conceive of traits that are any less "innate" than those produced and pressed into humans today. But these traits are nevertheless "*a priori*"; as casting molds, they preexist what one experiences, what one feels, and how one behaves, and as such, they "condition"[88] all these things. And since these "conditions" don't just predetermine *how* something is experienced, felt, et cetera, but also *what is* and *isn't* experienced, felt, et cetera, they are extraordinarily powerful and their range of applications extraordinarily broad. Those who have been shaped by them are no longer open to anything other than what broadcasts have prepared them for at home. They now only see, think, feel, love, and do what they have been preconditioned for. And this conditioning function is what broadcasts are meant to perform. But because none of these matrixes is allowed to disclose that it is a matrix, *these conditions must appear in the form of things and the matrixes as world-parts.*

This last point is of profound significance for our entire investigation, and there are two reasons for this:

1. The alleged "ontological ambiguity"—the spectral, phantomlike quality of broadcasts that occupied us at the outset—is now divested of its enigmatic nature. Because matrix-producers seek to deny that their templates are templates and their conditions are in fact conditions, they offer them as "world" to us and as "things." And this implies that they are offered as phantoms—*for phantoms also are designed shapes that present themselves as things.* The phantomlike quality of broadcasts thus turns out to be a desired effect, and their alleged "ontological ambiguity" nothing more than the outward form of something morally ambiguous: a deception.
2. This now adds a much-needed caveat to the definition of "idealism" provided above. To recall, we called any possessive attitude toward the world "idealistic"—any attitude through which the world appears to "be my world" because I am indeed claiming possession of it. There is, however, a fundamental difference between a conqueror (or Hegel's feeding animal) who is actively *making* the world his own and a situation in which the world is *made to be mine.* And it also matters *in what way* it is made to be

> mine. There are many things that can be "mine"—even the number branded into the arm of a prisoner in a concentration camp. When the world supplied to mass-humans is a totality made out of templates (as is the situation we are describing), then the world might just be replaced by a totality they picture, but this mental image is still only "theirs" because it is imprinted on them. "*My idea shall be what is world to you*"—thus speaks the will of the matrix-producer. This is how Hitler spoke. A follower of Hitler claiming that "the world is my idea" would have been unthinkable. And not because he, as a mass-human, falsely believed his idea to be his world, but because what he took to be his "world" was imagined and placed in front of him by someone else—it was, in fact, even delivered home to him.[89]

21. *How Needs Are Imprinted on Us. The Goods for Sale Are Today's Commandments—Commodities Are Thirsty and We Share Their Thirst*

What we are presented with are thus premolded objects that jointly purport to be "the world," and they are tasked with shaping us in their image. Yet this doesn't mean that the shaping process entails violence—or at least, where violence is deployed, it isn't necessarily felt as violence or even just as pressure. The formative pressure mostly remains as imperceptible to us as the weight of the ocean is to deep-sea fish. The more unnoticeable it is, the more certainly it will succeed. In the best-case scenario, therefore, the form-giving matrix is experienced as *a matrix that is actively desired.* If this outcome is to be achieved, desire must itself be molded into shape: Production processes are now tasked not just with *standardizing* goods but also with producing and standardizing our *needs* (to ensure that they thirst for standardized products). This largely happens automatically via the products that are supplied and consumed each day, for as we'll see in a moment, our needs adjust themselves to what they are offered and daily consume. But they don't quite fully adjust. A certain gap between the goods on offer and our needs always remains; demand and supply are never a perfect match. If this gap is to be closed, an auxiliary force must be called on. And this force is *morality.* But if morality is to become useful, it needs to be premolded, too: It must expose anyone who doesn't wish to acquire the things they are meant to acquire as "immoral" and nonconformist. Ideally, this exposure happens via the vehicle of public opinion—or, rather, via each person's "own" individual conscience (which acts as the mouthpiece of public opinion)—so that everyone is forced into desiring the right things. And this is the case today. The unspoken maxim we're all exposed to at any given moment reads, "*Thou shalt learn to need all the goods that you are offered!*" (or at least this is how this appeal to our "better self" would read were it to be translated into words). But although it is unspoken, it tolerates no dissent: *The goods on sale are today's Commandments.*[90]

Apart from a few surviving remnants of earlier customs, our "dos and don'ts" are now defined by what we are meant to buy. It is nearly impossible to excuse ourselves from the minimally required purchases offered as "musts"[91]—that is, as "must-buys." And anyone who tries to make excuses risks losing prestige, being branded an "introvert," missing out on employment opportunities, or being found out as someone who lacks the means. They are even in danger of making themself morally and politically suspect. The act of not buying is in fact considered a kind of sales sabotage—a threat to the commodity's legitimate demands. It's not just treated as a mere omission but as an active offense, similar to theft, if not even worse. Thieves, just like customers, affirm their loyalty to the alluring qualities of the commodity and its Commandment, whereas people who commit a nonpurchase dare to pretend to not be hearing the commodity's call—they insult the universe of goods. Although theft is, of course, not considered a desirable method of acquisition, thieves nevertheless prove themselves to be conformists; when caught in the act, they can easily be held responsible. Yet those who refuse to buy the goods for sale now even have the gall to dodge the long arm of justice by sanctimoniously relying on the alibi of negativity (of having done nothing at all). "Give me ten thieves over *one* ascetic" (as the Molussian saying goes).

The mere fact that I did not own a car (and could thus be caught redhanded in the act of a nonpurchase) placed me in the following tight spot in California in 1941.

Diary

Yesterday, as I was walking along the highway far beyond the outskirts of LA, a motorized policeman sped up to me and stopped. "Say," he called out, "what's the matter with your car?"[92]

"My car?" I asked somewhat incredulously.

"Sold her?"[93]

I shook my head.

"In the repair shop?"

I continued to shake my head.

The cop[94] stopped to think but couldn't quite seem to find any other explanation for my carlessness. "Why are you not using her then?"

"My car? . . . But I haven't got one."

This simple snippet of information again seemed beyond comprehension.

To help him out, I added, "I have never owned one."

I couldn't have gotten myself into any hotter water. Perfect self-incrimination. The policeman's mouth fell wide open. "*You have never?*"

"There you go!" I said, commending his comprehension. "That's the boy."[95] And with this I tried to resume my walk with a cheerfully innocent goodbye.

But this was out of the question. Quite the opposite. "Don't force me, Sonny . . . ,"[96] he now said, pulling out his notebook: "Stop telling stories!" The delight at having the clattering boredom of his job interrupted by the capture of a "vagrant"[97] now almost made him look endearing. "And why have you never owned one?"

Even I now understood that I was not allowed to say, "Because I have never been able to afford a car." And so I instead shrugged my shoulders and said as casually as possible, "Oh, I've never had any use for one."

These words seemed to cheer him up. "Is that so?"[98] he exclaimed, almost with enthusiasm. It immediately dawned on me that I had made an even bigger mistake. "And why has Sonnyboy[99] no use for a car?"

Sonnyboy[100] shrugged apprehensively. "Because he needs other things more . . ."

"For instance?"

"Books."

"Aha!" the cop said ominously and repeated: "Books."

With this he had now clearly heard enough to reach his diagnosis and said, "Don't act the moron!"[101] These words were meant to communicate that Sonnyboy[102] had been found out, that he was clearly "a highbrow pretending to be an imbecile," someone who was playing the idiot in an attempt to conceal his conscious refusal to recognize the command of the goods for sale. "We know your kind,"[103] he taunted, with a friendly nudge to my chest. And then, with a gesture that traced the entire expanse of the placeless horizon, he asked the question I had been dreading all along: "And where exactly are you going?" For although the road would reach San L. in forty miles, it led to nowhere else before then.

Had I tried to explain the aimlessness of my wandering, I would have exposed myself as a "vagrant" for good. Heaven knows where I would now be, had L. not rocked up at the very last minute in his imposing six-seater, as a veritable *deus in machina,* and beckoned me into his car with a "*Hallo.*" This arrival didn't just leave the cop baffled; it probably also did some serious harm to his "philosophy."

While overtaking us farther down the road, he looked at me and yelled across, "Don't do it again!"

What was I not to do again?

I was obviously never again meant to fail to acquire what the goods on offer command everyone to buy.—

Once goods are recognized as today's Commandments, one no longer needs to wonder why they are also acquired by people who wouldn't actually be able to afford them. For the prospect of failing to obey the Commandments and not making the purchase is something they can afford even less. Since when has the call of duty spared those without means? Since when has obligation spared the have-nots?[104] Just as, according to Kant, one's duty must be fulfilled even if it goes against one's inclination—yes, especially then—one must today also fulfill one's duty, even if one "cannot afford it"—especially then, in fact. The Commandments of the goods on offer are categorical. And when they proclaim their "you must," it would be pure sentimentalism to plead for exception because one's own "must-have-can't-afford-to-have" situation is too precarious.

This analogy is of course a philosophical exaggeration, but it exaggerates in the direction of the truth. It is unmetaphorically true that there is very little in the spiritual life of our contemporaries that has as much consequence as the discrepancy between the things one can't afford and the things one can't afford to live without. And it's just as consequential that this discrepancy becomes manifest as a "struggle": If today's humanity is characterized by a specific "*conflict of duties,*" then this attritional struggle raging wildly within the consumer and within the bosom of the family would be it. Yes, phrases such as "raging wildly" and "attritional" aren't out of place. The cause of the struggle may seem ridiculous and like some farcical parody of much nobler conflicts, yet this does not diminish its ferocity—it is certainly ferocious enough to provide the basic motif for a bourgeois tragedy about our times.

This struggle, as everybody knows, usually ends in a purchase—that is, with the victory of the "Commandment of supply." Yet this victory comes at a high price, for it marks the beginning of the enslaving obligation to *pay off the acquisition.*[105]

And once customers *have* a given object, regardless of whether it's paid in full or still being paid off, they naturally seek to enjoy their possession. And since they can only enjoy their ownership by putting the object to use, they will now begin to use the object *because* they have it—and, consequently, they will turn into its creature. Yet there's more. Now that they have the object, it would, of course, also be *morally* inconceivable to own it but not utilize it *to its fullest.* Failing to do so would, in principle, be no different from buying bread without eating it. So, if the television were only to be turned on occasionally and the radio only once in a while, one would voluntarily deny oneself something one has paid for, or for which a down payment has been made. This self-denial would also be to no one's benefit; these things *would simply go to waste,* which is, of course, unthinkable. So, if one resigns oneself to the nonstop deliveries these devices supply, if one lets oneself be molded nonstop, this is, at least in part, also for moral reasons.

But that's not all. *Once something has been acquired,* one doesn't simply use it—*one also comes to rely on it.* And as soon as this happens, the future is set on track: *One doesn't end up having what one needs; one ends up needing what one has.* Every new level of ownership psychologically congeals into the norm. The moment one no longer owns a branded product one has gotten used to, this product won't just leave a gap—it will generate *hunger.*— But something is always missing. True, most consumer products aren't actually perishable goods like bread and butter. But, for the benefit of production, all goods are now (designed to be) used up when used. It is thus the users and consumers themselves who ensure that this gap is generated: As soon as they own and use something, as soon as they have worn it out or used it up, they will need to buy this product again: *Need is generated in tandem with consumption.* In a certain sense, today's needs are modeled on "*addiction,*" which also means that needs now owe their existence and their specific form to the actual existence of certain goods and products.

The most ingenious commodities are those with qualities that create *accumulating needs.* Even in the United States, nobody would maintain that God or nature implanted in us humans a "basic need"[106] for

Coca-Cola, and yet thirst has now attuned itself to Coca-Cola over there. This has happened—and this is the crux of the matter—even though the hushed-up ultimate function of cola is not to quench thirst but to generate it: a thirst specifically for Coca-Cola.[107]— Demand is here the product of supply, and need is the product of the product; this allows the product to simultaneously secure its ever-accumulating production.—

This last example now shows that the "imperative" embodied by the "Commandments" of supply is not to be underestimated. It doesn't just issue from commands that are spelled out by the loud words of adverts: "Buy your Mozart underpants! Buy them right now! You must have them!" For with a bit of self-discipline, it is just about possible to resist this command, even though it already addresses us as the proud owners of the pants. No, the actual imperative issues directly from the things one owns. And although unspoken, their commands tolerate no objection. To stay useful, or at least not to be rendered immediately useless, *each acquired product demands the purchase of additional products; each thirsts for another—no, for many others* (this is, of course, also a question of prestige, for each product wishes to be surrounded by products of an equal standard). And their thirst for other products makes us thirsty, too: To buy products isn't difficult at all, but it's very difficult to have them. Once we own a given product, we must turn its thirst (for washing powder, for gasoline) into our own. And no matter how difficult it might become to feed the hungry mouths of our ever-accumulating things, we have no choice other than to turn their needs into our own. And before we even realize what is happening, this becomes a *fait accompli.* Anyone who needs *A* also needs to use *B*; anyone who needs to use *B* also needs *C*. There is not just a recurring need for *A* (as is the case with Coca-Cola), whole product generations are needed: *B* (which is demanded by *A*), *C* (which is required by *B*), *D* (which is sent for by *C*), and so it continues *ad infinitum.* With every purchase, we sell ourselves. With every product we, in a sense, marry into a product family that accumulates and breeds like rabbits, and which now demands our financial support. This, of course, brings a certain level of comfort, for we now scarcely need to contemplate how to live our life or reach our own decisions—we now simply respond to what this thirsty family calls on us to do, and so "time goes on."[108] This also means, however, that we are constantly being chased by a thousand family members who are keeping us in motion, putting us to work, and making our decisions. It means that we are living our life under a dictate, that our future needs have always already been chosen, and that we are never given the time or freedom to lodge—or even feel—any needs of our own.—

The naive will now warn us against ever letting ourselves get in with any of these "thirsty products." Yet such warnings are, of course, ridiculous—nonthirsty goods do not exist: It is not this or that product that thirsts for others, but the universe of goods as a whole. What we call "the thirst of things" is nothing other than the interdependency of production—the fact that all products draw on each other and are mutually dependent. It's as impossible to stay out of this cosmos of goods and production as it is *to be without being in the world.* And if some fool were to undertake the experiment of living independently by giving up even a few of the devices and forces that make up our world—electricity, for instance—this person would rapidly succumb. The system everyone born today is part of *nolens volens* cannot be allowed to have any gaps because these gaps would lead to the loss of the whole system.

The fact that every product offered to us as a "must" and bought as such now also harbors its own needs—needs that *become* our needs—represents *the climax of the matrix phenomenon: Our needs are now but the imprints or reproductions of the products' needs themselves.* What our needs will be tomorrow is written in neither the stars nor our hearts—not even in our stomachs; it is written in the freezer we acquired yesterday, in the radio bought the day before, and in the television brought home today. And it's their needs we will be listening out for tomorrow, hearts beating wildly in our chests.

22. *The First Principle of Economic Ontology: What Only Exists Once Doesn't Exist at All—an Excursus on Photography*

We just said that when not just our experiences are modeled but also our needs, then matrixes have reached their peak performance. And as long as we only have ourselves in mind as the subjects—no, as the victims—undergoing this procedure, then this is certainly true, as a deeper layer than that of our needs hardly exists. But this doesn't fully capture the power of matrixes.

Matrixes don't just mold us; they also mold the world itself. As this assertion merely seems to point to serial-production, it may seem self-evident. Yet, if we return to our original example, the phantom-production in radio and television, things are no longer as clear-cut. In this original context, our assertion now means that the artificial "world-models" reproduced and delivered to us as broadcasts don't mold just us and our worldview but the actual world itself. It suggests that the lie lies itself true—that conditioning has a boomerang-like effect; it signals *that the real turns into a depiction of its images.*

To understand the peculiar process by which this happens, we need to backtrack quite a bit.

Right at the outset, we had established how the real (or allegedly real) events we're supplied with turn into commodities (or rather into mass-products) when they are home delivered in countless copies. The relationship between event and broadcast thus represents a subset of the specific way that *models* relate to *the products reproducing them.*

But if one now asks which of these—the original model or the reproduction—is "real" in the economic sense, the answer must be: the reproduced commodity. Models only exist to be copied; they exist only for the commodity's sake: The more copies can be sold, the more real the commodity becomes. This means that the model is only real to the extent that its blueprint quality allows for the "realization" of the highest possible number of reproductions. If there were a fully developed economic ontology—that is, a doctrine of being as seen from the perspective of today's markets and outputs—its first principle would be: "*Reality is produced by means of reproduction; 'being' only 'is' in the plural—only when it constitutes a series.*"— Or the other way around: "*Once doesn't count: What exists only once doesn't 'exist' at all; the singular still belongs to the realm of nonbeing.*"[109]

This principle seems nonsensical and is indeed difficult to understand, for it counts neither "the general" nor "the singular" as "something that is," but only *the series.* As such, the principle is at odds with the classic alternative between "nominalism" and "realism" we are familiar with, but none of this prevents it from settling deep into our bones—especially among the least philosophical of us.

Anyone who has observed tourists visiting Florence or Rome—especially tourists from highly industrialized countries—will know how irritated they feel whenever they encounter *something unique*[110]—when they encounter one of those great historical objects, for instance, that just stands around there as an annoying one-off within our serialized-world. Yet these travelers now also carry an antidote against this irritation—a kind of syringe that promises instant relief. I'm of course speaking of the photo camera, a device that lets them instantly transform any one-off into a "subject" the moment its beauty or unclassifiable nature becomes a bit too much to take. The camera allows them to turn anything that is "all too specific" into something "commonplace," a reproduction that can assume a rightful place within the universe of reproductions.— Equipped this way, these tourists now swarm through the world as magicians who don't even need to touch their objects "*pour corriger sa nature*"—to fix the defect that, within a serialized-universe, anything unique represents simply by existing. They thus set out to "*take pictures*" [*aufnehmen*], to "*admit*" [*aufnehmen*] into the serialized-universe anything that has so far been excluded from it—an admission that happens by means of photographic reproduction. And once they have snapped away at anything unique and thereby fulfilled their quest, they feel relieved.

But besides "taking pictures" and "granting admission," another meaning of "*aufnehmen*" now comes into play; the word also means "*bei sich aufnehmen*," "*to accommodate in one's home*." By creating reproductions, these magicians also ensure that they "*own*" anything they reproduce. This shouldn't be misunderstood: It doesn't mean that they *only own an effigy*. The serial-products of their world—the things they live among, live off, and share their lives with—are reproductions across the board; all are copies of some prototype or model. These tourist-magicians hence already live their lives surrounded by effigies, and so they no longer know anything else. When owning something, they only ever "have" its effigy. With the photographs they now "have," they thus continue to enjoy the very form of ownership they are used to: *To them, reproductions are what is real*. In fact, they don't actually take photographs of what they see—for they only ever "see" to establish whether they want to take a picture, and they only take pictures of what is worth "having." Along the same lines, *the thing* they photograph is also not the "real thing"—to them, only the captured image is real; the picture through which the thing has turned into an example of a reproduction series, has become their property, and has been admitted into the serialized-universe. In ontological terms, our magicians have hence replaced "*esse = percipi*" with "*esse = haberi*."[111] To them, the "real" St. Mark's Square is not the one in Venice, but the one in the photo album at home in Wuppertal, Sheffield, or Detroit. *This also means that what counts is not being in Venice, but only "having been there."* And not just because this increases their social standing at home, but because only past moments make for reliable possessions. While the current moment is destined to remain an ever ungraspable, unreal, and unprofitable asset precisely because it doesn't *remain* (i.e., its fleeting nature means that it cannot be "owned"), the opposite is true for a past moment that has become an image and thus a thing. And this makes past moments the only thing that is real. Or, to put it ontologically, "*only what was, is*" [*nur Gewesen-sein ist Sein*]. If one of these magicians were truly aware of what their activity entails, if one of them did more than simply go along with it—which is, of course, highly unlikely since photography and philosophy seem incommensurate—they would justify a life spent taking pictures as follows: "Since I've transformed everything that was into a reproduction and thus also into a physical object, nothing in my life was futile; nothing was wasted or unprofitable. I've brought everything that was home with me—mostly in black and white, some of it in color. I even have some moving images. And so I can now continue to have it. Everything now *is* because it continues to remain; everything *is* because it is an image." For those who live like these magicians, "to be" means "to have been" and "to have been reproduced," "to be an image" and "to be property."—

To pursue the close link between technologies of reproduction and our memory (which has not without good reason been called "reproductive") would be too much of a digression, so I will only add that the link between memory and reproduction is ambiguous. On the one hand, photographs help us remember; on the other hand—and this is more decisive—these memories that turn into things have also stunted and replaced our own capacity to hold things in our memory and impacted our ability to produce the sentiments this creates. When contemporary man tries to gain an autobiographical image of himself, provided he still wants to consider himself as "a life," he assembles this from the photos he has taken. He no longer needs to produce any images from memory—they appear with the turn of a page; they don't need to surface, or if they do, it is only from the depths of the album. Just as is the case with St. Mark's Square, it's in the album—and only there—that his past can be found. To reconstruct it, he now relies solely on the (now unlosable) snapshots stuck into the album; if he keeps a diary, it is only kept in this album form.— That the life reconstructed thus consists almost exclusively of day trips and travels undertaken, and that nothing else seems to count as "a life," shall only be noted in passing.

It is essentially the museum principle that has triumphed as the autobiographical principle here: Everyone now encounters their own life as a series of images, as a kind of "*autobiographical gallery.*" This also means, however, that life is no longer encountered as something one has lived, for everything that was is now projected into the *single* two-dimensional plane of always available images. Time, where is thy sting?

If one were to offer Herr Schmid or Mr. Smith a journey through Italy, but on the strict condition that no photographs be taken and no memories set aside for the future, this Schmid or Smith would probably decline, thinking the trip a waste and thus an almost immoral imposition. If he were forced to undertake such a journey anyway, he would no doubt panic at some point along the way. For he wouldn't know what to do with all those sights "seemingly made to be photographed"; in short, he wouldn't know what to do with himself.— That travel agencies reject the slogan "Visit lovely Venice"[112] and opt for "Visit unforgettable Venice"[113] instead is entirely consistent. Venice is deemed unforgettable even before one has seen it. And one isn't encouraged to visit because of its beauty, but because it is unforgettable, just as one is meant to buy trousers because they cannot tear. It's also not its beauty that makes it unforgettable—because the unforgettable nature of Venice has been guaranteed, the traveler can also be certain of its beauty. For anyone traveling in this way, the *present moment* becomes nothing but the vehicle for something that "*will have been*": not worthy of attention for what it is, it becomes but a pretext for a future memory and is experienced only in the "Future Perfect." It has already become something phantom-like and unreal.

And needless to say, one does not just *travel this way.*

23. *The Second Principle of Economic Ontology: "What Is Unexploitable Has No Being"*

As little dignity as any unique item has in the eyes of our assumed economic ontologist, *the ontological dignity of nonmanufactured things* is rated just as low. I am speaking of the dignity of *natural objects,* those naturally unusable elements that are left over during the process of mass-production. Since these elements are economically speaking good for nothing, they are now treated as deadweight that deserves nothing better than to be annihilated and not "to be" at all. The second principle of economic ontology thus reads: "*What is unexploitable has no being or is not worth existing.*" And our epoch provides ample evidence that, depending on the economic situation, anything and everything can be condemned to this kind of worthlessness and liquidated into slurry, humans no less than radium-contaminated nuclear waste.

In comparison to the real existence of mass-produced ready-made goods, which are all destined to satisfy a need (or else "foresee" the needs they satisfy themselves), our economic ontologist doesn't consider nature as a whole something that is "meant to be." Despite its immense size, it lies outside of what he considers "Providence." Although nature has "some value" and "does exist" as raw material for products, he nevertheless considers it something incidental, mere *κατά συμβεβηκός*.[114] It only has value and being if it is invested with these qualities by the products it can be turned into. There are, however, also truly unprofitable things in nature—things producers not only have no use for but that can't even be liquidated. In the eyes of our economist, if he actually concedes that it exists, this unexploitable and unliquidatable cosmic excess—the Milky Way, for instance—represents a metaphysical scandal. It's a pile of material that was provided without any clear motive, which means that its existence can in a sense only be explained on account of some cosmic incompetence and lack of business acumen.

It seems likely that today's nihilistic complaint about the "meaninglessness of the world" (at least partly) gives voice to a *Weltschmerz of the Industrial Age—a world-weariness* rooted in the suspicion that this cosmic excess, "the much too much of the universe," will remain forever unexploitable, unprofitable, superfluous, wasted, and without purpose. (For the universe has clearly nothing better to do than metaphysically slouch around in the space it has inexplicably been given.)[115]

But I did qualify "if he actually concedes that these things exist." Normally, the existence of unexploitable things isn't even acknowledged; it is as unacknowledged as the fact of evil [*κακὸν*] is in classical systems of theodicy (in Plotinus's "evil is the absence of the good," for instance, where evil is "*given a place*" [*ein-räumen*] in the system and thereby stripped of its negative quality).[116] This analogy isn't meant to imply that economic ontology has developed into an outright theodicy, that it has formally uttered the doctrine that "there is nothing unexploitable." There is, of course, no such clearly stated doctrine, but in its actions—and these are more conclusive than any thesis—it seems positively fired up by a sporting ambition to outsmart nature. It is as if it were intent on "teaching nature a lesson," on confronting it with proof that its metaphysical laziness, its prudery and reluctance, and its pretense of independence from the cosmos of production will ultimately be to no avail. To use a somewhat awkward and sensitive image, this ontology seems motivated by a positive desire to rape nature, to give it children and forcibly extract fertility from it so as to make it publicly show that everything that is part of nature can be exploited—even if this necessitates the invention and production of the most absurd needs, even if these were attributed to its waste products. As Titan-like and unsurpassably self-assured as this nature-challenging attitude may seem, it isn't free of fear and trembling either. The Titan is also affected by the *Weltschmerz* of the Industrial Age; he, too, is afraid that he might not be a match for the overabundance he is challenging—that nature may inflict its revenge by supplying far too many beneficial effects (even if it were to remain frigid and tauntingly withhold its maximal fertility). Challenges that are accompanied by this kind of fear assume frantic traits, and since the violated party—that is, the world of nature—seems to reemerge from each violent embrace as a virgin, since it seems to act each morning as if it had not the slightest recollection of what had just been done to it, the struggle soon builds into a Sisyphus-like rage.

But let's move on from this image that mobilizes the language of myth in a not entirely convincing manner. What matters here is the underlying maxim: "*The existence of anything unexploitable is not allowed.*" Transposed into a positive imperative it reads: "*Make everything exploitable!*" In a certain sense, economic ontology is thus simultaneously an ethics that gives itself the task of bringing redemption to the world-chaos—saving it from its condition of raw materiality, from its "sinfulness" and "inauthenticity," by transforming it into something "authentic." It pursues the mission of turning chaos into a cosmos of products. In brief, it aims to bring about an *aetas aurea* of fabricated goods [*Fertigwaren*] so that "it is chaos itself, which, at the end of days, will stand transformed into a million ripe, fine and golden forms, an Apolline product, fermented into maturity and still radiant with its inner glow."[117]

The very language adopted here suggests that economic ontology also serves as a doctrine of justification: Once the contingent world turns out to be essential as material for production processes and ready-made goods, what once existed as a contingent, unfinished world is "justified and made proper." And so the existence of the human as producer is justified, too; for without human labor—conducted by the sweat of the brow—it would not be possible to transform the world and thereby save it. Our mission, as the economic ontologist sees it, is to make the world "wake up" and guide it to us, to its destiny: into the blast furnaces, factories, power plants, nuclear reactors, and radio and television stations. These are the "houses of being" in which the human tries to transform the entire world. This self-assigned task has such monstrous dimensions that the classic definition of the human as "*homo faber*" no longer seems to apply to humans gripped by this transformation fever. When *homo faber* created a world of his own, a world not foreseen by the world itself, he limited himself to only using *parts* of the world—and he left intact the parts he didn't need. It was in this limited act of world-creation that he recognized his freedom and his purpose. But for today's humans, the world is nothing but matter. They would rather force new needs onto themselves than leave anything that exists intact and unused; they want to process and transform the whole world and "finish it off." The pretension of today's human is thus no smaller and no less universal than religious pretensions or the pretensions of philosophical systems. Today's human is the smelter of being—or, at least, its shepherd.

It will probably be surprising to encounter this Heideggerian expression in a context that is so at odds with Heideggerian musings. There is an inestimable gulf between the "shepherd" and the "smelter"—between Heidegger, who makes language "the house of being,"[118] and economic ontology, which accommodates the world in the abovementioned transformation-, prison-, and slaughterhouses. But one commonality cannot be denied: the basic assumption that *being requires our help*. Heidegger and economic ontology thus share the highly peculiar belief that "being" has an intrinsic need of being "housed"—that it cannot manage on its own and wouldn't survive one second without our help. Both assume that being needs to find a stable or a home with us. By ascribing to the world the need to become *my* world (or ascribing the same need to being), these philosophies seek to furnish "idealism" (as it was introduced above) with a justification and root it in reality. They seek to ensure that the human is awarded a metaphysical mission and try to make the human believe in a task. Or, rather, they try to justify what humans do anyway by presenting it as a task. This desire is, of course, not incomprehensible. Both philosophies constitute desperate protests against "the human's place in the cosmos"[119]—or, to be more precise, against the fact that the human has *no* such place today, that naturalism has robbed it of any illusion of anthropocentric privilege by degrading it into a piece of nature, just one of a billion other pieces. By attempting to smuggle a special place for the human back in through the back door, by trying to define a mission and *a way in which the world cannot exist without humans,* these philosophies thus reveal how difficult it is to bear this lack of privilege and how difficult it is to cope with it. As the center of his flock, the "shepherd" is not himself a sheep. As the "shepherd of being" or the world, the human doesn't just "exist" the way the world exists; it has another special mode of being—and so the shield of its metaphysical distinction is shiny once again. The same is true for "the smelter of the world."— This makes both these philosophies examples of a new variety of anthropocentrism—an *embarrassed form of anthropocentrism.* Instead of claiming that the world is there for the human, they claim that *the human is there for the world.* The human is cast to the role of *cosmic altruist,* the *manager of the cosmos* who doesn't own what it provides and only wants the best for being and the world.—

As fascinating as it is to observe how two of today's philosophies, which diverge so strongly, nevertheless share basic tenets as *philosophies of today*—tenets they do not share with any previous philosophies—it is only economic ontology that is relevant to our topic. Or, to be more precise, only the belief that the world (in its current state) is not a finished, not a true world—which makes it a world that actually *is not yet,* a world that can only become true and authentic once it has been worked on, completed, and brought to market by us; a world, therefore, that is a world only once it is made to disappear *as* world.

For economic ontology, it is thus simply unbearable to think that there are unexploited, unprocessed events—things that are neither brought into circulation nor to market, things that anonymously appear and happen only to then vanish into nothingness. It is probably just as unbearable as it is for us to think that there are wheat fields or fruit plantations somewhere that are right now ripening toward their own decay without ever being harvested. That which merely *is,* is as if it did not exist. That which merely *is,* is wasted. If it is *to be,* it must be harvested. And this harvest—this harvest of what is happening and of what has happened—is now largely brought in by broadcasts. A dying man who is broadcast is saved; the defeat, when multiplied, is won; the solitary prayer, a million times reproduced, has been answered. It is only now, only in this format, that they *attain being.* But what these entities once were before becoming base and common is fading into empty semblance devoid of any essence.[120]—

24. Phantoms Are Not Just Matrixes That Shape How the World Is Experienced; They Also Shape the World as Such. Reality as a Reproduction of Its Reproductions

In the terms set out by economic ontology, neither the individual nor nature "actually exist"; the only thing that exists is the sum total of all goods that are part of production runs. And it belongs to the nature of these products that they attempt to be nothing more than the function they perform (i.e., need satisfaction)—nothing but the quality that makes them sellable and useful. No product can fully reach this target: Merely through its size, weight, and the operational demands it places on its user, each product also lugs around a set of sinful attributes that don't respond to any demand—attributes the buyer simply needs to put up with. In a bizarre sense, every product (just like the soul) thus seems embarrassed to be chained to a body and belong to nature. And each product pursues the ideal of compressing this bodily remnant to the most infinitesimally small size possible and reaching a quasi-angelic existence.[121]

This isn't just true for physical products, but for all kinds of products; not just for the physical matter products are created from, but for any kind of processed matter. It is also true, therefore, for *events*—the material that is processed to be turned into "broadcasts."

"Naturally occurring" events—events that happen as singular occurrences—are good for nothing. They lug around unusable sinful attributes; they are but raw material. In this state, none of them can "pass" the censorship of economic ontology. Before they can measure up, they must be multiplied, and since it would be pointless to multiply them in their condition as raw material, each event must in a sense first be "passed through" a "straining machine" and be filtered. It is only in this "processed" state that events are now deemed to be events. And as we are now dealing with products, the question of what the "actual" event was like before it was processed would not be relevant. When faced with jam that is ready to use, one also wouldn't ask what specific fruits had to be passed through the "straining machine" to come out as jam. No, the response "this is the real thing" sets in as soon as a product works, as soon as it proves itself in daily use.

Products and commodities thus embody a very particular concept of truth—they are true [*wahr*] only insofar as they *prove themselves in use* [*Bewährung*]. And this concept is perfectly upheld by the events that reach us in a processed state via radio and television. There is no sign of a deadweight dragging them down, nothing consumers might have to put up with, no distance to cover, no danger to face, no effort to be invested. Such is the level of perfection that broadcasts have reached that nothing is even left *after* they have been consumed: no pip or kernel, no hair or bone, not even the product itself (unlike books, for instance, which are still there after being read). While being consumed, the consumed product is absorbed like a pill, vanishes, and disappears. Apart from its invisible effect, which consists in transforming the consumer into a mass-human one more time, everything stays the same. Nothing needs to be cleared away or washed up, nothing has happened, nothing is left over, nothing remains—perfect inconsequentiality. The prospect of any unwanted accumulation of cultural goods has been stopped in its tracks. There is no risk of education.—

But this portrayal is still inadequate. It is not just our bread that is an artificial product; the wheat, the raw material it is made of, is artificial too. It may still grow, but its mode of cultivation and even the type of wheat it grows as are determined by what is most suitable for its use as a product. Cultivation (i.e., today's mass-production) no longer just consists of a retroactive reworking of the matter destiny has provided but already entails the actual design [*Lenkung*] of matter itself. Production processes now, in fact, try to intervene in the raw materials they rely on *as early as possible*. Ideally, this material isn't ever given any opportunity to just be "raw material." A great deal of labor is thus invested in inserting these materials directly into the fabrication process so that their growth or formation turns into the first stage of production. This also applies to the sector that produces "broadcasts," a sector that mostly relies on world events as its raw material. And so it happens that events are attempted to be cultivated and grown in a way that makes them fit to be mass-products. They are rendered *optimally reproduction-ready* as soon as possible, or even from inception; every effort is taken to ensure that they provide a substrate that doesn't pose any difficulties when being reproduced. *Reality—the thing allegedly reproduced—needs to be refashioned to ensure that it can measure up to the images it will eventually yield. It needs to be remade in the image of its images.* Current events need to anticipate their copies and obligingly take after them. There are already countless things that only happen the way they do so as to ensure that they can be used as broadcasts; some events even *only* happen because one wants (or needs) to broadcast them. In such cases, it has become impossible to tell where reality ends and the game begins. "When the judge and the witnesses and the lawyers . . . perform their functions before the microphones and batteries of floodlights knowing that perhaps ten million people or more are listening and watching their every move, the temptation to put on an act becomes irresistible." (Declaration by Judge Medina, as cited in *The New York Times*, September 3, 1954.)[122] In fact, the question of where reality ends and illusion begins is in itself flawed: Radio, screen, and the consumption of phantoms represent social realities of such consequence that they can contest most of today's other realities, so much so that they themselves determine "what is real" and "how things really happen." The lines by Karl Kraus, with which he already believed himself to be denouncing a scandal,

> In the beginning was the press,
> and then the world appeared[123]—

have already become harmless. Today they should read:

> In the beginning was the broadcast,
> and then the world happened for it.

This inverted (if not to say, perverted) relationship between model and reproduction is, of course, not entirely unfamiliar: Next to the thousands of projections for which they serve as models, the real film stars that actually walk around as flesh and blood in Hollywood count for nothing. They already seem like little more than the pathetic ghosts of their reproductions, ghosts that seem to be failing in their attempts to live up to their own close-ups.

And many of today's events—football matches, court cases, and even political protests—already bear a striking resemblance to these stars. When placed next to their broadcast versions, which are heard and viewed millions of times, they already seem inconspicuous and unreal—or at least they would seem this way had they not already been designed to be reproduced and broadcast. Instead of taking place for their original participants or for those who experience them in person, these events are thus *always already* meant for the millions who watch and listen to them via reproductions. Many of these events aren't actually important enough to warrant being broadcast; it's only the broadcast that makes them important; only the broadcast allows them to blossom into historical reality. Which also means that they are only organized to make a broadcast possible. *Theatrum mundi.*—

To a large extent, "the real thing" has thus become nothing but a pretext for its copies. Our contemporaries are as little interested in "actually" being present in such "original situations" as people reading books are interested in getting their hands on the matrix—that is, the typeset copy of the page; they are as little interested in this as the Platonic cave-dweller is tempted to get hold of the idea.[124]

So here we now sit, "born to see, employed to watch," the brotherhood of today's Lynceus . . . and we watch. Yet Lynceus doesn't seem to be our patron or our hero. And we aren't watching the way *he* did. Since we don't leave our home, because we lurk until our prey gets caught in our web, we watch like a spider. Our home has turned into a trap. Only what gets caught in it is world for us. Outside of it, there is nothing.

So here we now sit . . . and a small bite of world flies into our web and is ours.

But what just flew to us didn't actually fly but was flung to us. And it wasn't a bite of world but a phantom. And this phantom wasn't formed in the world's image but is a copy of a template, of a matrix.— And this copy has only become *ours* because it is meant to turn into a template for us, because we are meant to re-create ourselves in its image. So we won't call anything else "*ours*"; so we have no other world besides the one that is supplied.—

So here we now sit . . . in front of a copy that claims to be a phantom that claims to be a representation that claims to be the world. And we consume the copy and become like it.

But if there were still someone among us like Lynceus—someone who, in an attempt to tear free from this deception, were to set out into the world "to look into the distance" and "see what is close"—this person would soon abandon their quest and return as someone who was fully deceived. For all they would find outside would be the models for the images that were meant to serve as templates for molding their soul. They would find nothing but models modeled after their own images—nothing but the templates required for the production of templates. And when asked what the state of the real now actually was, they would respond that it now had the sole purpose of actually becoming real in the unreal image of its duplicates and copies.—

V. A Leap into General Perspectives

"So let me seem till I become."

—Mignon

"So let me become till I seem."

—V.

25. *Five Consequences: The World Is "a Perfect Fit"—the World Disappears as World—the World Is Postideological—Those Who Are Molded Have Always Already Been Molded—Existence in This World Is Unfree*

Let's recap the power of templates [*Matrizenleistung*]. As we have seen, they exert their formative power in two directions:

1. They shape real events: Since events now attain social reality only in their reproduced form, since they become "real" only as reproductions, they are from the outset configured to serve *as* a basis for their own reproduction.
2. This refashioned "real" now (acts as the "stamper"[125] that) molds the consumer's soul.—

If events now only ever happen the way they were molded to occur in advance, and if consumers, likewise, are always standing ready to consume, five consequences ensue that are decisive for the portrayal of our age:

I. *The world is now "a perfect fit" for the human,* and the human for the world—just like a glove fits the hand and the body fits into a pair of trousers (and vice versa).

Although the expression "off-the-rack item" is now often used to describe today's humans or products, our own comparison to clothing relates to something much more fundamental. It seeks to determine what category of object today's world belongs to.

It is an essential trait of clothing—and this trait also makes clothing its own category of object—that it is not encountered as a separate object "*in front of us*" but as something that "*fits us.*" And it fits so perfectly, *so frictionlessly,* that it is no longer actually noticed or experienced as an object.—

Dilthey famously presented the experience of "*friction*"—that is, "*resistance*"—as an argument in support of "the reality of the external world."[126] Since the human–world relation doesn't entail a neutral stance (toward something that, according to Descartes, could also turn out to be a phantom), but rather a collision and more or less incessant friction, it is extraordinarily important to foreground "resistance" as one of the world's characteristic traits.

This is all the more crucial because every human activity can be traced back to it: Every activity represents an always new attempt to reduce the friction between human and world to the smallest possible extent; every activity attempts to produce a world that might be a better fit, or perhaps even a world that "fits" as perfectly as a piece of clothing.

And it would seem that this goal is now closer to being realized than ever before. The human has been made to fit the world so perfectly and, likewise, the world the human, that the world's "friction" has become unnoticeable. And so:

II. *The world disappears as world.*— This new formula now reveals our reference to the object category "clothing" to be little more than a preliminary remark. For although it belongs to the nature of clothing to remain unnoticed as an object, it doesn't actually disappear when it is put to use. This is only true for one category of objects: *things meant for consumption* [*Genußmittel*]. These exist for one sole purpose: to be annihilated or, rather, absorbed. And the broadcast world belongs to this category.—

The idea of a world totally encompassed by this category is anything but new—as a materialistic *aetas aurea* fantasy, it is, in fact, ancient. This fantasy is called *Schlaraffenland—The Sluggard's Land.*[127]

In this land, one may recall, everything *can be consumed*—skin, bristle, and bones—precisely because nothing actually contains any "skin, bristle, or bones," any unpalatable scraps. In *Schlaraffenland,* the spatial or monetary distance that causes "friction" between consumers and consumer goods has likewise been annulled. This land contains things, "roast pigeons," that "transmit themselves" to their consumers by flying directly into their already open mouths.[128] Since they only exist to be ingested, eaten, and assimilated, these things have one sole purpose: *to lose their object character.* In the land of plenty, things only exist to not exist *as* world.

And this also describes today's "broadcast" world. When it flies into our eyes and ears, it should ideally be "*catchy,*" "*easy to grasp,*" absorbed without friction; it should become ours, yes, even "*us ourselves.*"[129]

III. *Our world is "postideological"*—in other words, it has no specific need for ideology.— Since world events today unfold as stage-managed performances from the outset, false worldviews (i.e., ideologies) that diverge from the world no longer need to be retroactively engineered. *Where the lie lies itself true, blatant lying becomes redundant.*

The postideological condition that *Marx* was hoping for, the one he foretold in his speculation on eschatology and truth, has thus, in a sense, been inverted. While Marx had calculated that philosophy (and for him philosophy also meant "ideology") would be brought to an end through the actualization of truth, a triumphant state of untruth has established itself instead. And because untrue statements about the world are now the "world," specific ideologies have become superfluous.—

The assertion that "world" and "worldview," reality and interpretation of reality, no longer count as two separate things will sound very strange. Yet this bewilderment dissipates the moment it is brought into relation with the corresponding phenomena of our age—the fact, for instance, that bread and sliced bread are no longer distinct entities either (bread is sold presliced today). World events that reach us in an already "sliced" (i.e., staged and preinterpreted) state are impossible to ideologically rearrange—as impossible as it would be for us to rebake and reslice such "ready-sliced" bread. And when things take the shape of "images" from the outset, we cannot then also "form our own image" of them at home. Yes, I am saying we *cannot:* It's not just unnecessary to come up with our own "version" but impossible.

This is an "inability" of a most peculiar kind; it represents an entirely novel type.

When we were unable to grasp or interpret this or that piece of world in the past, this was always because an object withdrew from us, or because it mounted some sort of unbreakable resistance. We've already seen that no such resistance exists today. What is surprising, however, is that it is this very *lack of resistance* that prevents us from grasping and interpreting the broadcast world. Or perhaps this isn't that surprising after all: whereas the smooth pill frictionlessly sliding down our throat is hardly noticeable, we cannot really miss the piece of meat that we have to chew. The broadcast and "easily digested" world is pill-like in this sense. Or, to use a different image, it has become a "*réalité trop facile*" (analogous to "*femmes faciles*")—it has become too easy, too accommodating, it "gives itself" to us the moment it shows up—and so we never actually get to "take" reality or even court it or its meaning.

IV. *Those who are molded have always already been molded.*— What is true of the broadcast world—that its widely assumed binary nature (reality vs. image) is already outdated—is also true of us, the consumers of this premolded world. It belongs to today's conformist situation that humans must be a perfect "fit" for the world, just as the world must be a perfect "fit" for them. The distinction between an initial state in which consumers are like a blank slate and a secondary process through which a worldview is impressed on them no longer applies. Consumers have always already been impressed by images and become receptive to role models and templates. To some extent, therefore, consumers always already have the form they are sought to be embossed with. Every single soul, in a sense, acts as an indented relief to its raised counterpart and thus sits perfectly on its matrix. Which means that the matrix stamp no longer actually leaves an "impression" on the soul, let alone cuts into it. The soul has already been cut to size for it, and so it doesn't actually leave an impression on the matrix either. The corresponding grooves are already there; they already fit perfectly.—

The back-and-forth between humans and world now entails two molded entities: an exchange between a matrix-made reality and matrix-made consumers. This is a highly spooky form of commerce, therefore: It's an exchange between ghosts that are interacting with ghosts (which were created by ghosts). But this ghostlike quality doesn't mean that life itself should now somehow be called unreal; it is real, and dreadfully real at that . . . if not to say really dreadful.

V. *It is dreadful because, in a world that amounts to a postideological Schlaraffenland, our existence is one of total unfreedom.*

It is, of course, true that thousands of things and world-parts our ancestors were excluded from today simply fly into our eyes and ears. We have even been granted the ability to choose what phantoms we would like to have flown to us. But since this delivery robs us of the freedom to approach things ourselves or arrive at our own position, we are nevertheless being deceived. The moment we receive the delivery, we're also surrendered (i.e., delivered) to it. This means that we're exposed to the same form of deceit that those gramophone records trick us with—those records that don't just play a piece of music but also include the audience's applause and impassioned reactions, an applause in which we're meant to recognize our own. Because such records don't just supply us with a thing but also with our own reaction to it, they *supply us with ourselves.*

The thing that happens so openly in these records may, of course, occur much more discreetly in other types of broadcasts; the only difference is the degree of noticeability, for this is what happens in every broadcast. There is no phantom that is transmitted that doesn't also immediately convey its "meaning"—what we're meant to make of it and feel—and since this "meaning" is an integral part of the phantom, it cannot be detached from it. There is no phantom that doesn't also provide us with our own reaction as a special discount.— We, of course, don't tend to notice this. The fact that we're so overfed with phantoms every day and every hour, with phantoms

> that present themselves as "world," prevents us from ever feeling any hunger for meaning and from arriving at meaning on our own. And the more we are stuffed with a prearranged world, the more categorically this appetite is unlearned.
>
> The fact that this unfreedom has come to seem so normal that we don't even *feel it as unfreedom*—or, if we do feel it, then only as something soft and comfortable—doesn't make this condition any less fatal. Quite the opposite: Because this terror moves silently on dove's feet,[130] because it definitively precludes us from imagining any alternative ways of being and any thought of opposition, it's in a sense much more fatal than any open and recognizable deprivation of freedom would be.—

We prefaced our analysis with a fable: that of the king who gave his son a horse and cart because he was traveling cross-country against the king's will. The king had accompanied his gift with the words "Now you no longer need to go on foot." What these words meant was "Now you are no longer allowed to," but their consequence was "Now you no longer know how."

And it would seem that we have now cheerfully reached this very state of incapacity.

26. *A Tragicomic Defense Mechanism: Contemporary Man Produces Frictions as Consumer Goods*

We said that because the world of plenty arrives as a consumption-ready processed good, we are prevented from processing it again ourselves.

This prohibition may make things comfortable, but it isn't easy to bear and accept. As creatures of need, it goes against our nature to live as if in a land of plenty; we are constitutionally ill-equipped for a world that is a perfect fit. On the contrary, we have a natural inclination to respond to our needs, to obtain the things we lack, to adjust these unfinished, unruly things *until* they "fit," *until* they have become "convenient" for us. *We are born with not just a need for sustenance but also the "supplementary need" to satiate this need ourselves.* It is unbearable to live without food; but it is also unbearable to live without the need to ever have to obtain it.

For the most part, we are, of course, not consciously aware of this "second need." But the moment we're kept from satiating it, when our primary needs are met without this being the product of our own actions, we feel cheated—cheated not of the "fruits of our labor," but of our "*labor for the fruit*." When this happens, we no longer know what to do with ourselves, for we expect a substantial part of our life to be taken up by trying to find a means to live. In short, a "second need," a "second hunger," now makes itself felt; a hunger not for prey but for effort, not for bread but for its procurement, not for the goal but for the journey that now itself becomes the destination.

That the "leisure classes" whose lives were exempted from exertion were afflicted by a positively *voracious appetite for effort* is well-known. But neither the foxhunter nor he who went fishing on a Sunday had any actual hunger for trophy animals—not primarily so, at least. They hunted not for prey but for the opportunity to be allowed to hunt. And when they killed a fox, stag, or perch, it was often because without a target it would not have been possible to obtain the pleasure of taking aim. The prize was nothing but a pretext for the toil and for covering the distance.

But this has now become a general condition. Improbable as this may sound, today, everyone belongs to the "leisure class" (and this includes every worker). This shouldn't be misunderstood; it only means that everyone now has their essential needs met by prefabricated products. Even the poorest cotton picker in the South buys precooked beans that are ready to eat. Yes, especially he. It is still as true as in the nineteenth century that workers have no share in the fruits of their labor, but now they equally do not participate in the work required to deliver consumer goods to them (and this especially applies to leisure products). And without this twentieth-century equivalent, the portrait of our age would remain incomplete: *Their lives—all of our lives—are doubly alienated: Life consists of not just fruitless labor but also fruits without labor.* "To eat fish," a Molussian saying goes, "one must go hunting rabbits; to eat rabbits, one must go fishing. There is no record that anyone hunting rabbits has ever eaten one."

This second alienation between labor and its "fruits" is the trauma that characterizes our situation of plenty. It is no surprise, therefore, that there is today such a voracious appetite for effort: a desire for homegrown fruit, or at least to try it once; the need to reach a destination by going on a hike; the dream to sit at a table that was built by one's own hand; the desire for resistance and the exertion of overcoming it.

Contemporary man now satisfies this craving, and this by artificial means. To ensure that he can navigate resistances and derive some pleasure from this, he specifically creates forms of friction, or rather, he has them custom-made for him. *Resistances have turned into products.*

This is nothing new, of course. Sport (which was not just coincidentally the twin sibling of industrialization) was already in wide use as a stimulant of this kind. To ensure that they *could* be conquered, the most unclimbable of summits were suddenly introduced as obstacles in our way (even though they could only be reached by traveling to them). Only this relabeling could ensure that their conquest could actually be enjoyed.

Yet the comparatively young, rampant hobby encapsulated by the slogan "*do it yourself*"[131] is much more characteristic. There are now millions who spend their free time putting obstacles in their own way. They give themselves technological challenges, renounce the "facilities"[132] the age provides, or else cobble stuff together that they could obtain at every corner. In 1941, I was already employed in a factory where "handweaving looms"[133] were being mechanically mass-produced for women who were craving the opportunity to finally be allowed to savor the pleasure of hard work in the evenings after working their job. Men, meanwhile, positively welcome any electrical faults in the house and every loose screw in the car, for these promise some bitterness and effort to sweeten their Sundays. And it is no coincidence that satirical magazines especially love the figure of the hobby watchmaker who de- and reassembles pocket watches; for when this poor child of our epoch also wants to have a go at making something himself, he's left with no other option than to start by taking something apart (his prefabricated world yields no other raw materials). Condemned to such demolition, he thus needs to first produce "raw material" from a finished product to then reassemble it in a second act of creation. This is the only way he can gain the small joy of having made something himself—or at least almost. The kinds of obstacles he puts in his way are thus identical to jigsaw puzzles, since all his act of creation amounts to is piecing together finished elements à la Hume. And the popularity of these puzzle games, which are also played by grown-ups, belongs to the same context.

Perfect happiness, however (and he would deserve to be happy, for how can he be blamed for what he is, the ill-fated epoch he was born into, or the pathetic nature of his attempts to somehow escape?), perfect happiness is what he hopes to find when he goes for a drive in the country on the weekend (to light a fire "all by himself" with the help of his brand new fire-spark gadget, the one "guaranteed to start fire in the most primitive way"). For as he traces Robinson's footsteps and barbeques the "frankfurters" he's kept fresh on dry ice (all by himself), when he puts up his tent as if he were a pioneer, and builds the table for his portable radio from ready-made blocks (yes, all by himself), true happiness cannot be far.

It will already be clear enough that this youth movement of grown-ups—this insatiable appetite to jump back to an earlier stage of production and get a break from being supplied with ready-made goods—is doomed to fail (though, as one of our epoch's few tragicomic traits, it would certainly provide an ideal subject for a play, a farce about our times). For in the hours after their toils at work, these millions are toiling entirely in vain. Industry (which had triggered this counter-movement in the first place) has now already taken possession of it, just as it has taken hold of every other movement that promises the sale of new products by generating new needs. So, even before the "do it yourself" craze reached its peak, companies had already started to make prefabricated "camping-gadgets"[134] to facilitate this trend. These items serve the paradoxical purpose of providing those hobbyists, who feel an urge to let off steam by creating obstacles and doing things themselves, with devices that can make this process as comfortable as possible. And since they are fully used to relying on whatever is advertised to them as the "most practical" solution (i.e., the most time- and labor-saving one), these customers cannot simply shake this deeply ingrained habit when transforming into self-makers [*Selbsthersteller*] overnight. And so, they really do buy all these apparently "practical" goods, which, of course, instantly ruins any prospect of deriving pleasure from doing things themselves. In just an instant—poof!—their pioneer tent stands there as if by magic, since everything required for its assembly suddenly appears in their hands. As if they were playing with model bricks, their own contribution is thus reduced to following a step-by-step instruction. And so they find themselves surrounded by the void again, and it's a true blessing that they've brought their radio along and they can summon back their phantoms in the customary way.— If this is not "dialectics," I have no idea what this word might mean.

The movement dedicated to "Creative Self-Expression,"[135] a movement that has already existed for a while (e.g., as "creative painting" or "creative writing"),[136] must also be included here. It encourages thousands to attempt to accomplish something themselves in the hours after work, on Sundays, or in old age (when one can no longer hope to be considered for a job anyway—also: "life begins at seventy").[137] This movement, hence, encourages activities in which "labor" and "the fruits of labor" are visibly connected.— All this, of course, also represents a countermeasure against the constant supply of ready-made products—or, to be more specific, against the never-ending supply of fully interpreted "world-pictures." As such, it must be understood as an attempt to smuggle some small consoling effort back into the hopelessness of existence in a land of plenty. Yet this movement is, of course, also doomed to fail. Let's not even mention that the followers of this movement—those who have suddenly become "creative" out of boredom, hygiene, or simply because "creative expression" is a "must"[138]—cannot actually have an oeuvre that matters to them, for the only thing they care about is having something to express. What is decisive here, rather, is that "being creative" is taught by means of mass-education or even via radio in the form of distant-learning courses ("how to get creative");[139] for this means that the elements of creativity are home delivered as a ready-made package too. In short, this tragicomedy is perfectly equivalent to that of our artificial Robinson. "Creative expression" is here, too, but an excursion into an antiquated mode of production and existence, undertaken by antiquated humans whose journey is facilitated by all the goods and ready-made luxuries the modern age can supply. Because both the mode of transport and the style of travel contradict the travel destination, this kind of journey can, of course, never arrive at anything.—

27. One More Time: The Real Is the Reproduction of Its Reproductions—the Metamorphosis of the Actress V. into a Reproduction of Her Own Image

It is likely that our concluding thesis amounted to the most bewildering claim of our entire investigation. It asserted that the real is already taking place with a view to being reproduced—for the sake of it even. It claimed that the real needs to facilitate its reproduction since its reproductions have now attained a more substantial social reality than the real thing itself, which turns the real into the reproduction of its reproductions.

To demonstrate that this isn't just some theoretical paradox, I will end by recounting a very concrete event. And at this point in our discussion, it no longer really matters that our example—the metamorphosis of the actress V. into a reproduction of her reproduction—is lifted not from the world of radio and television but from the movie industry. This narrow horizon was already pierced in the previous section—and not without good reason; treating the categories "phantom" and "matrix" (which are our sole concern) as exclusive to the world of radio and television from which we originally deduced them would have been misleading. Their range of application and the significance of our results are much broader than anticipated at the outset.

I quote from my Californian diary:

1941

Upon viewing V.'s screen test half a year ago, the producer M. remarked, "First, my dear, you need to become *more photogenic*,[140] then we'll see." His words had meant to say that "until you've more successfully used our phantoms as the matrix for your actual appearance—until you've remolded yourself in their image—there's just no way you'll ever be considered as a phantom to be seen on our screens."

Although V. had always been exceedingly proud of her wholly singular looks, her hunger for a career as a phantom proved to be much stronger than her pride. And so, she soon began the work of remolding herself, and she pursued this with an almost ascetic dedication. She sacrificed all her life's joys and all but depleted the last financial reserves of her soon-forgotten family and the ex-friends she had renounced. Since nobody can do this work alone, she sought the help of all those specialists (and here in California they already form their own occupational group) who regard the phantom as the ideal and actual humans as inferior, low-grade material in need of improvement. These specialists thus earn their living from the gap between the phantom and reality; their businesses are built on the fanatical desire of those who, like V., wish this gap to be surgically removed. And so, V. began running from the beauty salon to the masseur and from the masseur to the beauty salon. She surrendered herself to weight-loss clinics and wrinkle and crow's feet specialists; and—to the surgeon's great gain and her own ruin (or so I had sought to implore)—she even went under his knife. She let herself be worked on from the outside and the inside, back and front. She slept by the clock, and, by the sweat of her brow, she completed the prescribed hours—first in this position, then in that. Instead of enjoying her food, she started to weigh every salad leaf, and instead of smiling at me, she smiled at the mirror—and out of duty rather than joy.— In brief, never in her entire life had she worked so hard, and I doubt that even the initiation rites the young women of the Vedda[141] once had to undergo were more brutal than the ones V. needed to endure until she could be ceremoniously admitted into the phantom-world. So it's certainly no surprise that she soon became nervous, if not to say unbearable. As if already invested with the privileges of a phantom, she began to exact her revenge on the people around her [*Mitwelt*], treating us as nonexistent, as mere air—albeit air she had every right to breathe in and out as she pleased. She had lived this life for half a year (during which her old Adam or old Eve were so thoroughly reworked that there really was nothing left of them), when a new and unimaginably glamorous human arose out of her: the phantom. This epiphany happened about two weeks ago, and she immediately went to visit her phantom-broker a second time. Of course, it's not entirely accurate to say that it was *her* who set out that day. With her new hair, her new nose, her new figure, her new walk, and her new smile, she had already become a ready-made product, an indefinite article, someone else, "everybody else"[142] (since her hair was not really new but actually old, familiar from every movie, just as her nose and smile have been seen everywhere before). "So much the better," she said, and she was probably right. For as she told me after her second screen test, she had immediately thought it a good sign that the phantom-dealer did not recognize her, and this, apparently, had been an enormous boost to her "self-confidence" during the test (if the term "self-confidence" still applies here).— And today, fourteen days later, today's the day, the news has arrived, the unlikely thing has happened, the new test was accepted as OK, her life's dream is fulfilled, and this fulfillment is to be contractually confirmed. In other words, *she has climbed to the rank of being a matrix for matrixes;* she is allowed to serve as a matrix for the film images that will themselves serve as the matrixes for our taste.— She of course claims to be incredibly happy about this—but I'm not quite sure. The procedure of reshaping herself has caused so much serious damage that I find it difficult to say that it's *she* who is happy. The other her—the new one—may well be; but this is someone I do not know, someone I can do without. And because there is now only her, and because the girl walking next to me on the street is already moving the way she moved in the screen test (the way she will be expected to move in future films), she has already disappeared. She has already become the copy of her own image and the reproduction of her future reproductions. The final goodbye (which she has not yet spoken out loud but already administered anyway) is now probably just a few days away.

28. *It Is Not the Spectator, but the One Who Is the Spectacle That Commands Respect*

Although V.'s metamorphosis doesn't belong to the narrow scope of our original example, it is especially instructive—it demonstrates how the primacy of images over reality becomes a living motive when it is internalized, leading to a specific set of behaviors.[143] It also shows how the transformation of a person into an image matrix happens as an actual life process. The thesis advanced in our study—that being an image is today considered "ontologically superior" to "being"—is fully encapsulated by this case, and so we will continue to examine it.

It would be too easy to dismiss V.'s craving to become an image with words like "vanity" or "hunger for fame." The desire to be looked at and spoken about (vanity), and the hope that receiving this kind of attention will allow one to be something more or *be* at all (hunger for fame), don't explain anything. Vanity and the need for fame aren't explanations but philosophical problems in their own right, and highly obscure ones at that.—

Like thousands of others, V. grew up in a world in which only phantoms ("pictures")[144] were held in high esteem, a world in which the phantom-industry was (not without good reason) believed to be sensationally real. It was this world, and the prestige and power of these phantoms to act as templates, that she was molded by. From a young age, she had felt it to be an intolerable ordeal that she somehow "existed" within this world of images yet still only belonged to it as a "nonimage," as a "nonidol," and this soon developed into an unending sense of inferiority and invalidation. One needs to be clear about the etiology of the particular feeling of inferiority manifest here—it is a historical first. And although it hasn't yet been discovered by the discipline of Individual Psychology (which deals with nothing but feelings of inferiority), it is today's decisive variant of this feeling: *The idol-world that intimidates the insecure consists no longer of fellow humans but of human-phantoms*—and even things.[145] V. didn't feel inferior before the threatening example set by her parents, her siblings, her rivals at school or at the beach, but before the example set by reproduced images. Her neurosis was a symptom not of poor "social" adaptation but of poor technological adaptation to the image-world (as was already discussed in the introduction via an analogous case). If, for the petit bourgeois, it was a torment to live anonymously in an exclusively aristocratic world as a nonaristocrat "who doesn't count," it was now similarly unbearable for her to live in a world of exemplary phantoms.[146] She was constantly pursued by the feeling of being a *quantité négligeable,* if not even a nothing, and she was hounded by the fear that—should she not succeed in becoming a phantom—she might one day wake up and realize that she had never existed at all. In brief, the reason for her suffering was *a lack of ontological prestige.* When she began to fight for her career and to become a phantom, she was thus actually fighting to attain more *being*—yes, to even exist at all. Whereas Mignon sings, "*So let me seem till I become,*" her motto could have been this line in reverse: "*So let me become till I seem*"—till I'm *allowed to be* because I seem to be.[147]

And her craving for attaining *being* through *semblance* cannot be spelled out in any clearer terms than in the few throwaway remarks V. provided herself:

Diary

She had only just succeeded in her self-transformation when she suddenly exclaimed, "My God, what I used to be!" (and her contemptuous voice made it perfectly clear just how far up the ontological ladder she believed herself to have climbed in comparison to her earlier life). She had, of course, meant to say that she had once been *nothing* because she had only "been" and nothing more. Before her transformation, she had simply "*existed*" [*dagewesen*],[148] and this just as herself, and only ever once, only ever in the place she happened to be. Or, to restate this in the negative, because she was something unreproduced and unaltered, she had never been considered something to be looked at. She had never received any verification of her being because there were no consumers who confirmed receipt of her existence; no consumers who, molded by her *en masse,* could have confirmed that she was there. In brief, she had not been an idol to be emulated, nor a product for the masses; she had only ever been an anonymous "*who,*" never a "*what.*" And within the world she was part of, she was quite right to think this way: In the Hollywood-world, someone who is only a "who" is a nothing, and is hence simply not "there" [*da*] (when measured by the ontological status a "what" can enjoy).

V. didn't use this vocabulary, of course. But to her, these arguments would have sounded like simple "truisms,"[149] like trivialities that go without saying. And if one accepts the axiom of economic ontology that "unprocessed things have no being," that "reality is only produced by means of reproduction," then they really are. V. had simply applied these axioms, the validity of which she had no reason to doubt—in the world she was part of, these principles were considered binding, and they had always functioned smoothly.

Given the labor and effort she had invested, it certainly wasn't entirely fair that not only did I not stay quiet when she exclaimed, "But what was I before?," I even began to tease her. I jibed that she'd gained her "authentic being" [*eigentliches Dasein*] the very moment she'd "disowned" herself, just as she'd robbed herself of her own self. In the eyes of someone who, by the sweat of their brow, has finally managed to become a "what" instead of remaining a "who," someone like me—who is still hanging around as a "who" (and even deems this something to be proud of)—will, of course, only ever seem like a ridiculous gnome. And she taunted me accordingly: "*Here you go again with your 'self,'*" she sneered. "*Who cares about that?*" And since her last phrase had turned *the level of one's demand into the measure of one's worth* and the criterion of being, her question very effectively shut me up.—

The image-world, I just noted, had once made her feel like "air," like "nothing," like a petit bourgeois in an exclusively feudal world. And indeed, when I try to recall how she comports herself now—her new gestures, her new tone of voice, her new walk—the only image that springs to mind is that of a *snob* who has only just finally managed to belong to aristocracy but now exaggerates this affiliation. It is, of course, no coincidence that the Greek word for "nobility," *ἐσθλός*[150] (which derives from the root of the Greek word for "being"), denotes those people who "are" somebody, those who enjoy a higher degree of "being" than others. And V.'s ontological status is now likewise higher than the status of others, for she is now here as a mass-product, a prospective model for countless copies, whereas she had once, in an embarrassing past, only existed in the hapless singular and as unworked raw material—which meant she had lived in some obscure depths, far down among consumers, among the pathetic plebs.

Of course, it sounds strange that this aristocratic status was attained by rising to the rank of a mass-product. The mass and aristocracy contradict each other. Yet it makes no difference whether we say that this happened "by rising into the template-world" in which she transformed into an idol [*Vorbild*], or by rising "into the world of images" or "mass-products." Only idols turn into images, and they only become images by being multiplied and disseminated in countless copies.[151]

There is, by the way, another reason for the high esteem mass-products are held in today: A significant portion of today's products don't actually exist for us, it's us who exist for them; we are meant to sustain their continued production as consumers and buyers. But if our need to consume (and the lifestyle this need makes us adopt) is only created or at least shaped to ensure *that* there is a market for goods, then we are only a *means*, and as such ontologically subordinate to ends. Those who, like V., manage to soar from obscure depths to gleaming heights—to heights at which they may actually be "viewed" as consumer goods (instead of merely living off them)—are now held in high "regard" and thus belong to another, "more substantial" mode of being.

This desire to be considered, this process of turning into someone "substantial enough to be looked at," is especially plausible in V.'s case. By joining the motion picture industry,[152] she really did turn into something that was supposed to be *looked at and watched*.

Diary

Because she is now being considered for a role in which she will "be looked at," she of course can't acknowledge a gnome like me, someone who might at best be considered a potential consumer of her phantom-self (but only on the already-rare occasions he goes to the cinema). For a phantom, it is of course an outright *mésalliance* to go out with a real person; such ties between a product and a consumer are simply "unthinkable." I fear this means that V. will now have to look among her own kind for company, among phantoms. Well, she won't "have to" look, for the circle of phantoms is a world unto its own (although visible to all, it is nevertheless unreachable), a world to which she will now automatically be admitted. She will no doubt find someone there (or I should perhaps say "something"), someone who is also a "what," someone who also lives for being looked at by everyone; someone with a similarly larval heart beating in their chest and with whom she can now be one product-heart and one product-soul, and whom everyone will consider "a great match."—

If it were only a matter of formal intelligence, then V. would certainly have had no problem understanding me, for she doesn't lack intelligence. But understanding doesn't just depend on reason; it also depends on one's social standing. The aristocratic rank she now belongs to prohibited her from *still* understanding such things. If it was "beyond her,"[153] this isn't because it was "above her"[154] but because it was "below her"[155]—that is, she had already climbed too high to still *be allowed* to understand me. And so it would also have been unfair to accuse her of bad intentions or to become angry with her. It wasn't *she* who was acting; she was merely "going along." For had she swum against the stream, had she rejected what everyone around her deemed self-evident—*that becoming a product amounted to a promotion and that being consumed by everyone was proof of one's existence*—this would have been an act of downright arrogance.

3

Being without Time

On Beckett's Waiting for Godot

There are those obfuscating enlighteners who want to persuade us that it is more rigorous to discover religion in every important work of literature than to discover literature in every sacred text (as our fathers once did). These figures are extraordinarily industrious. Having only just set upon Kafka's great works with their premature piety, they are now already closing in on Beckett's clownery—intent on dressing it in false honorific robes. Perhaps it will this time be possible to put a stop to their skewed interpretations before it is too late. The following remarks will demonstrate just how little we will let the play be deprived of the honor it deserves.

1. *The Play Is a Parable in Reverse*

All commentators agree that it is a *parable,* but the argument about its meaning is in full flow. While bickering about *who* or *what* Godot might be, they are quick to provide answers as if this were Nihilism 101: "death," "the meaning of life," "God." But none of these commentators has racked their brain over the question of what mechanism all parables obey (and thus also Beckett's). This mechanism is called "*inversion.*" What does "inversion" mean?

When attempting to say that humans are like animals, did Aesop or La Fontaine actually show humans as animals? No. They swapped the elements in the equation—subject and predicate—and claimed that animals are humans. The curiously amusing estrangement effect fables achieve is rooted in this substitution. Brecht did the same thing a quarter century ago in *The Threepenny Opera.* To communicate that the petit bourgeoisie [*Spießer*] are robbers, he turned the subject into the predicate and vice versa; he introduced robbers as bourgeois. This, the fable writers' *quid pro quo,* must be understood if one is to approach Beckett's fable, for he, too, employs this same device. And this in a most ingenious manner: To tell the fable of a mode of existence that no longer knows any form or principle, and in which life no longer continues,

Beckett destroys the form and principle of the fable. The destroyed fable—the fable that no longer progresses anywhere—now appropriately turns into a fable about life that no longer progresses. If Beckett's "inversion" is translated back into its prior state, then his meaningless parable about the human turns into a parable about the meaningless human [*sinnlosen Menschen*]. True, this fable no longer corresponds to the classic formal ideal of the genre. Yet, since it is a fable about the mode of life that "no longer has any point to it"—the kind of life, therefore, that can no longer be condensed into the fable form—its formal defect and failure are the very moral of its story. It allows itself to be inconsistent because such lack of consistency is its topic; it takes the liberty of no longer narrating a "plot" or "dramatic action" [*Handlung*] because it is about the form of life in which no action is ever taken; it gives itself license to no longer offer a "story and character background" [*Geschichte*] because it portrays the ahistorical human, the human who has no story [*geschichtslosen Menschen*]. Nobody needs to deny, therefore, that the events and fragments of speech from which the play is assembled start and break off without any identifiable motive, or that they simply repeat themselves (often in such a perfidious manner that the characters don't even notice that repetition is at play). For this lack of motive is motivated by the subject of the play: life that no longer has any motives and no motor propelling it forward.

So, even though it is a *negative fable*—a fable in reverse—it is still a fable nevertheless (just as a photograph of a total solar eclipse is still a photograph, i.e., an image created by light)—even if for no other reason than the fact that, despite its lack of an extractable teaching, it still functions on the level of *abstraction*. Whereas the novels of the last 150 years were content to simply recount a formless life, *Waiting for Godot*[1] portrays *formlessness as such*. And its topic is not the only thing that is abstract, for its characters are abstract, too: At least on some level, the "heroes" Estragon and Vladimir are no doubt meant to represent "the human as such." In fact, they are "abstract" in the cruelest sense of the word: they are "*abs-tracti*," people who have been torn off, pulled away.[2] Torn off from the world, they no longer have any business being in it and no longer find anything in it either, which makes the world become abstract too. This is also why there is no longer anything on stage—nothing but the tree that stands center stage, a utensil that is indispensable for the fable's meaning. As an equivalent to the biblical "tree of life," this tree portrays the world as an instrument that is always there for one to commit suicide—it establishes the equation through which "to live" means "to not hang oneself." *The two "heroes are only alive but no longer in the world."* In view of the utterly unforgiving consistency with which this logic is executed, other portrayals of wordlessness (of which there is not exactly a shortage in contemporary literature, philosophy, and art) seem positively comforting. Döblin's Franz Biberkopf at least still stood right in the middle of the whirl of the busy world he was no longer part of; Kafka's surveyor K. at least still tried to enter his "castle"; not to mention the ancestor of today's generation, Michael Kohlhaas, who was certainly still grappling with this world, even if he treated it as if it were the Kantian kingdom of morality. All these figures, one way or another, still had "world": Biberkopf had too much world and therefore no definite one; K. had worldly hope; and Kohlhaas a false one. They had not yet arrived in the nonworld that only Beckett's creatures have reached. Only in their ears has the din of the world that still numbed Biberkopf become silent; only Beckett's creatures forget to even want to enter the castle of the world; only they abstain from measuring the world against the standards of another one. If this true worldlessness is to be represented by theatrical or literary means, it is quite obvious that it requires unusual devices. Since there can no longer be *a collision with the world* where there is no longer a world, *the resources of tragedy can no longer be mobilized* either. Or, to be more precise, what is tragic about this mode of existence is that it is no longer even granted any recourse to tragedy: It is always in its entirety a farce (instead of merely containing farcical episodes, as was the case in the tragedies of our ancestors). It can only be portrayed as a farce—*an ontological farce, not a comedy.* And this is what Beckett does.

How closely farce and abstraction are related is something Don Quixote had already shown us. Don Quixote, however, only abstracted from the way the world is, not from the world as such. Philosophically speaking, Beckett's farce is thus "more radical": He creates his farce-like comedy not by bringing humans into a world or situation they do not fit into and now collide with; he creates his farce by making his humans sit down without seating them anywhere. This turns them into clowns, for the metaphysical comedy of clowns entails a categorical confusion of existing and nonexisting things—the clown trips over nonexistent steps or treats them as if they were not there. But, in contrast to this type of clown (Chaplin is one of them)—a clown who, in a sense, collides out of principle and incessantly keeps busy to ensure that laughter continually reerupts—Beckett's heroes are *lazy or paralyzed clowns.* As it is neither simply this nor that object that is no longer there but the world as a whole, they no longer concern themselves with the world at all. With this, the type of human he selects as his *fabulae personae,* as the representatives of today's humanity, is decided: They can only be *tramps*—creatures who have dropped out of the world-plan (namely, out of the schema of civil society)—creatures who no longer have anything to do because they no longer have anything to do with it.

2. *The Motto: I Am Staying, Therefore I Am Waiting for Something*

Nothing to do anymore.— In the over twenty years since Döblin offered us in Biberkopf the portrayal of the human who is condemned to doing nothing and thus to worldlessness, a set of diverse historical developments has made the status of "doing" even more suspect. Not because the number of unemployed people has increased, which is not the case, but because millions who are in effect still doing something nevertheless feel like they are "*being done*"—that is, "*acted upon*" while doing so. They are active without ever deciding the objective of their work or being able to know this objective; they may, in fact, even work in the full knowledge that their work is suicidal. In short, *today, one is bossed around to such an extent that even actions have become a form of passivity. And even in situations where actions make one feel like death because of the effort they require (or when they bring death even), they still feel like a form of doing nothing or good for nothing.* Estragon and Vladimir, who both do absolutely nothing at all, are representative of millions of actively employed people—nobody will be able to deny this.

What makes them so representative is, of course, only the fact that, despite their inactivity and the pointlessness of their existence, they want to "*carry on*" all the same; they are only representative because their portrayal has not turned them into tragic candidates for suicide. Estragon and Vladimir are as far removed from the noisy pathos of the desperado heroes found in nineteenth-century literature as from the hysterics of Strindbergian figures. They are more genuine: just as unemotional and inconsequential as regular mass-humans happen to be, who aren't pushed into calling it quits by meaninglessness either; even the nihilists among them still want to live, or at least "don't want to not live." But even this antivoluntarist formula might still be too much of a doctrine. The Estragons and Vladimirs live on precisely *because* they are living a pointless life, because the habit of doing nothing—nothing by their own initiative—has already stunted the choice of not doing so, the freedom to end things.[3] Ultimately, no actual motive may actually be required: They live on because they happen to be here, and because anything other than existing is not presented as a life option.

Beckett's play chronicles this mode of "life"—the life of the human who carries on because it happens to exist. But the play does this in a manner that fundamentally diverges from all previous literary portrayals of despair.

The following motto could be put into the mouth of every classical desperado figure (Faust included): "*We have nothing more to expect, so we won't stay.*" Estragon and Vladimir, however, deploy "inverted variants." They seem to say, "*We are staying, so this means we are waiting,*" and "*We are waiting, so we also have something to wait for.*"

Their slogans seem more positive, more optimistic, than their ancestors' motto. But they only sound this way. For it cannot be claimed that the two are waiting for anything definite. So much so that they constantly need to remind each other that they are waiting and what they are waiting for. They are actually waiting for nothing at all, but since they quite evidently exist and continue to exist day in and day out, it is also impossible for them not to conclude *that* they are waiting. And because they are "waiting" every day, they cannot help but conclude that they are waiting for *something.* Just as we can't help but conclude that the people we see standing in the rain late at night at a bus stop are waiting for something—something that "won't keep them waiting." It is thus pointless to ask who or what Godot might be: *Godot is nothing but the name given to the fact that a mode of life that pointlessly continues is one that miscomprehends itself as a life that is "waiting," "expecting something."* The positivity of the two characters amounts to a double negation—to the *inability* to acknowledge *the senselessness* of their position—and not to anything positive, therefore. And in saying this, we are in effect only reiterating what Beckett has himself disclosed about the title of his play: that it is not Godot who matters to him, but only the "waiting."

3. *Beckett Does Not Show Nihilistic Humans, but the Inability of Humans to Be Nihilists*

French commentators turned to the Heideggerian expression "to be thrown into the world" [*Geworfenheit*] to characterize the play's portrayal of the mode of life in which one continues to wait because one happens to exist. This is wrong. With this term, Heidegger acknowledges and openly signals the contingency of one's own existence (only to then stubbornly reconceptualize this arbitrariness in order to turn that very existence into one's own "project" [*Entwurf*]). But the two heroes of Beckett's play do neither one nor the other—just like the millions they represent: They neither acknowledge their existence as contingent nor attempt to offset their own contingency by converting what was "thrown to them" into a positive "project." They are far more unheroic than Heidegger's *Dasein*, far more believable, far more "realistic." As little as they would be able to think, while standing before a house or chair, that these objects are exclusively "there," or "there for nothing," they would also not be able to entertain the thought that their own existence is "nothing" or void.[4] What they are, therefore, is "metaphysicians": *They are unable to give up on the concept of meaning.* Heidegger's expression explicitly dethrones the concept of meaning; *yet Vladimir and Estragon*—whose experience of being makes them deduce that they are waiting for something—*are the lord keepers of the great seal of meaning in a situation that is manifestly meaningless.* It is not just wrong to say that they are "nihilists" (as they have been portrayed); this is the exact opposite of what Beckett wants to show. Because they do not lose hope (and are unable to lose it), they are in fact naive and hopelessly optimistic ideologues. *Beckett is not staging nihilism but the inability of humans to be nihilists, even if the situation is of unparalleled hopelessness.* The abject sadness the play exudes is thus not so much generated by the hopeless situation the two heroes are in as by the fact that as people who just continue to wait, they are so absolutely unable to face up to this situation. It is rooted in the fact that they are *not* nihilists. It is this inability that also makes them so powerfully comic.

There is nothing funnier, as comedy has established in the course of over two thousand years of praxis, than total trustfulness that is totally unjustified. This is evident in its affinity for the figure of the cuckold—that is, for the man who, despite all evidence to the contrary, is in a sense constitutionally prevented from ever suspecting anything. Estragon and Vladimir are very much his brothers: They resemble the "*maris imaginaires*" from the French fairy tale[5]—those "men who were born as husbands" and who, every night, would once again begin to wait for their wives to come home, even though they lived alone on deserted islands and had never married.— In Beckett's eyes, we are probably all like them.

4. *Proofs of God's Existence* Ex Absentia

No, not with a single word did Beckett imply that they will come, that "Godot" exists, or that he will come. That the name "Godot" conceals the English word for "God" is as certain as it is definite that the play is not actually about God but only about *the concept of God.* It is thus no surprise that God is evoked only in vague terms: What it is that God does, or so it says in the play's theological passages, was not known; as far as one can tell from what has been heard, it is assumed that he does nothing at all. As for his messenger boy (a figure who is the brother of Kafka's Barnabas), the only thing he has to say as he arrives every day is that Godot sadly won't be coming today but will surely come tomorrow. With this, Beckett makes it clear enough that it is precisely Godot's nonarrival that sustains the waiting for him and the belief in him. "Let's go."— "We can't."— "Why not?"— "We're waiting for Godot."— "Ah!"[6]

The similarity to Kafka cannot be missed; it's impossible not to think of "the message of the dead king."[7] But it is irrelevant whether this is a case of direct literary influence; both Beckett and Kafka are *enfants du même siècle* and thus sustained by the same preliterary source. Whether it is Rilke, Kafka, or Beckett, their religious experience paradoxically always stems from religious futility—the fact that they do not experience God. Paradoxically, it thus stems from an experience they share with nonbelief. In Rilke, it stems from the unreachability of God ("First Duino Elegy"); in Kafka, from the unreachability during the quest (*The Castle*); in Beckett, from the unreachability while waiting. They all share the proof of God's existence: "*He isn't coming; therefore he exists*"; "*Parousia isn't happening; therefore he exists.*" The negativity we know from "negative theology" seems to have affected religious experience itself, and in this way, it is incredibly intensified. In negative theology, God was only described *by means of missing qualifications*, yet it is now *the absence of God as such* that is turned into the proof of God's existence. It scarcely seems possible to deny that this applies to Rilke and Kafka, and the motto Heidegger borrowed from Hölderlin—"*Where there is danger, grows the saving power also*"[8]—belongs to the same category of proof *ex absentia.* Beckett's characters now also belong in this company. But only his characters, of course—not Beckett himself. His is a special place, for he doesn't go along with the conclusion he has his creatures reach (that Godot's nonarrival is proof of his existence); in fact, he even portrays this conclusion to be absurd. *His play is thus not religious but at most about religion.* "At most": In a sense, it only portrays a faith that believes in nothing but itself. And this is not a faith.

5. *Life Becomes a Pastime*

When one asks oneself how such a disintegrated (yet nevertheless not abandoned) life "takes place" *in concreto,* one is asking about its temporal *movement.* I say *movement* because the impossible (as our use of language knows) no longer "goes" anymore—or, expressed in positive terms, because time only moves forward in a life that is pursuing a goal, setting out for something. Which is what the life of Estragon and Vladimir no longer does. That is why Beckett's play rightly marches on the spot. That's why the events and conversations begin to go around in circles (just like actors who are playing passersby exit the stage on the left only to reenter as allegedly different people on the right). The before and after become like left and right, their temporal character neutralized. After a while, the circular movement creates the impression of being stationary, time appears to be standing still, and—if it is permitted to re-create Hegel's term "bad infinity"—turns into "bad eternity."

Beckett executes this in such a logically consistent manner that he simply repeats the first act again, if with subtle variation (which is probably without precedent in the history of theater). Against expectation, he thus never presents anything unexpected; this doesn't just allow him to achieve an absurdly astonishing effect, he also stirs in us the specific type of horror that any encounter with *amnesia* instills. For, with one exception, his figures are not in the slightest bit aware of this repetition; and even as their attention is drawn to this, they remain unable to recognize in any way that what they are experiencing or saying is nothing but the recap of what they had experienced or said yesterday or the day before. And the inclusion of this amnesia is entirely consistent: For where there is no time, there is also no memory. But time is not as "stone still" [*versteinert*] as it often is in Kafka.[9] Since Beckett still retains a rudimentary amount of activity—and it will shortly become clear what kind of rudiment this is—some minimal sense of time remains. Time may not actually flow, yet its substance can still somehow be pushed back, shoved aside, and turned into "the past"; it has, in a sense, turned into a stagnating *time mush* [*Zeitbrei*]. It can, of course, only be set in motion for a few seconds or, at best, minutes. The moment the hand keeping time in motion is withdrawn, even if just for an instant, then everything slides back into itself (and nothing remains that might reveal that something had happened). But, at least for a passing moment, one was able to produce and to enjoy "time."[10]

The rudimentary activity that, for a passing moment, was improbably able to set this time mush in motion no longer really amounts to an actual "activity." For this action has no other aim than to propel time forward, and in a life that is active in the "normal" sense, such movement is the consequence of action, not its aim. The activity amounts to nothing other than *killing time* [*Zeitvertreib*]. Once the sole aim of being active becomes that of achieving a consequence in the sense of "temporal succession," it has given up on any other kind of consequence. When Vladimir and Estragon play at "going away," they stay; when they play at "helping each other," they hardly lift a finger. Even their good mood or outrage vanishes so abruptly that their sudden nonexistence feels like an explosion in reverse—like *a negative explosion*. And yet they continually "become active" once again because such activity keeps time moving—it pushes a few meters of time behind them, which plays them closer toward the alleged Godot.—

This goes so far—and the play here takes on genuinely heartbreaking tones—that the two even suggest feelings and emotions to each other. They even fling their arms around each other, since affective motions are also a form of movement and hence promise to push the sludge of standing time forward a few additional meters. "Well?" Estragon notes, what "if we gave thanks for our mercies? At least there's that?"[11] And as Vladimir asks what "At least there's that" might mean, Estragon responds, "That much less," by which he means to say: This will shorten the stretch of time that separates them from Godot. And indeed, such activation of togetherness, the continual reinvestment in the possibility of enduring meaninglessness together, is the most consoling remedy for bearing this windless lull. Without their touching and desperate mutual attachment, without the idling engine of their conversation, without their bickering, without them leaving each other and finding each other again—all of which takes time—they would be truly lost. It is thus not just for formal technical reasons that Beckett presents us with a pair of heroes—not just because a play about a single person waiting alone, about a *Robinson of expectation*, would congeal into a mere image. The pairing is also intended to show that each is the other's pastime and that conviviality helps the meaninglessness of existence be endured, or at least helps cover it over. Such conviviality, of course, does not represent an absolute guarantee that time will flow on, either; it merely provides some help here and there. And when the partner responds to the question "What have I done with my pipe?" with the words "Charming evening we're having,"[12] these monologic cues and responses resemble the sword thrusts of two blind duelists, who are trying to persuade themselves that they are fighting a duel by stabbing around on their own in the dark.

Nobody will deny that "killing time" is something that also exists in the context of "normal existence," during the intermezzos that free time and leisure represent. The common expression "pastime" in itself indicates that we try to make time go away, or at least try to make it move forward, by playing at being active—that is, by playing games—otherwise it would be at risk of stagnating. One will now object that we only do this in our free time—for in our own lives there is a clear demarcation between serious things and play. Yet the seriousness of Estragon's and Vladimir's life (or the thing, at least, that constitutes its misery and, measured by reality, its abstruse unreality) consists in the fact that they must incessantly keep time in motion—that they must incessantly play.— But can this distinction between us and them really be justified? Is there really still a clearly identifiable line between our real life and our play?—

I don't think so. The reason Estragon's and Vladimir's pitiful struggle is so impressive is probably just that their attempt at pretending to be active holds up a mirror to the fate of the mass-human. On the one hand, machine-mediated labor (robbed of any visible aim and goal) is so far removed from what is illusionistically called "*human action*" that it has itself become a kind of pretend activity: There is not the slightest difference (neither structurally nor psychologically) between "real" work and the most pretend pseudowork that one is driven into out of distress. Yet, on the other hand, this kind of work now makes humans lose their balance to such an extent that they feel compelled to invent "hobbies" to help them regain some sort of equilibrium, to help them "relax" and "kill time." Paradoxically, therefore, they now use their free time to set themselves what appear to be real goals; they try to enjoy their free time by playing at working. One way to make this happen is by explicitly regressing to what, in comparison to their salaried work, are obsolete modes of production—for instance, by working with the coping saw or at the allotment.— They only do this, of course, if the patronizing nature of their daily work hasn't already so utterly ruined them that they have lost the ability to take their free time (their "play," their way of "killing time") into their own hands—only if they have not already become fully dependent on the conveyor belt of radio to make time pass. Yet the strongest evidence (stronger than any theoretical comparisons or equations) that activity and inactivity *coincide* today is provided by the fact that the two now converge in millions of homes and workshops—wherever a workflow and the continuous flow of radio programs converge to form *one single* stream. In short, it is only because free time and work, activity and doing nothing, seriousness and play, are today so hopelessly entangled that the ludicrous earnestness with which Estragon and Vladimir attempt to maintain the appearance of being active seems so terribly serious and strikingly topical.

Any activity (and this includes any pretend activity) is in effect equally well suited to making time go away, but all these pretend activities are also equally difficult to get started with. The ability to do something without meaning it (or to pretend to be doing something) requires precisely the kind of freedom that has already been paralyzed by the passivity of an Estragon- or Vladimir-like life. It is thus entirely consistent that Beckett lets them play but play in vain, for he lets them play in a way that no longer allows them to be a match for the task of "filling their free time."

And this all the more so because, unlike us, they cannot yet avail themselves of designated and recognized types of leisure activities. To make time pass, they can turn neither to sports nor to Mozart sonatas but are forced *to invent* their own forms of play, which they must do by reaching into the arsenal of everyday activities. Whereas we have the privilege to play football and, when finished, start again, Estragon plays the *da capo* game "boot off–boot on." He doesn't do this to reveal himself a fool but to fool around with *us,* to show us—using the recipe "inversion"—that our games (whose meaninglessness has already been covered up by the public recognition they receive) are no better than his. The inversion in the scene "Estragon plays boot off–boot on" thus means: "We are also playing a 'boot off–boot on' game," something phantomlike, an activity that makes it seem as if we were doing something. The scene is ultimately not just an inversion but amounts to a full rotation: "Our *actual* boot off–boot on" (our *daily routine*) is nothing but play—that is to say, it is clownish, inconsequential, merely motivated by the hope of "making time pass." And, "just like Vladimir and Estragon, we have been cursed to the luxury and misery of inconsequentiality."—The only difference is that these two clowns *know that* they are playing, while we don't. And with this, it is the two heroes who have become the serious people, and *we* have turned into the comedians. Such is the triumph of Beckettian "inversion."

6. *The Antagonists Enter the Stage*

It is quite obvious that to Estragon and Vladimir, the lot of those who don't have to keep the time mush moving themselves (or who keep it moving as a matter of course because they wouldn't even know what else to do) must seem enviable. And these antagonists are now encountered in the duo *Pozzo-Lucky*.

The attempt to decipher these two—to establish who they might be and what they might mean—has occupied commentators no less than the question of Godot's identity. But all these attempts at decryption have gone awry because *this duo is already a decryption in itself*. What does this mean?

It means that Beckett used Pozzo and Lucky to translate something back into the sensual realm that had previously only existed in conceptual space—that he freed this something from its speculative existence as a cipher. It means that he didn't turn a sensual reality into an abstract cipher. *What?*

Since the early 1930s—when Hegel's dialectic and Marx's theory of class struggle first started to cause excitement in the literature and philosophy of the youth of France—the famous pairing of "*master and slave*" found in *The Phenomenology of Spirit* has dug itself deep into the intellectual spirit of those born around the year 1900. So much so that one can probably say that it now occupies the place that the image of "*Prometheus*" once held in the nineteenth century—in short, this famous pairing has now turned into *the image of the human as such*. Sartre is the key witness for this transformation. With "Orestes" in *Les Mouches*, he resurrected the typical Prometheus figure for one last time (in the figuration that was customary from Goethe onward, via Shelley and Byron, all the way to Ibsen's *Brand*), but he only resurrected it to then replace this emblematic figure with Hegel's image for good. The decisive aspects of this new symbol are "*pluralization*" and "*antagonism*"; put otherwise, "*the human*" is now represented by means of *a human duo*. The individual (who, as a metaphysical self-made man, had once fought a Promethean struggle against the gods) has now been replaced by *humans* fighting for power *with one another.* It is no longer the singular but *the plural* that counts as *the real:* Power, and the struggle for power, are now what "is." And with this, we arrive at the decisive point, for it is only power and the struggle for power that act as the engine *of time:* Time is history, and in the eyes of dialectical philosophy, history owes its movement solely to antagonism (the antagonism between human and human or class and class), so much so that if this antagonism were to come to an end, history would find its end at that very moment also.

This Hegelian thought-image—the principle of a continually advancing history—now moves across the stage in the figures of Pozzo and Lucky, a stage on which, so far, only "being without time" had been operative (or actually hadn't been operative at all). That the appearance of this new character duo is intriguing for the audience is understandable for several reasons. There are aesthetic reasons: The stagnation that the audience had initially rejected as an unreasonable imposition but then accepted as the "principle of the Godot-world" is now violently disturbed by the incursion of two figures who are genuinely doing something. It is as if an unmoving image were transforming into a film before our very eyes.— There is also the complexity of the allegory itself; it differs from almost everything we know from tradition. While the allegories we are familiar with (virtue, freedom, fertility, and so forth) decorate the abstract and dress it in a sensual reality, allegory is in this case tasked with undressing the abstract of its decorative elements; it is made to denounce the cruelty and misery that the philosophical "master and slave" formula doesn't make evident enough, so as to show it in its fullest possible nudity. *The aim pursued while allegorically dressing things up is thus to undress them; the function of this theatrical illusion is to disillusion.* The performance that stages dialectics is itself dialectical. And when this feels so intriguing—and not just for us but also for Estragon and Vladimir, who never quite manage to overcome a certain shyness toward Pozzo and Lucky—there is sufficient reason for this.

But as shy as these two may be vis-à-vis the new duo, there is one thing that Estragon and Vladimir cannot conceal: that they envy Pozzo and Lucky. It is no longer necessary to explain why, as the interpretation now comes full circle all by itself. Now that we have traced the image of hell an atemporal existence amounts to, it is clear that the governors of time—even the most hellish ones—must seem privileged to those condemned to "being without time." Pozzo, the lord master, is enviable because he doesn't need to "create time" by himself or make it progress; he does not have to expectantly wait for Godot either, for Lucky is pulling him forward anyway.— Lucky, the slave, is enviable because he isn't just "able" to trot forward but *must* do so, as Pozzo is always behind him and always after him, ensuring that this is so. And although this duo moves past the two timeless ones without realizing that they had already moved past yesterday (they are, in a sense, moving as if they were still "*blind history*" that is as yet unaware of its own historicity), in the eyes of Estragon and Vladimir, they are both already blessed creatures. For regardless of whether they are being pulled or chased, both are always being kept in motion by the other. So it is certainly understandable that they suspect Godot in Pozzo (although the latter has never heard the name of this dynasty [*ihres Geschlechtes*] and mispronounces Godot's name as a matter of principle), for under Pozzo's whip their waiting could find an end. It is no mere coincidence, therefore, that Lucky, the pack animal, is called "the lucky one": Although he is burdened with everything, although he may only ever carry around sand-filled bags his entire life, he is nevertheless fully freed from any burden. If Estragon and Vladimir were permitted to stand in his place, they would no longer need to tread on the spot but would be able to move on because they would have to—hell would have lost its sting, and once in a while, they'd even be tossed a bone.

After all this, any attempt to wring any positive or even comforting features from this image of the human and the world would amount to mere reassurance. Yet Beckett's play nevertheless differs from almost all those other nihilistic documents in which the present pronounces itself in literary form; it differs in *its tone*. The tone of these other documents is usually either serious (because it does not yet know the human warmth of humor) and rightly called "animalistic," or else it is cynical (because it no longer cares about the human) and is thus again inhuman. But the clown—and we have shown just how clownish the play is—is neither animalistic nor cynical; the clown is of a sadness that makes the hearts of all humans join in solidarity since clowns mirror the sad lot of the human as such and bring relief by means of this solidarization. It is no coincidence that no other figure was met with as much gratitude as the miserable figure performed by the early Chaplin. The farce seems to have become the last refuge of love of humanity; the complicity between the sad is our final consolation. And even if what is sprouting there on the inconsolably barren ground of meaninglessness—that is, the mere *tone* of humanness—is but a tiny consolation, and even if consolation doesn't know why it is consoling us and what Godot it is consoling us for, it nevertheless proves that warmth is more important than meaning; and that the final word does not belong to the metaphysician, but solely to the friend of humankind.—

See Günther Anders, "Sein ohne Zeit," *Neue Schweizer Rundschau,* January 1954. [The version of "Being without Time" Anders published in 1954 corresponds almost exactly to the text republished in *The Obsolescence* in 1956. Besides some minor differences in individual sentences, the 1954 version also has a slightly extended introductory paragraph that comments on the impact of Beckett's play after its first performance in January 1953.—Eds.]

4

On the Bomb and the Roots of Our Apocalypse Blindness

> Courage? No, a lack of imagination!
>
> —Molussian saying

> *L'imagination assez pleine et assez étendue pour embrasser l'univers comme une ville.*
>
> —Montaigne

1. *Trivialized Objects Call for Exaggerated Portrayals*

Anyone who is cast onto a *terra incognita* cannot immediately begin to chart this unknown world and create a cartographic record. They have to surrender to chance, take things as they come, and roam about. Whatever catches their eye first—a tree or a mountaintop—will be what they head toward. That they will reach this point by the most direct path is highly unlikely. Since the path is pathless, they will inevitably get stuck at some point and be forced to change direction at another, they may even have to turn around. And even if they manage not to forget their destination, it will certainly be lost from sight.— Yet such wayward wandering will still not be without gain. Bit by bit, it will allow them to become familiar with the land; this is the only means by which terrain features can become visible from more than one side, and it might even lead them to a place from which a first all-encompassing panoramic view opens up. This point could, of course, have been reached by a more direct route—had there already been a path. But such a path will only be forged through the detour.

As a philosophical terrain, the *bomb*—or, more precisely, *our existence in the shadow of the bomb*, for this is our topic—constitutes such uncharted territory. An immediate cartographic mapping isn't possible. We will have to allow ourselves to roam and be content to just observe and mark out the details that catch our eye. Their arrangement will at first be arbitrary, and the connection between them obscure. But as we forge our way, a way is forged, and the way we proceed will change. As some things begin to reveal themselves, some even from more than one side, and others form into preliminary connections, it will increasingly start to feel as if we are moving on a path, or even something more methodical—although the precise point at which this happens cannot be clearly identified. In short, the terrain will gain contour and begin to take shape. But it will only *begin* to take shape; I make no claim to having reached a vantage point from which a valid cartographic image could be produced.

Yet, even if the overall image may still seem blurry, I've tried, from the very first step, to bring the shape of individual features into as much focus as possible; I've tried to formulate things in as pronounced a manner as I could. And this overemphasis and exaggeration requires some further explanation, lest it be misunderstood.

It doesn't stem from a desire to deliver punch lines—given our topic, this would be appalling. On the contrary, my exaggerations are exclusively motivated by the peculiar invisibility our object is cloaked in. This object, which should be arresting our gaze nonstop with its menacing and fascinating presence, is instead *the very center of our negligence.* It's the business of the age to look away, not to hear, and live as if it didn't exist—humanity seems to have conspired not to mention it at all. An object of this kind cannot "simply be described." If it is a defining trait of an object that it is only ever encountered in situations in which it is being actively obfuscated, trivialized, or repressed, then its portrayal must correct this obfuscation. It must *exaggerate* its contours, or else the resulting image wouldn't be true. And it must be exaggerated to the same extent that it is usually "*downplayed.*"—

In other words, what makes our object so difficult to talk about isn't just that it is a "*terra incognita,*" but also that it is systematically kept incognito. The moment one even simply mentions it, the ears one tries to reach don't listen.[1] If there is to be any prospect of reaching the ear of the other, one must thus intensify one's speech as much as possible. This is why my formulations are overpronounced. The age of artlessness—the happy age in which one can afford not to exaggerate—has not yet been reached.

Indirectly, this also signals that this essay isn't predominately academic in nature (or, at least, not in its main parts). There are topics one already fails to address by addressing them the wrong way. Since the bomb hangs not just over our university buildings but over all our heads, it would also be entirely inappropriate to adopt academic jargon to philosophize about the potential apocalypse to members of a specific discipline. Besides, academic philosophy also appears the least "interested" in this "topic"; before academic philosophy gets involved, one must usually wait until the victims of the blows reality inflicts are not just dead but also forgotten—only then might it take the trouble to turn these blows into "philosophical problems." In books of academic ethics, the termination camps and infrastructures of liquidation are even today still an undiscovered future subject.

In light of this, nothing is more important than finding a language that can be widely understood: Philosophy must be conducted in a popular tone. This introduces our second difficulty: There is no such thing as "popular philosophy."

Mass-friendly motorways that provide solutions to philosophical problems exist as little as "royal roads"; to conduct philosophy means "to make things difficult." And I am as aware of this as those professional philosophers whose line of work consists in being unrelenting in this respect.

But such reservations no longer count. Other things are at stake today than defending the honor of things that are difficult.

At any rate, I won't even elaborate on how this philosophical ideal is in itself a sign of contempt for one's fellow humans—a contempt for all those who make things too easy for themselves by living as "they" do (and who would want to let their ontological mind be troubled by the misfortunes of those condemned to such an "average and pre-chewed" existence?).[2] But when addressing today's humanity, a humanity faced with apocalyptic danger, the ideal of making things difficult turns into a mere gimmick and distraction. People who are in danger don't need things to be made more difficult for them; they are only served by words that might help bring them to their senses. And the moment this course of action becomes *necessary*, the question of whether this is even *possible* is wrong because it has become inadmissible. To make things difficult or not—as a moralist one must somehow muddle through, one must somehow find a language that isn't just understood in certain buildings—that is, in universities. This task is of such importance that it will be of little consequence if the resulting observations aren't deemed "philosophical." All the less so (to cite a Molussian saying) because "the defining trait of the true philosopher is that nothing is less important to them than how their reflections are classified."

Whether words such as "morality," "moralistic," "ethics," and so forth still apply to the reflections we are about to embark on is entirely unclear. In view of the monstrous magnitude of our object these words sound feeble and entirely inadequate. Until today, moral questions have always been questions about *how* humans ought to treat humans, *how* humans ought to relate to humans, *how* society ought to function. And apart from a few desperate nineteenth-century nihilists, there has barely been a moral theorist who didn't take it for granted *that* humans will continue to exist and *that* they ought to exist (and until very recently, it would also have been meaningless to debate this). But because of the bomb, and because of our position or lack of position toward it, this question *has become* acute.

This means that all those questions asking "*how*" we might best live have been replaced by one single question, one single "*if*": *if* humanity *will* continue to exist or not. This "if" sounds clumsy, and the contemporary—afflicted by apocalypse blindness, fearing fear (their own and that of others), nervous about making themself and other mortals nervous—doesn't want to acknowledge it at all. But it's true all the same, for the nuclear device is nothing but this question.

We must not fail to hear this dreadful "*if*." Ominous and full of menace, it hangs over these words—as if it were a "blood moon" (or so the ancients would have said). And I hope that at least in the hour spent reading, readers will not be able to forget the thing looming over us.—

I. Opening Shock Observations

2. *We Are the Infinite—Faust Is Dead*

If there is anything in the consciousness of modern humans that is still deemed absolute or infinite, it is no longer the power of God or the force of nature, let alone the alleged authorities of morality and culture: It is *our* power. *Creatio ex nihilo,* the mark of omnipotence, has been supplanted by its counterforce: *potestas annihilationis, reductio ad nihil,* a power that is now in our own hands. The Promethean dream of omnipotence has at long last come true, albeit in a very different form than hoped for. Because we have the power to inflict on one another an absolute end, *we are the lords of the apocalypse. We are the infinite.*—

This is easy to say, but the situation is so horrendous that all of history's ups and downs seem to have become insignificant; all earlier epochs seem to be shrinking together into mere "prehistory." We do not just belong to a new historical generation of humans. Although we are anatomically unchanged, our completely changed relation to the cosmos and to ourselves *makes us beings of a new species*. We now differ from the previous type of "human" no less than, say, in Nietzsche's eyes, the "superman" [*Übermensch*] would have differed from the human. In the affirmative this means *we are Titans* (and this is not a metaphor); or this is true, at least, during the brief or long interval in which we are omnipotent without yet having made definitive use of our omnipotence.

In the short time since our supremacy began, the gulf separating us Titans from yesterday's humans has become so wide that the latter are already beginning to feel alien to us. The prime example of this is *Faust*, who desperately longed to be a Titan, the figure in whom the last generation of our forebears recognized "themselves." This may sound like a sacrilege against Goethe and his work (which has already been sacrilegiously diminished into a cultural asset anyway), but my suspicion that *Faust* has become a figure we can *barely comprehend* is not a criticism of Goethe. What had so tormented "Faustian man"[3] when he lamented that he "could only ever be finite" is something we can now barely intuit. Because of the "infinitude" we hold in our hands, the infinite longing for the infinite, which for almost a millennium had been the source of the greatest sufferings and spurred the greatest achievements, is rapidly fading. So much so that we now only seem to "*know*" that this longing once existed. The thing most characteristic of our parents, the last humans, has already become invalid for us, their children, the first Titans. Our parents' innermost feelings are already unknown to us, and the alternatives through which they understood themselves and articulated their own existence have already become obsolete.[4]

A sort of "infinite longing," of course, also exists today, even one that could easily turn into an epidemic. But this isn't some past longing that is *still* alive today; it is a longing that is only just beginning to erupt. I am speaking of the boundless nostalgia for the "good old" age of honest finitude—the desperate Luddite desire to be allowed to be human again, just as we were in the "golden age" of yesteryear. It is a desire to discard the Titanic quality that we acquired (or had imposed on us) from one day to the next— this is a most dangerous and dubious desire for as long as it remains mere sentiment, it only weakens those who feel it while also strengthening and lending substance to the position of those who actually wield this omnipotence for real.

Back then, when the Luddite uprising started as a protest of hungry workers against competing machines, who could have foreseen that this revolt would assume dimensions that can only be captured in the language of myth? For how else if not with the formula "*the Titan who desperately wants to be human again*" could today's longing be put into words?—

Omnipotence, curious as this may sound, only seems to have become properly dangerous as it passed over into our hands. In earlier ages, there was always a Noah, always a Loth. Every previous form of supremacy proved to be merciful, regardless of whether we believed it to be natural or supernatural (and even this distinction now seems secondary). All these other supremacies had only ever posed a partial threat; they only ever annihilated individual entities: "only" humans, "only" cities, "only" kingdoms, "only" cultures—but *we* continued to be spared (if this "we" is understood to mean humanity). It is no wonder, therefore, that the idea of total danger didn't actually exist, notwithstanding a few natural philosophers who toyed with the idea of cosmic catastrophe (e.g., a total freeze) and a small Christian minority who are *still* waiting for the end (and are thus not doing so *again*).

It is doubtful how much this has changed and how conscious humanity has become of the apocalypse—*but it certainly should be.* There are few reasons to believe that we, cosmic parvenus and usurpers of the apocalypse, will be as merciful as the superior powers have been up to now—whether out of kindness, indifference, or simply by chance. Because the men who are now actually the lords of the infinite are as little a match for their possession—imaginatively and emotionally—as we its prospective victims are, there is probably only minimal ground for hope. Perhaps there is no hope at all, for these men are incapable and will necessarily remain incapable of seeing in their device anything other than a vehicle for finite interests, if not even for most limited party-political goals. Because we are the first humans in charge of the apocalypse, we are also the first to live permanently under its threat. Because we are the first Titans, we are also *the first* dwarfs—or whatever we should call ourselves now that we are collectively finite beings, not just mortal as individuals but also as a group; beings whose existence is granted only until it is revoked.—

3. The Premise "All Humans Are Mortal" Has Today Been Replaced with "Humanity Can Be Killed as a Whole"

We have just lived through a time in which death from natural causes was an unnatural death—or, at least, not an everyday occurrence.

During the years of war, people who simply passed away were deemed enviable. They must have maintained some secret channels to unknown cosmic powers beyond the grip of tyrannical terror (for how else could they have attained their knowledge of how to escape the collective fate of being killed?), and it was only this select few who could afford the peacetime luxury of dying an individual death even in apocalyptic times. And even if at times there was a different understanding—one in which this natural death was viewed as a testament to freedom and independence akin to stoic suicide, making the person who died naturally seem like the last sovereign human—here, too, it was the violent, unnatural death that was actually considered "natural," as being killed was the primary way to die. The model for our experience of finitude was Abel, and not Adam.—

What was true for the everyday experience of war was even more true in those installations in which the final front lines of terror were taking shape (just as ordinary military front lines were beginning to disappear). I am speaking of the extermination camps, where the machineries of death were operating with such absolute reliability that no uneconomical remnant of life remained. In these camps, the respectable premise "All Humans Are Mortal" lost its meaning once and for all and already became an object of ridicule. Had the sign above the entrance of the gas chambers read "all humans are mortal" instead of "Shower Room" (a sign meant to facilitate efficient work), this would have provoked jeering laughter—a laughter in hellish unison in which the voices of the condemned would have been joined by the laughter of those escorting them in. For this old adage had already been displaced by a new premise, one that would have had to read "*all humans can be killed*."

And since then, the truth resides in this new statement. Much may have changed in the past ten years, but because of the bomb and the threat we live under, it still resides just as comfortably in its new home. If there was a change, it was only for the worse, for it is now not just "*all humans*" but *humanity as a whole* that can be killed.

This shift has propelled history into a new age. Its individual ages would thus have to read:

1. All humans are mortal.
2. All humans can be killed.
3. Humanity can be killed as a whole.

4. Solomon's Saying "Whatever Is to Come Has Been Already" Will Be Replaced with "Nothing Ever Was"

It is thus only today that our mortality has truly become mortality.[5]

In the past, of course, there was nobody who could exclude themselves from being mortal either, yet everyone still experienced themselves as mortal within the human-world—that is, within something bigger than themselves. This world may not have been considered immortal, yet it was deemed something lasting—something that survived one's own death. All those strange efforts to outlive oneself, to gain immortality through glory, thus owe their existence to a space one was able to die in and die into—a space whose own death or deathlessness was never questioned. These efforts were, of course, not very successful. It is not exactly a sound metaphysical investment to try and make oneself known among the unknown and attain immortality among beings who are mortal themselves. And so even the famous were only ever reminiscent of the ship passengers in *One Thousand and One Nights,* who may have enjoyed the highest renown on deck, but still remained entirely anonymous since nobody on land knew about the ship.

And yet, in comparison to today's situation, this was still comforting: Tomorrow's glory, which was called "eternal," was still holding the weight of eternity in check. Status *on* the ship was still able to cancel out the status *of* the ship—but it is this ship itself that has now been lost.

Today, as fear of death concerns the ship itself—humanity as a whole—it is unclear with whom today's mortals should deposit their memories to ensure that they are known in the future. If humanity were to be engulfed by the waves, then all who remember, and all potential guardians of memory yet to be born, would be dragged down into the darkness too—so that they will not have existed. And this would not just affect this or that mortal; there would be nothing that wouldn't have been for nothing: no people, no human, no language, no thought, no love, no struggle, no pain, no hope, no solace, no sacrifice, no image, no song. In short, there would be only that which "once had been" and nothing else.

Even for us today, still alive in a world that still exists, the past that "was"—the past that is nothing more than "what once had been"—seems dead. But now, even *this* death would be dragged down and forced to die a *second death,* so that everything that *was* would *no longer even be that.*—For if nobody is left to remember anything that was, how would that which *was* differ from that which never has been? The future, whatever is to come, wouldn't be spared either; it would be dead before its first hour of life. So that the bleak tense of Solomon's sermons, the all-consuming future perfect "whatever is to come has been already"[6] would be replaced by the even bleaker "*nothing ever was.*" And with this, this nothingness would commence its unwitnessed (and hence legitimate) reign.

5. *How Prevention Is Prevented*

Let's assume the bomb were to be deployed:

It would be inappropriate to call this an "act." The process by which this deed would be set in motion would be so mediated, so nontransparent, it would consist of so many steps and so many partial steps taken by so many authorities (of which none would be *the* decisive step) that, in the end, nobody would have done *it,* as everyone would have merely done some other thing.—

To mitigate any residual risk of exposure to a guilty conscience, entities have been constructed, oracle machines—*automatic electronic consciences*—to which responsibility can now be delegated; and our cybernetic computing machines are nothing other than this. As the epitome of science (and therefore progress—which renders them unconditionally moral), these machines now whirr and hum and assume responsibility while humans stand by with as much gratitude as triumph, washing their hands in innocence.

To the machine-operator, the question of whether the "aim" that is activated with a combination of switches is a responsible one (or even just meaningful) becomes irrelevant the moment the machine begins its computation—the very question is, in fact, forgotten. If he were to doubt the machine's result, he would be doubting the scientific principle. And where would this lead if such a precedent were set? That he has *substituted "responsibility" for a mechanical "response,"*[7] and that the delegation of responsibility onto an object (he considers "objective") has transformed the moral course of action into whatever the machine deems factually correct and the prohibited into whatever it deems incorrect—this instills no fear in him at all. Yet this shift is decisive. The oracle's sole capacity consists in calculating what measures might pay off—or, rather, what losses might still be advantageous in a situation defined by the parameters *A, B, C, . . . N.* And since only *finite* numbers can be entered into such computations, *our life is ipso facto entered as a finite number, too;* it becomes a variable that can also be annulled. The very method of calculation has hence already "annihilated" us even before we are annihilated for real.

But even if there were no such robots, the mere fact that the labor process behind this monstrous deed is a thousand times subdivided and mediated—the mere *complexity* of modern organization—would make the deed much *easier* to execute. This sounds paradoxical, but it is not. These highly convoluted organizational pathways weaken or fully absorb any obstructing energies resulting from moral scruples; they thus perform a similar function to electrical resistors.

Once an organization is operative, the smoothness of its operation replaces the very idea of moral action. As long as a scheme has a "satisfactory" organizational structure and runs cleanly, its purpose also seems clean and morally sound. And this not just because the whole operation works so well, but because the operation is itself never visible *as* a whole. The countless specialized workers engaged in the process only ever see the step they are themselves working on, and each of them is considered conscientious as long as they perform their task conscientiously—and thus nothing immoral can be seen anywhere. Or, to put it another way, since their attention is only ever fixed on the task in front of them, nothing immoral can be happening anywhere else, since there isn't "anywhere else" for them. "Dirt divided up a thousand times is clean" (Molussian proverb).

The conscientious work of all these conscientious workers thus adds up to a monstrous lack of conscience.[8] But because these individual workers are never able to intuit the relation between the particular step they are working on and any others, they never even become aware of this. This "lack of awareness" could, of course, be called "a lack of conscience." But if this old phrase were to be employed, it couldn't be used to suggest that these workers act against their conscience—this possibility would, in fact, still be reassuringly human, for it assumes an entity capable of having a conscience; no, "a lack of conscience" could now only mean that these workers are *excluded from the possibility of having a conscience* in the first place. They don't just lack morality but also immorality. To accuse them of a lack of conscience would be as pointless as accusing a hand of cowardice. That is to say, it would be pointless because the subject being referred to cannot possibly be a candidate for this kind of statement.

There is thus nothing in the way of the production and use of the thing: It is precisely the great number of participants and the complexity of the device itself that prevents any such prevention.

II. What the Bomb Is Not

6. The Bomb Is Not a Means—the Absolute Has Been Reached, the Comparative Is Redundant

Everybody somehow knows, of course, what the bomb "is." But most only "*know*"—that is, they know in a most hollow and superficial sense without truly grasping what it is they know. In a certain sense, their knowledge even consists of *falsehoods*. I'm not trying to imply that they are misinformed or only inadequately informed. This may be true for specific details, but not for the main point. And it is not just us, we who are to varying degrees well- or misinformed, who "know falsehoods"; this is also true for those with full clearance and actual power over the bomb. What am I trying to say?

That even when we simply think of the bomb, we already place it in the wrong category.

As we speak of the bomb (or even merely picture it in thought), it is, of course, inevitable that we adopt some sort of conceptual terminology. And so it is also only natural that we mindlessly revert to the categories we are already used to and which we rely on to navigate our world. Yet we must pause to consider whether this is entirely justified. It will certainly do no harm to consider the possibility that the bomb is an entirely abnormal object, an object *sui generis*—that is, the only one of its kind. It will also do no harm to brace ourselves for a task that hasn't existed in this form before—with the possible exception of theology, which, by its very definition, deals with something "singularly unique." To prevent itself from attributing ontologically inadequate qualities to God, theology (as "negative theology") limits itself to only listing what its unique subject is *not*. And the same might now apply to us—and the analogy is, of course, exclusively of a methodological kind—we might also have to limit ourselves to identifying what the bomb is *not*. Only this will ensure that we do not misclassify it from the outset simply by classifying it.

When we think of the bomb, we especially reach for the conceptual coupling of "means" and "ends." And it is this categorization—the term "means" to be precise—that I had in mind when I said we use the wrong categories. *The bomb is not a "means."*

What makes this so difficult to grasp is not just the fact that the bomb is produced as a weapon (and weapons are obviously means), but also that this category has attained a level of universal applicability it has never previously enjoyed. We have in effect come to view the world we inhabit as "*a world of means,*" a universe in which nothing but means exist—so much so that, paradoxically, even "ends" are now pushed into the background as secondary (since unlike "means," they seem to lack an "end"). One can hardly expect, therefore, that today's mass-humans—and this also includes those with direct power over the bomb—comprehend that this world now also includes something that is *not* a means, even something that is manmade.

Why is the bomb not a means?

The expression "a means to an end" inherently assumes that the "means" is absorbed by its end while "mediating" it, that the "means" ends in the end, just as a path ends in its destination—that is, the "means" is assumed to disappear as a separate "variable" as soon as the destination has been reached.— But is this also true for the bomb?

No.

It doesn't disappear as a "variable."

It is *absolutely too big.*

What does "absolutely too big" mean to imply here?

That if the bomb were to be used, even its smallest effect would already be greater than any human end, be it political or military, regardless of how great such an end may be. "*Effectus transcendit finem.*" Not only would the effect be bigger than any end, but it would probably also put an end to any further articulation of ends, and thus to any further use of means. It would *wipe out the means–end principle as such.*

To call an object of this kind a "means" would be absurd.

But this is not all. Let's for a moment assume that someone were pursuing the herostratic aim of global annihilation (which isn't the case). Even under these circumstances, the further production of nuclear weapons wouldn't amount to a production of "means," or at least not today. Quantity has already turned into quality: The virtual violence stacked up in nuclear arsenals—all experts agree—is already absolute. It is already perfectly sufficient to realize any such "cosmically herostratic" end. Which is also why the continued attempts at increasing the size, explosive power, and number of weapons seem devoid of sense. The absolute danger already wielded, the effect that can already be triggered, cannot be increased. No improvement of the means would result in anything new, only in novel comparatives and superlatives of the adjective "dead."

This is a first in the history of production. The "spirit of industry," possessed by the obsessive idea that every technological product has a duty to improve and become comparatively better, simply cannot comprehend that such a comparative addition might be possible while also being pointless—that the caliber and effective range of the thing can be improved but not its effect. "Once upon a time," a Swiftian text of today might read, "there was a senior administrator in nuclear weapons production, who was charged with ensuring that the explosive power of hydrogen bombs was daily increased. Having tried to fend off this thought over long months, this man was no longer able to close himself off from the insight that no improvement of his product would improve its effect, and so his mood began to darken—yes, he even became clinically depressed. 'Don't you understand, doctor!' he cried, 'Don't you understand! A means that can be improved, while its effect remains unchanged! Has anyone ever heard of such a thing? I tell you, only ever the same effect, only ever the same end of the world! Is this not enough to drive one crazy? Is this not a scandal? Is this not truly the end of the world?"—

But enough Swiftian tales; reality is fantastical enough. Let's memorize our central point:

If someone were to make use of the bomb in the foolish hope of enforcing some finite end, they would achieve an effect that wouldn't bear any resemblance to their aim. The path wouldn't end in the destination—or, more precisely, the means wouldn't lead to the end, but the other way around: The end would meet its end in the effect of the alleged "means." And not just in *one* effect, but an unfathomable chain of events, in which the end of our life would probably just be one small link.—

To speak of a "means" would thus only make sense if someone were genuinely pursuing the herostratic aim of bringing about the "end of all things."

7. *Sequel—the Means Justify the Ends*

The process by which the conceptual coupling of "means–end" has degenerated has been long in the making.[9] But regardless of what the distinct phases of this process might have been, means and ends now seem to have taken each other's place: *The production of means has become the end of our existence.* It now often happens (in all countries, the trend is universal) that things that used to count as "ends" are justified through an appeal to how useful and effective they are as means. This applies to hygiene products [*hygienische Mittel*], for instance, which also promote and facilitate the production or purchase of additional products. (The same applies to leisure, love, even religion). The title of the small American book *Is Sex Necessary?* (although, of course ironic) is symptomatic.[10]

Anything that cannot identify itself as a means—as some sort of remedy, instrument, product, or mediating device[11]—is barred from today's thing-cosmos. This is also why ends—precisely because they are not means—are deemed pointless and lacking in ends themselves. They are deemed to lack purpose so as to ensure that they, too, can prove themselves useful as devices. Ends, to use the language of mathematics, are to become "means to the power of means"; they are to turn into devices for mediating devices, by making them sellable, for instance. Today, *ends serve the purpose of being a means in the service of means.* This is a simple fact, and if this formulation sounds paradoxical, it is only because the situation is itself paradoxical.

This "mediating" function is especially apparent in the context of ends that are "*retroactively assigned.*" I am speaking of ends that are assigned to things after the fact—"to make them honest," so to speak, to secure a rightful place for them within the commonwealth of devices [*Mittel*]. When a chemical process yields a derivative, this creates the task of finding some purpose for it (or else inventing one), for this is the only way this compound can have the opportunity to rise to the rank of a means. There are indeed only very few cases in which it isn't possible to find or invent a purpose for something by forcefully remodeling demand. These retroactively made-up ends thus serve the purpose of facilitating the transformation of matter. Born as a "*good-for-nothing,*" matter is told to "man up" and make something of itself so it can prove itself within the community of devices. No "thing-in-itself" is allowed to exist—only means and devices, or, at least, potential devices.

There is, of course, no one quite as obstructive to the interests of those who fabricate means as the type of critic who rejects not only a fabricated device [*Mittel*] but also its purpose. The allegedly unrestricted freedom of criticism is, in fact, highly restricted: It is only ever meant to critique the suitability of some given means, remedy, or device; this is the only form of criticism that exists; there is no critique of ends. For *the critique of an end* would likely *disrupt the production of the means serving this end,* a situation that would set a most perilous precedent. Ends exist, after all, for the purpose of providing *a raison d'être* for means and devices. If a critique of ends were now to question this *raison d'être,* this would amount to an attack on the sacred principle of means-production itself. In other words, *the means justify the ends.*

The means justify the ends. This formula isn't meant to be a joke, and not even a "philosophical exaggeration." This inversion of the notorious formula is the secret motto of our age. It describes the setting in which the bomb came to see the light of day as an alleged means among other means and in which we, the stunned world, for the first time witnessed its blinding light.

If the bomb is a means, what is true for all other means must also apply to it.

To remain legitimate, it must have an end, a purpose.— Issuing warnings against its purpose (and warnings that might even lead to the ban of nuclear weapons) would set an untenable precedent, paving the way for warnings against the ends of any number of means. There would be a risk of rapid proliferation, until the "entire world" of means (and its fundamental principle) was undermined. It is thus clear that the mere thought of such a warning already amounts to a nightmare for the managers of the means-cosmos. As absurd as this may sound, when warnings against the bomb are suppressed, it is because these managers fear that such warnings will put their "*entire world*" at risk. They hence feel a duty to overlook that the real danger lies in *not* opposing the bomb, and in failing to do so successfully—for this really does put the "entire world" at risk (the real world: all of our lives), yet this is precisely what they seek to divert attention from. That's why it belongs to their typical (if certainly mostly unconscious) *tactic* that they successively *minimize their object,* trivialize its novelty, make light of its anarchic character, and introduce it as just another means among many others.— In short, they try to deny that it is their own product that is threatening and exploding their time-honored cosmos of means. The favorite trick by which they attempt this is to simply *pin* the anarchic element (which cannot quite be denied) onto those who issue warnings. They portray them as anarchists and revolutionaries, denouncing them collectively as *traitors.* So much so, that the word "traitor" is now used for those honorable people who are revealing the truth to the deceived, a humanity that has been sold out and betrayed. As the first blame tends to fall on those who call something out and never on the perpetrators, it won't be surprising that this maneuver is proving successful.

But this success will only be temporary. In the long run, it won't be possible to conceal that it is the bomb, and not those warning against it, that is the anarchic element. Humanity senses, even though it cannot usually articulate this, that it is confronted by an object uncannily at odds with the "world of means" that is usually presented to it as "world"—that is, that this object is a *unicum.* Humanity intuits that "the bomb is not a means."

This insight, which cannot be repeated often enough, isn't just negative—it is the positive and fundamental ascertainment of a negativity. And the riposte "If not a means, what else might it be?" doesn't make it any less valid. At any rate, those who are being threatened do not owe those who threaten them a definition of the threat.

There isn't anyone, in fact, who could provide an answer to the question "What else might it be?" The question is devoid of sense. If it were possible to answer, this would imply the existence of other objects of the same type—objects the bomb would merely be an example of. Which is precisely not the case. The bomb cannot be classified. It is an ontological *unicum,* and this constitutes its anarchic character.

Entities that could not be classified used to be called "monstrous." That is to say, the word "monsters" referred to all those entities that, despite having no identifiable being, were somehow *here* all the same, wreaking havoc with their troubling nonessence—the kind of entities, therefore, that simply laughed back at the question when asked what they are.[12]

The bomb is an entity of this kind. Although it is essenceless, it is here all the same, and its nonessence is keeping us on edge.

8. *Not Just the Murderer Is Guilty, but Also the* Moriturus

As is the case with most other means and devices, this nonmeans simply fell into the lap of those who now control it. It cannot be denied that the existence of the bomb was planned,[13] but, given the lack of human imagination, the same cannot be said for the situation its existence has brought about.

Indeed, those who are now suddenly in possession of this power are also "only human." Which means that they are no less limited than us. And because they are just as preconditioned for finite things and as attuned to them as we are, they remain just as unable to grasp that the thing at their disposal no longer bears any resemblance to a "means."

This means that the bomb cannot be clearly attributed to anyone and that the moral situation remains entirely unclear. Which only makes the device more dangerous. Moral problems can only yield unequivocal solutions if they are the type of problem where the innocent and guilty can be clearly told apart.

The only thing that seems clear is who is *not* to blame. For the kind of sentences that we have also used on several occasions—"We are hurtling toward the self-made apocalypse," "We are no longer a match for the objects we ourselves have produced"—the kind of sentences, therefore, that broadly assign responsibility to *us* ("us humans") are, in fact, incorrect. To suggest that it was "*us*"—that is, the majority of humanity—who wanted this apocalyptic device to exist, or that it was us who planned and produced it, would be absurd. And not just absurd but also dangerous, because such sentences are extremely convenient for those who now have the misfortune of being the actual "subjects" of the bomb—those who are actually able to preside over its production and use. Talk of "the suicide of humanity" creates a convenient and hence ideal impression of broad responsibility. It provides a wonderful alibi by presenting everyone as virtually complicit and nobody as being to blame right now.

But this is not the case—there are people who are to blame. Regardless of how ambiguous the question of responsibility was until today, the real question of guilt starts now—now that we know what the bomb signifies. Regardless of how innocent one might have been so far, one now instantly *becomes* guilty if one doesn't open the eyes of those who do not yet perceive, or attempt to make oneself heard by those who do not yet comprehend. *Blame, responsibility, and guilt* [Schuld] *do not lie in the past; they can be found in the present and the future. The potential murderers are not alone in being guilty; we, the morituri, are guilty too.*

9. *The Tests Succeed, but the Act of Testing Fails*

We opened section 5 with the words "Let's assume the bomb were to be deployed."

This opening sentence was wrong. It was too optimistic.

The bomb *is* being used—it is used *all the time.*

I am thinking not of its first two deployments—the destruction of Hiroshima and Nagasaki—but of its use in the immediate postwar period, during which it was employed in two distinct ways.

1. As a *means to pressure others.*— This applies to the first eight years after the war, up to the explosion of the first Russian H-bomb in the summer of 1953.— Yet the assertion that "it was used to pressure others" is, of course, also entirely misleading. To exert pressure on others, the monopolist owners of the bomb never had to utter the sentence "If you don't do *X*, then . . ." The kind of action that Hitler—in obscure, preatomic times—took against the Prague government with his coercive threat had become entirely redundant. *The bomb's mere existence*—having mere possession of it, the mere ability to put it to use—*automatically already amounted to an ultimatum:* The bomb is blackmail by its very essence—it's an act of extortion in the form of a thing, regardless of whether one wanted to use it as such. It would have been blackmail even in the hands of holy St. Francis.— Of course, this doesn't deny that some powerful groups and important individuals more or less officially made use of the bomb as a means of exerting pressure. And this is still the case today—which is easy to understand, and not just for psychological reasons, since it's almost unbearable to merely "have" an instrument of power, and such a powerful one at that, without ever putting it to use. A sense of socioeconomic duty must also be factored in, since the nonuse of a device—especially one developed at such immense cost—would be considered wasteful and thus morally unconscionable from an economic point of view.— During the postwar period, while America enjoyed a monopoly over nuclear weapons, the bomb was never physically deployed but *only* ever used rhetorically. It would be highly unfair not to mention this, but just as wrong to praise it. If not committing murder where murder would be possible were classed as a morally positive act, where would this lead? Yet there is little doubt that those men who limited themselves to blackmail, rather than using the bomb in its "ideal" function—that is, as a weapon—self-righteously believed that they acted

"idealistically." The practice introduced by Hitler, where one presents oneself as a "savior" by listing as "saved" all those one could have killed but did not, is continuing unabated today. And if the definition of life as "not yet having been murdered" introduced in the concentration camp is still valid, we have these self-righteous men to thank, who congratulate themselves for not committing murder.

This form of blackmail now boasts an additional feature. Whereas conventional weapons and honest criminal blackmail once confined their threats to those they were directed at, the bomb now always also holds humanity as a whole at ransom. The bomb isn't just a "total weapon" in its strategic use (the way strategic bombers were deployed during the last World War), but also by the very nature of its physical composition. The "total threat" it represents is a sign not of some special "maliciousness" but of an inability: *an inability to aim.*

I would, of course, never doubt the precision of the bomb's ballistic trajectory or its accuracy on target—these are undoubtedly admirable. I only mean to say that, because its effect is so vast and the zone of its spatial and temporal fallout so unlimited, it must necessarily *overshoot its target. Its omnipotence is its defect.* It creates a set of options that are absolute. It can either extort everyone or no one; it can throw the baby out with the bathwater or throw nothing out at all. Ever since the bomb was constructed, humanity has been living in a state of blackmail, with the population of America under as much coercive pressure as any other, if not more. For the staging of rehearsals for nuclear war has already made its civil population "H-bomb conscious."[14]

This form of blackmail, by its very nature, blackmails everyone, which also makes it a form of "*self-extortion.*" With the exception of the poison-gas intermezzo during the First World War, this is also a historical first. The frantic "either you . . . or I will kill us both," which was only known from most private settings, is now being most gruesomely impersonated on a global scale.

2. But there's another reason why we claim the bomb *already is* in use. And when speaking of its "inability to be aimed," we already came close to introducing it.

The phrase "Let's assume the bomb were to be deployed" implicitly accepts a boundary that no longer applies: that is, the distinction between "preparation" and "application," between "*test*" and "*the real thing*" [*Ernstfall*], *a distinction* that was until yesterday valid in the realm of experimental physics and technology. It is a characteristic of the bomb that it erases this very distinction.

The insularity of the test environment was a defining aspect of technological trials: the trial process was always contained, always separated from its environment; reality was always left untouched. Every experiment represented an isolated system, a process that occurred within a constructed microcosm with no or only negligible effect on the macrocosm, and this remained true regardless of how far-reaching the conclusions drawn from this microcosmic model might have been (for instance, by yielding facts that could establish "laws"). And the sealed-off and self-enclosed microcosmic playground was just as foundational to the concept of the *machine*. The "sealed environment" was, in fact, the very prerequisite of scientific work: it didn't just (negatively) restrict an untold number of incalculable and uncontrollable worldly determinants, it also made it possible to (positively) define a limited set of preconditions. And since these preconditions could be controlled, the same was true for their effects. The successful development of technology was a result of this method of constructing artificial—that is, hermetically sealed—microworlds within the world that totally remodel and reconfigure reality. This insulation also made every experimental procedure appear somewhat "*game-like*": The enclosed nature of the playing area— whether we think of a chessboard or a football pitch—is a characteristic of every game; it is as constitutive of its logic as of its opportunity for escapism. That's why some experiments were even called "beautiful": "beautiful" like a flawless game of chess or an entirely self-contained and consistent artwork that stands apart from the world. Sealed off from reality during their experimental stage, experiments only became historical realities through their technical utilization and industrial applications.

None of this counts anymore: *Nuclear "experiments" are no longer experiments.*

That these tests are carried out in an "insular fashion" on coral atolls or somewhere else in the world oceans—that there is an effort to salvage the principle of isolation by resorting to the geographical ur-meaning of "insularity"—is still quite typical. But these final attempts at isolation—these desperate attempts to continue conducting "experiments"—remain futile. Regardless of how perfectly *these tests succeed, the act of testing fails* since it immediately turns into something else—something more than a test. The effects are so immense that the instant the experiment is conducted, the "laboratory" becomes coextensive with the globe. It thus no longer makes sense to distinguish between "trial" and "implementation": Every experiment has become "the real thing." And the various "tests" that have already been conducted have indeed had such effects. Nobody will be able to claim that the Japanese fisherman Aikichi Kuboyama—the first victim of the hydrogen bomb, who died in August 1954—only died an experimental death.[15] Just as nobody can deny that the Japanese trawler *Fukuryū Maru,* contaminated by fallout at a distance of 130 kilometers from the center of the blast, was in the area *experimenti causa.* But to name individual cases is almost misleading. The contamination these "experiments" cause is already universal: Even now, the air, sea, rainwater, earth, flora, fauna, human-world, and all sorts of foodstuffs are already affected and infected—it is apparently not yet even possible to estimate to what extent. And since certain "half-life values" (the time required until half the original radioactivity is emitted) are fatally long when measured by human lifespans, the effect of these alleged experiments is also temporally unlimited. What is "played" experimentally today can become a "real" genetic "emergency" for children, grandchildren, or great-grandchildren—maybe even the ultimate one. Geiger counter readings of thyroid glands (the organ the radioactive isotope iodine-131 accumulates in) conducted in the widest area around the testing ground (1,500 miles) have already established a hundredfold increase in radioactivity.[16] The role that *syphilis* until recently played as a fatal contamination for future generations (in Ibsen, for instance)[17] has now been taken over by *radioactivity.*

These "experiments" are thus actual parts of our reality—our historical reality. Whether we say that "experiments aren't experiments anymore" or that "they are historical events" amounts to the same thing.

History knows no joking around—no "just once won't hurt." History refuses to revoke anything; the word "trial" isn't part of its vocabulary. There is no experimenting, since anything that introduces itself with the self-effacing assurance of being only an experiment immediately takes place "once and for all" and "for real."— This doesn't mean that "nothing should be attempted anymore"—that would be absurd. But one needs to be clear about what one is doing: A process one initiates without being able to seal it off from the world does not take place in the conditional mode of "what could happen," but in the indicative: "It is happening." Such processes don't just happen *in* the world—they happen *as* world-events; they are not just experimental but actual—they *have lasting effects* and lasting consequences. And this is as true for our most intimate, private histories as it is for history in the wider sense. Just as there are no test marriages that could simply be erased from the lives of couples "who just wanted to try," there are also no "only experimental" attempts at making a country fascist. The attempted turns into fate; it becomes an indelible part of history. Just as there can be no experiments in mass-sterilization—should the test subjects survive, they will be maimed for life—atomic explosions can likewise, though perhaps to an even lesser extent, never "just" be set off "once," *experimenti causa,* or "just a few times." The things we once used to describe with the small word "just" no longer have any equivalent in the real world. Nothing is just "just." Nothing is "just in the past." Every "just once" immediately becomes a "once and for all"; every once becomes a *nunc stans.* And the differences between the ages have likewise been erased: There is no longer a past that isn't also a present, no present that isn't also a future, and no future that doesn't harbor the future perfect "we will have been" as a deadly germ within. As short-lived as any one of us may be, we are all "*bigger than ourselves.*" The products we create, the effects we set in motion, are so long-lasting that

> they won't just affect us but also our grandchildren and great-grandchildren—who, one day, will be facing them with raised palms, shouting "Halt!" in their direction.

Thus, when we claimed that these experiments had broken out of the trusted hermetic seal of isolation (which they had still occurred within in our fathers' days) and into the dimension of historical reality, we were still highly conservative. The force of these experiments, stripped of their experimental nature, is so elemental that the historical world itself risks breaking apart the moment they break into it. *These experiments are not just "historical" events; they are "historically supraliminal."*

10. *The Expression "Historically Supraliminal"*

That there are millions of facts that are "historically subliminal"—so insignificant that they would never be factored into the plans of politicians or taken into account by historians—is plausible enough. The afternoon nap I took three years ago today may have been part of the historical situation, but it belongs to the nature of history to omit such aspects of itself *as* history and to omit them from the record as it gets on with things. History only *exists* insofar as it *sifts* itself.

Yet there are also events of such incalculable size that they extend beyond the dimension of what we can even consider a historical state. This becomes clear in the context of natural disasters. The sinking of Atlantis (assuming it happened) wasn't a historical catastrophe or a catastrophe *within* history. As an endpoint *of* history, it wasn't actually something that could be recorded by history anymore. It was "*historically supraliminal.*"

The same holds true for nuclear experiments, and even more so nuclear war. The preparations leading up to it may still belong to history (its initiators, after all, hope to accomplish some future historical aim). But the instant they commit to its actualization—the moment such a war would commence—history would be done for. On the day of the first explosions, the dimension of history would collapse too. "At the end of the path, the specter of general annihilation looms ever more distinctly" (Einstein, in his appeal to Italian nuclear scientists).[18] What would be left would not be a historical situation anymore, but just a debris field, beneath which everything that was once history would lie buried. If the human were to survive regardless, it would not survive as a historical being but only as a pitiful scrap—contaminated nature within contaminated nature.—

If there ever was anyone with the competence to assess the situation, this person was Einstein. And this was how he assessed it. And we read about it in the newspaper.

And how do we react?

Just as we react to any other newspaper report: We don't react at all.

How so? Is it courage? Stoicism?

"Courage?" a Molussian saying goes, "No, it's a lack of imagination!"

It is a lack of imagination because we are "*apocalypse-blind.*"

The next section attempts to explain why this is so.

III. The Human Is Smaller Than Itself

11. *We Live in the Age of the Inability to Fear*

This is our situation; this is how frightening it is.

But where is our fear?

I see none. Not even a medium-sized one. Not even the kind of fear that arises amid the danger of a flu pandemic. I find no fear at all.

How is this possible?

Not a word is allowed to be uttered against the legitimacy of labeling the time of tyranny and war, the time of the camps, occupations, and burning cities, "The Age of Fear" [*Zeitalter der Angst*]. But in the ten years since the final catastrophe, this expression has experienced a strange and most unreal career path. If one went out looking for fear, for *true* fear, in Vienna, Paris, London, New York—or wherever else the expression "*The Age of Anxiety*"[19] is *en vogue*—one would return with a most meager haul. The printed word "anxiety" would, of course, be found everywhere. It can, in fact, be found in whole swarms, in whole bales of publications, of which hundreds are daily consigned to the wastepaper basket to be replaced by hundreds more the next day. Anxiety has long become a commodity: today, everyone is talking *about* anxiety—but very few are speaking *out of* fear.

For in the time since the war, enough time has passed to remember that the word "anxiety" has a highly philosophical and literary past—that it was even elevated to the status of a philosophical term by

Kierkegaard. Inspired by Kierkegaard and Heidegger, this word has now become prime journalistic merchandise. It cannot just be found in the handcrafted pages of every artsy magazine; because nothing is currently more exoteric than the esoteric, it is just as much at home in the sleeping pill adverts of provincial newspapers.— In short, to give oneself the feeling of being *up to date*,[20] of "belonging" (be it to this or that), anxiety is all one needs.—

Today, in the age of the bomb, the designation "The Age of Anxiety" has thus become not just untrue but also untruthful. The pitiful speck of fear we can muster now and then, and usually only when pushed, isn't anything to feel smug about. I am yet to see a contemporary be totally overwhelmed, left numb beneath the threat crashing down on them like a giant wave. The most I've seen are some who appear startled. Yet they weren't actually startled by *their fear*, but rather by the sudden realization of just how *unafraid* they actually are. And then there are also those who try and make themselves afraid but get stuck in the run-up, only to feel embarrassed by how easy it was to let go as they put the newspaper away. (To be more precise, they feel embarrassed about not being able to do anything other than let go and return to familiar dimensions and their immediate everyday concerns.)

No, measured against the amount of fear we should be feeling and should be able to muster, we are but *illiterates of fear*. And if our age deserves its own designation, it should be called "*The Age of the Inability to Fear.*"

I admit, the bomb made its entry—if the analogy can be permitted—at the dramaturgically worst possible moment that could have been chosen. It stepped onto the stage in the very final phase of the war, just as the immediate fear of tyranny and war was for the first time beginning to wane. It appeared just as millions, for the first time after many years, dared to go to bed without fearing the police or bombs in the night—just as, in the less affected areas of the globe, "good old life"[21] began to resume. Could it now really be expected, amid this collective sigh of relief, that one brace oneself for a new danger? A danger that was, allegedly, incomparably greater? Or at least that one be open to the *possibility* that such a danger might exist? This was rejected; no such thing was possible: Any threat that wasn't an imminent danger anticipated for the coming night was deemed laughable, and so it wasn't registered at all.— And what didn't register that first instant couldn't be learned to be perceived later on. And one year later, the danger had already become something familiar—something one had read about a hundred times. It was already boring. Today, it has already become a "good old danger,"[22] a cherished facet of our postwar life.—

In his list of essential freedoms, Roosevelt also included "Freedom from Fear,"[23] and he thereby spelled out the incommensurability of freedom and fear for the first time in this way. Yet, even though nobody had come up with this formula before, he was already too late. The clock of world history had already turned to five past midnight; his petition was made just as it was becoming invalidated by the new task looming on the horizon—a task almost diametrically opposed: the task, as it reads in *Grimms' Fairy Tales*, of "*learning to fear.*"[24] For what we are lacking most of all is the "freedom *to* fear"[25]—in other words, the ability to *feel the amount of fear that would be warranted;* the fear we would actually have to produce if we really wanted to attain "freedom from fear"[26] and free ourselves from the danger we are in. What it comes down to, therefore, is this: *to fear so as to be free,*[27] or even just to survive.

The fact that we who are living alongside the bomb lack such fear doesn't prove that we are especially courageous people.[28] This defect of ours has other roots. Many others, in fact, and very diverse ones at that. And if we are to overcome the complacency with which we are drifting toward self-annihilation, these roots must be unearthed. These, the roots of our apocalypse blindness, will be exposed in the next sections—or at least some of them will be. I make no claim for a comprehensive list.

We will separate these roots into two groups. We'll first track the philosophical-anthropological root—that is, we'll start by examining a characteristic that has to do with *the fact that we are human.* We'll then turn to the *historical roots* of apocalypse blindness—to those facets that have to do with *our current state of being.*

12. *The Main Root of Our Apocalypse Blindness: The "Promethean Gradient"—the Human Is Bigger and Smaller Than Itself*

The simple fact that the first root of this blindness was already mentioned in the introduction signals that we believe it to be the more important of the two. It consists of the "*Promethean gradient.*"— What does this expression mean?

It means that our various faculties (such as our powers to make, think, feel, imagine, take responsibility) differ in the following respects:

First: *Each faculty has its own unique relationship to size and scale.* Their "*internal volumes*," their "holding capacities," "performance parameters," and "effective ranges," *vary.* Example: *Today, we can easily plan the destruction of a major city, and we can also execute this plan with the help of the means of destruction we have ourselves produced. But our capacity to actually comprehend and imagine the resulting effect is entirely inadequate.*[29]— Yet, *the little we are able to imagine*—the indistinct image of smoke, blood, and ruins—*is still a lot when compared to the tiny quantum of feeling and tiny sense of responsibility we are able to produce as we picture the annihilated city.*— Each faculty thus has its own limit, beyond which it ceases to function and beyond which it no longer registers any further increases—there is no congruence between their effective ranges. "We can murder thousands," as we said in the introduction, "but we can only imagine maybe ten dead people—and we can at most cry over and feel remorse for one." And what is true for tears or remorse is true for emotion as such, and thus also for fear: It cannot live up to the performance of our other faculties. And when our fear attempts to align itself with them and behave "appropriately," this ends in failure. The anxious human can only master much smaller tasks than the producing human, and in this sense "*the human is smaller than itself.*"

"This differential," one will object, "is well-known."[30] Aren't we here simply dealing with the difference between "abstraction" (which moves from the singular case to the general) and perception (which is always directed at something concrete and bound to the narrow scope of its field of vision)?— There's no reason to deny a certain link between the latter differential and the specific "differential" we have in mind. It is even likely that the capacity to abstract can itself only be understood with the help of a fully developed theory of the Promethean gradient.— But the two differentials aren't identical: The annihilation of a city (which we evidently have the power to bring about) is not an act of abstraction; and the act of imagination we are incapable of performing is not an act of perception.—

Viewed in isolation, the "differential" between our various faculties cannot be deemed a defect, of course. Quite the opposite; it would probably be biologically disadvantageous or even nonsensical if our various faculties were to all achieve as much as each other, if they all had the same capacities and "effective ranges." Their divergence is thus certainly rooted in the specific functions they each perform in life.

Yet the differences between our various capacities can grow to such extremes that our faculties lose sight of each other. When this happens, they become incapable of referring to the same objects, and the bonds between them are torn for good. This is the case today. As proof of this, we can listen to a bomber pilot who, upon returning from his mission, was asked by an interviewer what he was thinking of during his flight. In response, he perhaps naively, perhaps cynically, stated, "I tried and tried but I just couldn't stop thinking about the $175 still missing until the fridge in my house is paid off."— This is how far our actions and our conscience have been torn apart; this is how separate their respective objects have become.—

That our reason is "limited" and what this limitation signifies is something Kant has taught us. But we've routinely overlooked the fact that the same applies to our imagination and our capacity to feel—that they, too, are limited and incapable of surpassing their limits. Yes, even our feelings, which were always considered positively "unrestrained and excessive," are hemmed into the narrowest of limits. Feeling appears to share a similar fate to reason; it evidently also has a restricted capacity, and although it may to some extent be elastic, it is limited nevertheless. And this is true of all emotions, not just fear.

And so, we arrive at another variation of the already-familiar formula that serves as the leitmotif of our reflection.

We are incapable of feeling regret or crying over ten murdered people; our capacity to feel remorse cannot "expand" itself sufficiently to "comprehend" ten murders.

We might be able to imagine these ten dead—when pushed.

But kill? We can kill tens of thousands. Easily. And this capacity for killing can be increased further without difficulty. Or:

We can be afraid of our own death.

Yet it is already too much to empathetically intuit the mortal fear of ten others.

But at the thought of the apocalypse, the soul goes on strike; the thought remains but a word.—

All this should be investigated by a *Critique of Pure Feeling*—not in the sense of reaching a moral verdict, but in the Kantian sense of tracing the limited performance of our capacity to feel. What should upset us today is not, like Faust, the fact that we are not omnipotent and omniscient; rather, inversely, it is that our capacity to feel and our power to imagine are too small in comparison to what we can know and produce;[31] that we—as feeling beings—are smaller than we actually are.—

And it is now quite irrelevant what part of what specific "level" we isolate when observing this "gradient"; its structure is always the same. The gradient between "doing" and "feeling," for instance, is no less important than that between "knowing" and "comprehending": that we "know" what consequences a nuclear war would have cannot be denied; but we only "know" it. And this "only" indicates that our "knowledge" remains in the closest proximity to ignorance, or at least to incomprehension, to which it is much closer than comprehension. In the end, we perhaps only proclaim, "I know, I know," so that we, for heaven's sake, don't have to know *even* more—so that we can use the cover of the alibi of stated knowledge to retreat back into ignorance again.

There are, of course, those who know more about the bomb than others, even some with far greater knowledge. But regardless of how much one knows, none of us "knows" with the degree of comprehension that would actually be required: The commander in chief knows as little as the infantryman, the president as little as one's pal. The gradient between knowledge and comprehension exists, irrespective of personal status and differences in rank; nobody is immune. Which also means that there are no competent people here—the power to decide over the apocalypse categorically rests in the hands of the incompetent.—

When I chose to describe the widening gap between our faculties as a "Promethean" gradient, I did so in view of its basic motif. The name reflects the differential between our "Promethean accomplishments"—the products we have the power to make—and all our other powers and accomplishments. It signals that we cannot live up to the "Prometheus in us," and is thus also meant to link our reflections back to the topic of the first essay and the phenomenon of "Promethean shame." The name reflects the close connection between the two essays.

Second: *Just as each faculty has its own relationship to size and scale,* each now also has its own *degree of elasticity* or, rather, its own *level of rigidity.* We are not only able to produce, do, and think more than we can imagine, let alone feel; our capacity to make and think can also be *increased ad libitum,* whereas our imagination seems much less expandable. Yet, in comparison to our almost inflexible and rigid capacity to feel, it still stretches very far.

Although this second point is probably just a shift of emphasis regarding the incongruity between our faculties, this new accentuation is important since it signals that each faculty has a different relationship to *history*. That we primarily think of our changing deeds, products, and ideas when considering what is part of *history* is no coincidence. For the speed at which their effects, functions, and performance parameters change is much greater than the speed at which our other faculties transform, so much so that these faculties are now lagging behind our products as something that hasn't been historically synchronized. Inversely—and we will return to this shortly—it's also no coincidence that there is a history of "*res gestae*" and a "history of ideas," but no analogous "*history of feeling*." In fact, the speed at which our emotional life changes is so slow that it was even possible to attribute it to "nature" and treat it as a constant, ahistorical aspect of human existence. We will soon see whether this was justified.

IV. Development of Moral Imagination and the Plasticity of Feeling

13. *On the Spiritual Exercises That Are Required Today*

Since our faculties each have their own scale and relationship to history, we are now the most inhuman, most torn, and most internally malapportioned beings to ever have existed. While our capacity to feel is still stuck in the artisanal phase of cottage industry—we can at best feel remorse for one person slain by our own hand—our capacity to kill, even to engage in corpse-production, has already reached the proud stage of industrial mass-production. The performances of our hearts—our inhibitions, fears, foresight, and regret—are developing inversely proportionally to our other accomplishments: They are growing smaller as our other capacities increase. In comparison to the rift that affects us today, all the older spiritual conflicts that we needed to put up with—for instance, the conflict between "mind and body" or "duty and inclination"—were perfectly harmless. Such conflicts, regardless of how dreadfully they may have raged within us, still became manifest *as a conflict*; while clashing, these opposing forces still remained attuned to one another—their incongruence was still human. Even when reduced to nothing but a battlefield on which these two opposing forces met, the existence of the human was never in dispute: There was never any doubt that it was *the one same human* who was struggling with itself via two opposing forces. And because these forces never lost contact during the struggle—duty never split from inclination, nor inclination from duty—their intuitive connection, their shared identity, was still authenticated and mutually confirmed; the human was still there.

Not so today. Even this minimal guarantee of identity has been lost. What makes today's condition so utterly dreadful is precisely the fact that there's no sign of a struggle anywhere. Everything seems peaceful and perfectly in order—a collective smiling[32] now obscures the situation. Our faculties have moved apart and can no longer see each other; and, having lost sight of each other, they no longer bicker and clash.[33] And since they have ceased to fight, they have also stopped causing each other any grief. In short, the human *as such* no longer even exists; there is now only the one who acts or produces over here, and the one who feels over there. And only these specialized human fragments are real. What filled us with abject horror ten years ago—the fact that the same human could work as an employee in a termination camp *and* be a good family man; the fact that these two fragments didn't impinge on each other because they no longer even knew each other—this *abject harmlessness of horror,* did not remain an isolated case. We are all the descendants of these—in the truest sense of the word—*schizophrenic* beings.

If this is the case—and if everything is not to be lost—*then today's decisive moral task must consist of the training and formation of our moral imagination.* That is to say, we must try to overcome our "disproportion" [*Gefälle*]; we must attempt to increase the capacity and elasticity of our imagination and feeling until these are able to match the size and scale of our own products and the unanticipated magnitude of what we can bring about. An attempt must be made to coordinate our self, making the part of us that feels and imagines commensurate with the person we already are in our actions.—

An entire generation—the generation of those aged fifty today—was enthralled when Rilke spoke of feeling as something "*to be performed,*" "*to become accomplished in*"; when, in his final poems, for instance, he spoke of feeling as "not being performed" or as something obscurely futural that was "not yet being accomplished"—such as in "the accomplishing of love" in the *Duino Elegies.*[34] Rilke's petition was, of course, rooted in other experiences than ours, and what he was trying to say has nothing to do with our condition today (our inability to imagine and emotionally live up to our own products). But the notion that one *should also be able to feel* couldn't be found anywhere in the literature of his day. And if we, too, were fascinated by his vocabulary, this was because we sensed (without being able to identify why) that his words about insufficient feeling had captured one of the decisive defects of our time. And because we also sensed that he had identified a most necessary task.—

At present, it is unknown whether this state of "disproportion" can be overcome; it is an open question whether the task we need to perform is even "accomplishable" at all, or whether the performance of our imagination and feeling can be voluntarily increased. Our capacity to feel may be limited a priori; it may be rigidly fixed, or at least not expandable at will. Should this prove to be true, the situation is hopeless. Yet the moralist cannot simply accept this prevailing assumption; it may be the product of laziness or else be based on an untested theory of emotion. So, even if they believed it entirely improbable that these limits can be breached, they would still have to demand, at least of themselves, that the attempt be made. Only the actual experiment can determine whether this is possible or impossible, and whether these limits can actually be breached. The question of whether there are any past examples of such voluntary expansion of feeling, or even instances of entirely new feelings, can be picked up later—and it will be. But at present only the actual experiment matters—the attempt to perform "moral stretching exercises," that is, "hyperextensions of our usual capacity for imagination and feeling." In short, we must perform *spiritual exercises* [*Exerzitien*] in order to transcend the putatively fixed "*proportio humana*" of our imagination and feeling.—

The writer of these lines is fully aware that, at this juncture, his text no longer represents a "text" in the normal sense and that he is no longer simply addressing the reader as "a reader." Then again, there probably is no text whose *raison d'être* isn't that of being more than text, and we've now reached the *raison d'être* of ours.—

The writer is also aware that his request amounts to a form of violence. In fact, the demand that humans ought to voluntarily expand their faculties bears a striking resemblance to the violently excessive demands he so vehemently rejected in his discussion of "human engineering."[35] But I see no other way. The weapons of the aggressor determine those of the defender. If it's our fate to live in a world (produced by us) that in its excess eludes our imagination and feeling and thereby places us in mortal danger, then we must attempt to *reel in* this boundless excess.

Yes, "reel in"—and this makes it clear enough how our experiments differ from those of "human engineering." Whereas "human engineering" tries to make us "*sicut* gadgets"—that is, make everything about us conform to the machine-world—we hope that our experiments will allow us to "catch up with" and "reel in" the machine-world, just as one "reels in" a cast line. We hope they will allow us to *bring* it *back*.

Concrete instructions for conducting these experiments don't seem possible to me, nor even mere descriptions of what they entail; they elude communication. The last thing that can be shared is what happens *on the threshold*, immediately before one takes the actual step and attempts the experiment. I'm talking about the moment at which one tries to visualize the goal by "*telling oneself*" what one hasn't been able to feel and imagine; when one tries to tempt "one's inner bastard"—one's unwilling imagination and lazy emotions—into action and forces them to complete the workout. The expression "to tell oneself" signals that the exercise entails *a call*. But because it is a call directed across the Promethean gap, as if the faculties stranded there were people, this call is not comparable to a call of conscience, which is primarily *heard*. It is a call one is actively calling oneself, and it is the faculties left on the far side of the gradient—our imagination and feeling—that are meant to listen or be enticed into listening.

With this, really everything that can be put into words has been said. What happens after stepping onto the threshold—the actual awakening of the faculties, their cautious attempts to feel beyond their boundaries, their efforts at measuring up to the objects placed in front of them as their task; in brief, their actual self-expansion—cannot be communicated anymore.

There is no denying that this account is reminiscent of religious practices, and I don't pretend otherwise. I wouldn't even object if this self-transformation were compared to the practices so extensively documented in the history of mysticism. Yet, in our context, the word "mysticism" shouldn't be understood in its customary (i.e., vague) sense; it must be read as a specific reference to the attempt to reach altered states of consciousness with the help of techniques of self-transformation. It would have to evoke actual attempts at gaining access to states, regions, or objects that one would otherwise be excluded from.— I will return to this problematic in detail in the appendix below.

This, of course, doesn't mean that our exercises amount to true mystical practices. Despite the resemblance in type, the difference is fundamental: Whereas mystics seek to open up metaphysical regions and view their lack of access to them as something metaphysical in itself (believing it a consequence of their own metaphysically inferior position), our own efforts seek to grasp objects that are under our control, even objects, such as the bomb, that we have produced ourselves. These objects, therefore, cannot be said to be unreachable by *us*, but only by us *as beings who imagine and feel*. What is to be bridged is thus not a transcendence; it is at most an "*immanent transcendence*"—that is to say, "*the Promethean gap.*"—

Yet, regardless of what we call them, these attempts to bridge this gap elude description. The only thing that matters is to actually make the attempt.—

V. The Historical Roots of Apocalypse Blindness

14. One Believes in No End; One Sees No End—the Idea of Progress Has Induced Apocalypse Blindness

If one considers how even insignificant historical occurrences (or mere omens) were once enough to stir huge waves of eschatological excitement, fear as much as hope—a mass-excitement in the absence of means of communication—then the current "eschatological calm" becomes that much spookier. It persists even though the end has actually become possible and nearly limitless means of communication are available. Apart from the excitement in scientific circles, not the slightest panic can be detected anywhere (this, by the way, is a historical first—fear of the apocalypse affecting the nonreligious). True, the organized mass-actions mobilized against the use of nuclear weapons do signal the acute nature of the threat, but these protests certainly don't prove that millions are holding their breath in fearful anticipation of the end. It's not even possible to claim that eschatological excitement and readiness have lapsed—this century's revolutions didn't lack eschatological ambitions. Quite the opposite, in fact; it was their positive ambition to "annul and sublate" [*aufheben*] history, to establish a posthistorical condition, and to realize a classless society or "*Reich*." The powerful symbolism of declaring the onset of a new age—a gesture borrowed from the French Revolution—or the qualification of the Reich as the "thousand-year" Reich leave no room for doubt. Yet mythologizing words such as "revolution" and "upheaval" shouldn't blind us to the fact that these movements were, in a sense, "only" ever transfixed on the kingdom of heaven. There was no notion of an apocalyptic ending, of a

Judgment Day that would also be a day of judgment for their eschatological adherents—only the state that would arrive *after* the end.

This situation is most peculiar. In times past, eschatological hope was automatically supplemented by apocalyptic fear. But the apocalyptic element of this equation has now been obscured if not entirely erased. People put their trust in "saviors" like Hitler but didn't believe any prophecies of doom. And when these "saviors" actually brought this catastrophe about, their followers weren't even able to retroactively recognize or acknowledge this evil as an experience they had lived through.

What prevented this?

The belief in progress.

The generations-old belief in the apparently automatic progression of history has robbed us of our ability to anticipate the "end." The belief in progress has instilled in us an attitude toward time and the future that hasn't yet lost its shape, and even those of us who have lost our faith in progress remain affected.[36] There is thus also a "gap" between our attitudes toward time and our new ideas: Our attitudes haven't yet been able to synchronize; we *are* still a reflection of yesterday's beliefs.—

When it comes to the necessary task of learning fear, it seems unlikely that any generation predating that of the eighteenth century (before theories of progress began their triumphant rise) would have been as ill-equipped for this as we are now. Belief in progress, by its very definition, has transformed history into something *without an end*—into a happy calamity, a steady, unstoppable process of continual improvement. Its idea of eternity is the child of optimism and comparative improvement. As such, it could never lead to a day of judgment (let alone a hell).[37] It couldn't even lead to a heaven: As the best of all possible states, heaven is the enemy of the *better* state and would hence suspend the process of improvement. For the disciples of the idea of progress, there was thus neither "good" nor "bad," only the comparative—only things that are better. Or, to be more precise, if anything were bad, this was only because it hadn't become better yet, only because it still needed to be overhauled—something that would inevitably happen soon. The bad was hence treated as something transient and fleeting, already a thing of tomorrow's past. Belief in progress thus turned the notion of the negative into something just as unreal as it is for adherents of theodicy. And indeed, anything perceived to be "negative" was relegated to the past, whereby any lingering traces of hellishness that had stuck to it were likewise erased. In short, *one was not prepared for a "bad ending" because there was neither anything bad nor an ending anywhere.* The idea of a "bad ending" was but a blankness in the soul; it was as inconceivable as, say, the idea of a spatially finite universe is for us.—

The immense impact of dialectics on the nineteenth century quite understandably makes our suggestion that it lacked a notion of the negative seem quite strange. Yet, if we compare the meaning of the negative in dialectics to the notion of negativity implied in the idea of hell, its changed meaning is inescapable: The negative is turned into something that renders things "positive"—it becomes a *catalyst that aids fermentation.* Since its power now lay in provoking new life and new movement, its "pure negativity" was, in a sense, negated in advance. Goethe's portrayal of Mephistopheles—that he always wants the bad but creates the good—also applies to Hegel's version of negativity.—

If the "*History of Feeling*" introduced as a desideratum above were to be written, the loss of this fear of the apocalypse (and of hell) would need to play a significant part. The profound changes this loss effected in our ancestors—and which, in turn, also altered us—were no less fundamental than the transformation their own ancestors underwent during the "Copernican Revolution." Without this "loss," the confidence of modern man would never have grown to the dimensions it has actually assumed, and our current inability to fear would remain incomprehensible.

This isn't meant to suggest that fear of the apocalypse and hell should be artificially reinstated. I am only trying to indicate that it was the anticipation of judgment and hell that once taught us humans how to fear, and that the torments awaiting us in hell provoked much greater fear than any "inner-worldly" danger, including our own death. This means that the fear required today bears a much closer resemblance to that now-lost form of fear than anything we may have endured during the recent terror regimes or carpet bombings.[38]—

15. *Even One's Own End Is Denied*

The idea of progress thus creates a mentality with a very specific image of "eternity"—the image of an endlessly improving world. Or, to be precise, a mentality coupled to a very specific defect—the inability to even *intuit* an end. Regardless of how this is viewed, it means that anyone who believes in progress, in effect, regards not-ending as a basic law. And this is the decisive point: For as such, the notion of eternal improvement is universally applicable, even to one's own life. Which means that one doesn't face, *cannot* face, one's own death either—*one denies one's own death.* As the idea of progress cannot possibly be squared with the humiliation of dying, this avoidance is logically quite consistent. The fact that people continue to die cannot be changed, of course, but death's sting can be blunted and the shame of dying hushed up. Stated in the positive: A world can be created that is so neatly sewn up with positivity that no embarrassing questions about death show through the seams—a world in which there are no reminders of this ultimate humiliation, so that only a few (on as few occasions as possible) are reminded of death.[39]

Two randomly chosen examples can illustrate how this denial takes place:

First: In the United States, it has arguably already become impossible to find death anywhere. Only those things that are successively becoming better are deemed "to actually exist," and because of this, one isn't quite sure what to make of death. Unless, that is, death can itself somehow be tagged on to this universal law of increasing quality. And this is what is happening. The unstoppable advances in cemetery design have already been extensively covered by Evelyn Waugh; and it's undeniable that one dies much more pleasantly today than fifty years ago. And since the dead are made to look pretty, if not even alive, it's not actually the dead that are buried in these cemeteries, it's death itself.— It is, of course, not entirely possible to suppress the fact that the dead have experienced some sort of change, and so they are simply sent away: The saying "*partir c'est mourir un peu*" has turned into "*mourir c'est partir un peu*,"[40] for they are sent to a place of such picturesque beauty and easy accessibility that it is described as if it were an attractive holiday resort. And even before the fact, the cemetery ads in subway trains and along the highway had already provided the *moriturus* with ample opportunities to become familiar with its tempting amenities. But "*moriturus*" is precisely the wrong word to use: The living person is precisely not considered a *moriturus*—that is, someone who will die. The dead are simply treated as if they were continuing their life elsewhere after some sort of "change of residence."[41] The expression "to continue life elsewhere" now identifies the type of "immortality" developing in the optimistic world of progress: It doesn't consist of "eternity," but of the endlessness of one's current existence.—

Second: In places where death is acknowledged (its existence cannot fully be suppressed)—in scientific contexts, for instance—it is detoxified by being declared an aide in the ascent of life. Darwinism already provided a classic example, for it understood death—even the death of entire species—as the "sifter of life," so to speak. It hence put death in charge of the office responsible for facilitating the awarding of mandates to fitter (and "thus" more legitimate) forms of life, a duty it dispensed by letting the weaker (and "thus" life-unworthy) forms of life fall through the sieve. By recoining the *negativum* "death" into the *positivum* "ascent of life," Darwinism fulfilled the function of providing a naturalistic theodicy that also introduced progress into nature.—

These illustrations of the role that death (or, rather, its denial) plays today were selected quite at random. Yet, to show just how comprehensively the philosophy of progress has been exercising its rule, it didn't seem misguided to choose two examples that lie far apart in time and geographic location while also stemming from very different spheres of existence.—

Given that humanity's view of death was still so inadequate and opportunistic, it would indeed be rather unfair to expect it suddenly to be capable of comprehending this "ending" in an apocalyptic sense or even brace itself for its arrival.—

16. An Additional Task: The Voluntary Expansion of the Temporal Horizon of the Present

Although the technological "utopias" dreamed up by the disciples of progress did pre-envision the truth, and although the future was the only time that truly mattered to them, their middle-class (as opposed to their eschatological-revolutionary) followers largely remained blind to the future. In a sense, they didn't need to look to the future because it was coming to them "all by itself," with a new and always improved version arriving every day.[42] When compared to Christians' anxious anticipation of the apocalyptic end, their futurism was thus not nearly as intense.[43] As fast as we may have been running into the world of progress, we were running as if shortsighted. Our *Zukunftshorizont*—our perceived horizon of the future—was most provincially narrow. The day after tomorrow was already no longer considered the future anymore.

This paradox tries to convey the simple fact that just because things lie ahead does not automatically mean that they count as "the future." The year 1967 is clearly treated as "the future." The same, however, cannot be said of the year 2500; and we are also unable to see the humans of the year 2500 as our great-grandchildren. Their place in time seems hidden somewhere in the fog; these people seem to be "none of our business." Not to mention the year 10,000 CE, which seems as archaically remote as the year 10,000 BCE.— So, even in the age of progress, life simply went from one day to the next, even if this day was one of daily rapid change.—

But these happy days are now over.

The future isn't "coming" anymore—it's no longer considered something that "arrives"; *we are now making it.* And we are making it into something that also contains its own counterpossibility: the possibility of its own end, potential futurelessness. Even if this end doesn't happen tomorrow, our actions *today* have the power to make it happen the day after, in the generation of our great-grandchildren, or in the "seventh generation."[44] Because of the *lasting effects* of our actions, we are, in fact, already inhabiting this future. Pragmatically speaking, it already is *the present*; it's as "present" as an enemy who may in some superficial sense be distant, but who is already within range of our weapons and can, therefore, be struck.

We thus have power over a time span we do not usually consider and cannot consider to be "the future." Our actions do more than our perception comprehends. *We are throwing further than our shortsightedness allows us to see.* So, we again find ourselves before a "Promethean gradient," faced with the familiar task of having to overcome it: the task of "not lagging behind ourselves." It is up to us to ensure that we do not remain "smaller than we are," that we "catch up with ourselves," and that we don't bring about our own demise because we're no match for ourselves. All these expressions will already be familiar from the section in which we discussed the task we are required to complete today (section 13, above).

Montaigne's phrase "*to embrace the universe as one's town*" served as the epigraph to this chapter.[45]

This Montaignian motto must now be translated into the temporal sphere: We must reach beyond ourselves to the most distant point in time; we must fully grasp its existence and ensure that it remains evident to us. "Greet the unborn as your neighbors" (as they say in Molussia).

It is thus, again, the "expansion" of a faculty that is at issue: the expansion of our ability to intuit time [*Zeit-Vermögen*]. Yet, rather than being asked to predict this or that occurrence in the manner of a prophet, what is asked of us is to look into time as if we were looking from an airplane or a mountaintop. We are being asked to attempt to comprehend this expanded temporal horizon as *our own* time.

Just as radio and television can pick up even the most spatially distant things, channeling them to the *one* place we're in, we must now pick up temporally most distant, most futural events so we can synchronize them with the *one* moment we're in. Since these futural events are bound to the present, they are happening right now. Which makes them our business and concern because our actions today make us something that "concerns" them.—

It is no doubt a very unusual relationship to time that is being postulated here. The future, allegedly, no longer lies "*ahead of us,*" but is present right here; it is "*with us*" as our captive. We certainly won't be able to acquire this novel relationship to time overnight, but let's hope that there is enough time left for us to learn.—

17. "The Things I Can't Do Anything about Are None of My Concern"

But let's pause a moment here to gain some methodological clarity about what we are doing.

Our problematic is set out in a most unusual manner. Whereas any half-reasonable research focuses on an actual performance or achievement (say a specific cognitive process) so as to examine its mechanisms and conditions of possibility, what we are searching for are the roots of *a nonaccomplishment*—the roots of our failure in the face of the apocalyptic situation we are in. Our focus is thus *something nonexistent.*

This is absurd. And, indeed, as long as it is solely undertaken as an epistemological study, the examination of the question "What prevents us from perceiving the apocalyptic situation?" makes no sense. If we were to start asking "Why does this or that thing *not* exist?" there would be no end to the questions, since nonexistent things can be thought up at will. Our problematic only makes sense because it is a moral question—that is, because we are considering something nonexistent that actually ought to exist, a *desideratum.* As strange as this combination of words may sound, our study is *a moralistic epistemology*—in other words, *a moralistic theory of knowledge.*

There is, however, a second reason we are using this strange term.

The factors responsible for "apocalypse blindness" are in themselves moral in nature.

Because *the moral situation* in which we encounter a given object (a specific issue or situation) also *determines whether we actually perceive it*. This means that our ability or inability to perceive something is contingent on whether an object "concerns us" [*angehen*]; nothing actually "sinks in" [*eingehen*] that doesn't concern us.[46]

Angehen, eingehen—with this play on words nothing has of course been resolved. On the contrary, this is where the problem actually begins. After all, we have no patent, no secure criterion by which we could decide what concerns us and what doesn't—that is, what is our business and what isn't. At any given moment, there are countless things that leave us cold and fail to concern us, even though they "ought" to be our concern: such as, for instance, the "apocalyptic situation" (but not just this, of course). There are in fact thousands of *issues* we *cannot or are not allowed to concern ourselves with* and which we cannot possibly involve ourselves in because our concrete life situation (work demands, property relations, peer pressure, political violence, and so forth) precludes us from doing so. We lack the opportunity; we lack the *freedom* to regard them as "our concern." When I isolated the moral situation as the decisive factor in our comprehension or incomprehension of an object or state of affairs, it was this I had in mind. If we are robbed of the possibility to somehow participate in the deciding of something, we will soon no longer concern ourselves with this thing at all—unless, that is, we are actively trying to resist it. It's thus not just true that "ignorance is bliss" but also that *"the things I can't do anything about are none of my concern."*[47] We remain unable to perceive whatever we've been forced to "disregard." This becomes evident in the context of totalitarian states. Once doing something about certain measures comes to be considered absolutely unthinkable, humans will soon no longer take notice of these measures at all. This is also why, for instance, the question of whether thousands in fact knew about the termination camps and liquidation installations is inadequate. It's probably true that many thousands did know, yet they took no notice because taking action was ruled out from the outset. And so they just carried on with their lives as if they didn't know. Just as we do, even though we "know" about the bomb.

To fully comprehend our own apocalypse blindness, we must thus also view it in relation to the moral situation we are part of. It must be viewed through the lens of what we are permitted to do and what we are prohibited from doing—what we can and cannot do, what we should and should not do. A portrayal of our moral situation—or as much of a portrayal as is necessary for shedding light on our defect—thus needs to be attempted.

18. "Mediality": We Are No Longer "Acting" but Only Participating—Our Acts Have Been Stripped of Their End Goal; and So We Live without a Forward-Looking Perspective, without a Future; and Hence without Any Comprehension of the End of the Future. This Makes Us "Apocalypse-Blind"

There is no disagreement: The nature of our activities—in other words, our approach to *work*—has fundamentally changed. Except for some uncharacteristic remnants, working today means "being a coworker" in a labor process that is administered by an organization [*Betrieb*] and fully coordinated with it.[48] Individual solo work has, of course, never been the main form of human labor—which is why I am foregrounding "the organized character" of today's work. For today, what we are dealing with is not primarily collaboration with other workers; work amounts to a "cofunctioning" with an operation. And although this operation is untransparent, workers must comply with it nevertheless, and their other coworkers, likewise, only belong to it as cogs in the machine.

This is a trivial point. But what applies to our work (and this is far less trivial) now also applies to our "*actions*"—or, perhaps we should better say, to "*our activities*," as it already sounds like an exaggeration to say that we are "acting," and such dissonance must be taken seriously. With the exception of a few sectors, our "activities" now take place within the framework of some prestructured organization that is as untransparent as it is impossible to avoid, and as such, they amount to conformist coparticipation [*Mit-Tun*]. Any attempt to determine to what extent a given participatory behavior is "active" or "passive"—where "being engaged" in action ends and "initiating action" begins—would remain entirely inconclusive; as inconclusive as any attempt to establish to what degree a given machine interaction is active or merely reactive because we must inevitably conform with the machine's functionality. The distinction between the "active" and "passive" has become secondary; today, human existence consists neither primarily of being "*a driving force*" nor of "*being driven*"; neither of acting nor of being acted upon.

Instead, it has attained the neutralized character of being *active-passive-neutral*. Let's call this a "*medial*" existence, a mode of being in which our existence corresponds to the capacity of the devices and intermediaries we rely on.[49]

This "mediality" prevails everywhere—not just in countries that enforce conformity through violence, but also in places where this happens gently. But it's much more noticeable in totalitarian settings, of course. And so, I will choose a typically totalitarian behavior as an example of "mediality":

During the trials prosecuting Nazi "crimes against humanity," a clear pattern could be observed: The defendants repeatedly reacted with hurt, bewilderment, or even outrage at being addressed as "persons"—that is, at being held personally accountable for the abuse they had inflicted and the murders they had committed. It would be totally wrong to consider these defendants as arbitrary examples of inhumanity, callousness, or obstinacy. When they failed to exhibit remorse, shame, or any other moral reaction, their responses weren't missing "despite" the fact that they had participated in these atrocities, but mostly because they had "*only*" participated. In some cases, these responses were, in fact, missing *because* they had participated. Such defendants equated "moral behavior" with full "mediality," and because they were people "who participated," they had a clean conscience. "If only we knew what you want from us!"—they could have said, spelling out their moral obstinacy—"Back then *we were* perfectly decent (or 'moral,' if you prefer). After all, it's not *our* fault that the organization we satisfactorily participated in has now been replaced by another! Today, it is 'moral' to participate here; back then, it was 'moral' to participate there. So what's the problem?"—

This attitude made possible the most horrific crimes.— Yet anyone who gasps at these crimes as erratic events of our epoch closes themselves off from actually comprehending them. When viewed in isolation, these crimes seem devoid of reality—or at least of a reality that can be understood.

The only way they can become comprehensible is through a classificatory lens—by asking oneself what type of actions these atrocities represent and what activity they are modeled on. And this reveals that their perpetrators—or most of them, at least—committed these crimes in a situation in which they, in principle, behaved no differently than they were used to behaving in the workplaces [*Betriebe*] that molded them.

This sounds shocking, and it's perhaps inevitable that this claim will at first be misunderstood. After all, there are no workplaces where workers are taught how to commit mass-murder or torture (or at least none we refer to with words such as "factory" or "office"). What is meant here is something different—something much more trivial: a circumstance that is mundane yet nevertheless rarely thought through to its final consequences. It is the fact that *the conformist principle of "mediality"*—of participation that is "active-passive-neutral"—*prevails in every workplace.* This is a principle that no one could ever be held accountable for introducing and which has effects that are usually neither dreadful nor immediately visible. And this approach to work is regarded as entirely obvious and legitimate everywhere, whether one is in Detroit, Wuppertal, or Stalingrad. It's now a key feature of the average workplace—or at least the large corporations that are decisive today—that they demand full commitment (regardless of their goal) and that workers are expected to "actively go along with the workflow." Yet, although a worker's sole purpose is that of contributing to the goals of a given organization, they have no idea what these goals might be. In analogy to the fundamental problem of Marxism, we could say that the worker doesn't even "own" the envisioned goal; these goals are none of the worker's business. Once this is the case, once workers *do not know* the goal—don't need to know it or aren't supposed to—they evidently no longer need a *conscience* either.[50] The kinds of "actions" that are examined or even dictated by one's own conscience have thus been suspended and replaced by the conscientiousness of one's contributions to the workplace (which are "active-passive-neutral"). The only thing still resembling a "good conscience" while at work is hence the paradoxical satisfaction of having successfully switched off one's own conscience, an achievement that might even make one feel proud. If a factory worker or office employee were to refuse to participate in the machinery of work because its product went against their conscience or even a universal moral law—because they feared its immoral use or the possibility of this—they would, at best, be deemed a fool and would immediately be made to suffer the consequences for their unusual behavior.

Since work is considered "moral" under all circumstances—and this is one of the most fatal traits of our age—any goals and outcomes of this work are perceived as "morally neutral" while the work is being done. *The work product,* regardless of what one is working on, remains "*beyond good and evil.*" And any other characterization of this situation, any nonnihilistic portrayal, would amount to a cover-up. *Non olet:* "work"—and this is where the calamity reaches its climax—"never stinks." Psychologically, the idea that a product one is working on might infect the activity of work doesn't even arise, not even if the product is most reprehensible. Morally speaking, the link between the product and its production has been fully severed. The moral status of a given product (poisonous gas, for instance, or a hydrogen bomb) casts no shadow onto the moral status of the workers participating in its creation. In order to do the work they are doing they need no conscience, regardless of whether they know what this work is.[51] As already mentioned, this "lack of conscience" is ingrained in today's workplace [*Betrieb*].—

The "medial-conscienceless" human is thus produced in the workplace; this is the birthplace of the conformist. The humans molded there seem innocuous, but they only need to be reassigned to a different range of duties, to a different "operation," and they suddenly appear monstrous and fill us with abject horror, even though they haven't really changed. The moment this happens, the suspension of their own conscience (which had already been a *fait accompli*) now appears as naked unscrupulousness; the suspension of their sense of responsibility, as pure "moral insanity."[52] As long as we refuse to face up to this, refusing to acknowledge that the coordinated human is forged in today's workplaces—that work is itself the model for this synchronization—we remain incapable of understanding today's conformist contemporaries. In other words, we remain unable to comprehend those "morally obstinate" men, who, in the trials just mentioned, refused to feel any remorse or responsibility for the atrocities they "participated" in.

This should not be misunderstood. The very last thing on my mind is "*comprendre pour pardoner,*" understanding to forgive. It's mere chance that I, unlike others like me, didn't fall victim to these men.— What I want to show is that these atrocities, because they are rooted in the "medial nature" of today's working practices, are intimately connected to the nature of today's epoch. This makes them much more dreadful and disastrous than was assumed during the (all too fleeting) attempts to face up to them (e.g., in the discussions about "collective guilt"). They are more dreadful and disastrous because their precondition—today's approach to work—remains unchanged; and because we don't even know where to turn for a remedy, as we are unable to even imagine an approach to production and work that would differ from the one that currently prevails.[53] That's why we consider it pure self-deception to view these atrocities as "erratic events" that only arbitrarily strayed into the realm of history, and only on this one occasion, with nothing to suggest that they might be repeated.— Since "mediality" and conformism are more prevalent today than they have ever been, it is, in truth, entirely unclear what might prevent such repetition. Should a Herostratus one fine day decide to commit some "genocide" or another, he would have no reason to doubt the reliable participation of his contemporaries. He wouldn't need to lose any sleep. They wouldn't let him down. They would be on call, motorized, rearing to go, for *participation* is what has been drilled into them *as* workers. The conscientiousness they've procured in place of a conscience (or that our epoch has imposed on them) is like a sacred vow. It amounts to a pledge to never see the outcome of their activities, and when they cannot avoid witnessing it all the same, they must immediately forget it and not let it sink in.— In short, they vow to "*know not what they do.*"[54]

With this, the dreadfulness of today's moral dilemma has been defined. On the one hand, we expect today's human to participate 100 percent, since such participation is a precondition of work, or at least work ethic; on the other hand, we would like to demand (and we genuinely feel that the world should be the kind of place in which this can be expected) that, "outside of the sphere of work," humans behave "as themselves"—that is, "nonmedially," in brief: morally. And this is an impossible situation, for two reasons:

1. In the situations that really matter, there is no "sphere outside of work." That is to say, it has already been ensured that any decisive tasks awaiting today's humans only ever appear as work tasks; and that humans, even when killing, aren't "acting" but are simply doing their job. Dreadful as this may sound, *the employees in termination camps weren't acting; they were working.* And since the aim and outcome of their work was none of their business and concern, and since work *qua* work is always "morally neutral," the same applies to what they got done.—
2. Today's human (and I am now speaking of the average human) is expected to simultaneously embody two utterly distinct modes of being. They are expected to behave as "conformists" at work but as "nonconformists" whenever they are "exercising their agency." They are hence expected to live a schizophrenic

> life and bear this mode of existence, a life shaped by a hierarchy [*Gefälle*] between two contradictory modes of action that can never be leveled out. This *schizophrenia by decree* is so monstrous that all previous demands placed on humans by moral principles, even the most impossibly challenging ones, shrink into amicable incentives and suggestions.—

We said that medial humans take a "secret vow" "not to see, not to know what they are doing." They pledge to never set their sights on the *eidos* or *telos* of their activity; they vow to remain "*telos-blind*" (by analogy to the expression "apocalypse-blind").

As the word "pledge" assumes a form of freedom that medial humans no longer possess, it is, of course, only a metaphor. We only use it to convey that such humans aren't *supposed* to know what they are doing; this ignorance serves business interests. And it would be a mistake to assume they want to know. Indeed, *while at work*, having a clear sight of the *telos* (or even of *its telos*—the use to which the work product is put) would be of no practical use. Such knowledge might even interfere with their work. By the time the medial human gets involved, this *telos* has already been decided. The worker thus never arrives at the idea of paying any attention to the *eidos* of their work. Often this has also become *impossible*: While completing their (conveyor-belt-like) work, workers only ever deal with one small step in the labor process—with tasks that continually appear and disappear again. Their work never finds completion in a finished product. It is only ever superficially "stopped" by the sound signal, which cuts into the endless game of repetition at the end of the workday.

This points to a radical transformation of the human. From the early Platonic dialogues all the way to Heidegger's analysis of "relevance as constitutive of being,"[55] human activity has been characterized as the pursuit of an *eidos* that is sought to be realized as an action is performed. It is this *eidos*—the *eidos* pursued through an activity or its product—that is "removed" from medial actions; they are *eidosless.*— Aristotle distinguished between two types of human activity: those that pursue a goal, a *telos* (such as cooking), and those that are entirely self-contained and thus carry their *telos* within (such as going for a walk). Yet, in the context of today's work culture and related spheres of action, these categorizations have become superfluous, since *telos* and *eidos* have both been removed. As we work at a machine or participate in a coordinated activity, we are as devoid of a destination as when setting out for a leisurely stroll.

What implications does this account of the medial human have for our actual topic, the roots of apocalypse blindness? In what sense is "medial existence" one of these roots?

The connection is so close, in fact, that every aspect of mediality provides a valid point of departure; every aspect leads to apocalypse blindness on an equally direct path.

1. Since medial humans are "active-passive-neutral," they remain *lazy and indolent,* despite the immense significance work plays in their lives. They only ever "go along with it," "surrendering themselves" to the current state of affairs. As such, they count on things continuing as they are (whatever the costs may be)—a continuation they need not take responsibility for.
2. Since their activities never find completion in an actual *telos*—since these activities are only ever interrupted by external factors—medial humans lack a genuine relationship to the *future.* Whereas people who act and make plans create a temporal space—a future—into which they project their plans and actions, "medial humans" only ever tread on the spot. Through their work they are, of course, trying to secure their own future, but there is no contradiction here, for their work only ever consists of the obstinate repetition of an always-identical step. Medial humans, in effect, live an "atemporal" existence—"your future is taken care of."[56] That is to say, they exist without a temporal horizon in which the end of the future—the possible apocalypse—could appear. Since they know no future, they are also incapable of grasping the acute threat of its end.
3. Since medial humans are used to activities for which no conscience is required—and isn't actually desired—they are *conscienceless*; and they are conscienceless in all good conscience. This means that they never feel any scruples with respect to the wider implications of their work—all the more so because they consider their own work as categorically "immune" to any such contamination.
4. Medial humans also believe that each product is "morally neutral." The assertion that an absolutely immoral product could

exist—even one for which a whole dedicated world of production has been established—inevitably sounds ludicrous.—

In short, all aspects of medial existence have conspired to prevent medial humans from noticing what the bomb represents. And so, they continue to work toward their own end, as hectic as they are lazy, without any sense of what the word "end" actually means.

VI. Annihilation and Nihilism

19. *One's Deeds Are Not a Reflection of One's Character; One's Character Is a Reflection of One's Deeds.— Today's Categorical Imperative: "Own Only Such Things Whose Inherent Maxims You Would Will to Be the Maxims of Your Own Actions"*

Regardless of whether we assign blame to those who see, the morally decisive factor, of course, is not apocalypse blindness, but the *bomb itself.* The decisive factor is that we *have* it. And since, regardless of whether those who have the bomb want this or not, "to have" here automatically means "*to do*" (section 9, above); the moral issue we are dealing with here is that of *the bomb as a deed.*

In the context of a deed of such magnitude, its moral character no longer reflects the character of the person responsible for it—it doesn't depend on their intentions, whether good or bad, or on their level of awareness or convictions. It is determined solely by its *effect.* There is a threshold beyond which all psychological differences, even the crudest ones, must be ruthlessly suspended—

beyond which it becomes an academic luxury to hold onto any distinction between intentional criminality and some minor offence [*crimen veniale*];

beyond which a perpetrator, regardless of their alleged intentions, must be judged *as if* the threat posed or effect caused was their actual intent;

beyond which, since the existence of the human is at stake, the seemingly inhuman principle is valid: *Through their actions "a person can only ever be as good or as bad as their deed,"* rather than judging deeds based on the good or bad nature of the person responsible for them.

Of course, this principle is not allowed to be applied to normal deeds: The distinction between intentional and unintentional acts remains sacrosanct. And even in our case, its suspension can only be justified as long as it is accompanied by tireless activism that attempts to educate—only as long as the ceaseless attempt is made to confront the unperceiving perpetrators with the effects of their actions, and only when no opportunity is missed to try and prevent them from becoming guilty in the ultimate sense.—

Whether this kind of activism can succeed—or how long it might take before it has an effect—cannot possibly be predicted. At least it has begun and appears to be gaining more support each day. But our moral judgment of the perpetrators cannot be adjourned, pending its outcomes. The principle, therefore, must be: "*In dubio contra reum*"—guilty until proven innocent. Although this principle may again sound inhuman, its sole concern is for humanity.— As long as *they don't abolish these devices,* as long as they threaten us simply by *having* them, as long as they continue with the activities they falsely refer to as "tests," they must *be deemed guilty.* And because the effect of their activities amounts to *annihilation,* they are guilty of *nihilism*—nihilism on a global scale.

With this, we have arrived at our final thesis: *The masters of the bomb are nihilists in action.*

This sounds strange. We typically picture "nihilists" as figures of a very different kind: We usually associate this term with either the anarchists of prerevolutionary Russia who engaged in pitiful individual actions or the radically skeptical European intellectuals of the same period. Or else it makes us imagine those dictators and abhorrent lackeys who could only affirm their own existence through the screams of their victims—only through the principle "I annihilate, therefore I am."—

Do the men with power over the bomb really fit into this picture?

Not in any immediate sense; psychologically, they resemble none of these types. Some of them will scarcely have heard of nihilism, if at all, then only in sayings such as "overcoming nihilism" or "the crusade against nihilism."[57] Most of them, in private, are no doubt good-natured, decent, bourgeois people, and thus capable of neither the melancholy nor the cynicism that even the most average nihilist couldn't forgo without dishonoring their profession. No, the men in question are "positive" deep down into their bones. Their bank accounts are healthy, and their principles no less so.

But they are nihilists all the same. Even if they profess the healthiest philosophy and the most positive religion, and do so without any pretense, their character, philosophy, religion, and entire personal integrity is nevertheless nothing but surface decor and hypocrisy before the tribunal of the "objective spirit." For regardless of whether they know it, whether they even want to know it, they follow a very different kind of philosophy and morality—the philosophy and morality of the thing attached to them and to which they are attached. Before the tribunal of the "objective spirit," "*everyone has the principles embodied by the object in their possession.*"

This is a third strange formula, one that strains our habitual ways of thinking even further. So, first, let's start by explaining what it *doesn't* seek to imply.

It isn't meant to have any Marxist connotations; it has nothing to do with social theory—that is, with the now commonplace assumption that "consciousness" (morality, philosophy, mentality) is shaped by "existence" and "circumstances." It *doesn't* suggest, therefore, that what has made these men so "bad" are the bad "circumstances" they found themselves in (circumstances which happened to include their unfortunate authority over the bomb). It *doesn't* suggest that their "evilness" is morally invalid or unreal because it was not originally born out of malice but is simply a product of "circumstances." Nothing along these lines.

We don't believe that those who have the bomb are in any decisive sense shaped by the fact that they "possess" it, rendering them totally evil. We even consider it likely that their private, subjective convictions—their feelings, mentality, and behavior—are still as harmless, decent, and virtuous as they were before. For whereas the social theorist argues that "bad as they are, they are ultimately not so bad after all, since only the circumstances are to blame for their crime," we are saying that:

They are not bad at all, but their lack of malice is not a virtue; it is exclusively a consequence of the "Promethean discrepancy"—the fact that they are limping behind the thing they "have," that they have remained too small to be shaped by the monstrosity they hold in their hands. We are saying that they are incapable of becoming as monstrously evil as the bomb and its deeds. In other words: *Their virtue is nothing but a symptom of their failure,* and, from a moral perspective, this makes it null and void.

And, second, *their virtue* only concerns their behavior within a sphere of a life in which the bomb doesn't yet even "exist," which makes it *inconsequential,* null and void.

No, genuine virtue would today consist of something very specific.

Only those who fully realize that their moral character exclusively depends on their "possessions" could be called virtuous—

only those who spare no "strenuous effort of the concept" [*die Anstrengung des Begriffs*][58] in their attempt to actually get a hold of their possessions, take ownership of them, and "reel them in" (in the sense specified earlier in section 13);

only those who take every one of their "possessions" to task and force them to confess their secret maxims;

only those who examine their conscience before every one of these maxims, to assess whether they would wish them to also be the maxims of their own actions—if not even the principles of "a universal law";[59]

only those, finally, who are resolved to eliminate any "possession" that didn't pass this test . . .

only someone who takes all these steps can today be called "moral." Their imperative, the categorical imperative of the atomic age, would hence read:

"Own only such things whose inherent maxims you would will to be the maxims of your own actions."

Today's humans are, of course, not used to "examining the conscience of the things in their possession." No matter how accustomed they may be to treating people like "things," it has never crossed their mind to react to "things" as if they were "people." But this is what is called for now. The "things" that are decisive today (those which constitute our modern world and decide its fate) are no longer just "things" anymore but maxims that have materialized into objects; they are congealed modes of action.—

We have already established (section 6, above) how we try to reassure ourselves by calling all devices "means," without making any distinction between them. Let's now add that this move is precisely what makes us "immoral."

20. *The Janus-Faced Formula of Nihilism*

As friendly and decent as the masters of the bomb may be in those aspects of life where the bomb doesn't yet "exist," *as* masters of the bomb they follow the maxim of the thing they possess; and next to this, all their other qualities shrivel into nothingness. The bomb's maxim, we said, is "*nihilism.*" What does this mean? *What exactly is "nihilism"?*

To answer this question, we must briefly turn our backs on the bomb and examine the situation from which "the classic form" of nihilism erupted. By saying it "erupted" (as opposed to "developed" or "matured"), I am already highlighting an important aspect of nihilism: It originates from a *situation of shock*; it was the effect of a *catastrophic event*.

The event in question was the collision that suddenly exposed Russia to the "*West*" (and hence also to the *natural sciences*). Overnight, and without any forewarning, Russia encountered a conception of the world from which God had been entirely removed—a world that was blind, devoid of rights and purpose, unredeemable, and in no need of redemption. In short, it collided with the concept of a natural world, entirely composed of matter, of which humans were now allegedly also part, and this to the exclusion of anything else. And since this exposure happened suddenly, without the period of transition Central Europe had been granted during the nineteenth century, Russia was denied the opportunity to varnish life in this newly godless and meaningless world with even the semblance and gloss of a positive *modus vivendi*—through the application of "morality" or "culture," for example.[60]

Anyone who has read *The Brothers Karamazov* will be well aware of the total upheaval in Russia's emotional and spiritual habitus that was caused by the explosion of the theocratic world, triggered by its collision with the natural sciences.[61] Russia found itself in a historical constellation that demanded something it couldn't possibly cope with. It was asked to accept that the world that had an exclusively religious meaning the day before had overnight become solely a matter *for physics*. In place of God, Jesus Christ, and the saints, it was suddenly asked to accept the "*law of nature*"—*a law without a lawgiver*, simply there, meaningless, yet nevertheless utterly "unbreakable." Since this law exclusively determined what *is* (and determined things to their very core) while also remaining totally silent about what *ought to be* (because *the ought* is not an empirical entity), the situation was all the more untenable: Suddenly, everything was "*nature*"—that is, *everything was the same*. One should listen deeply to this formula (which used to be put into the mouth of Death, who, without regard for a person's standing, devoured all), for it reveals the essence of nihilism. The formula is "Janus-faced"; it has two contrasting meanings:

On the one hand, it says that "everything is of *one kind, nature*"—and, as such, the formula represents the foundation of "*metaphysical monism.*"[62]

On the other hand, *by acknowledging only the laws of that which exists*, the formula also means that "there is nothing that ought to be; nothing matters; everything is permitted"—and as such, it also represents the programmatic foundation of radical *amoralism*.

Formulated in the wake of Schelling and Weisse, *Lotze* famously asked, "In what properly consists the fact—how it comes about or is made—that there is something and not nothing[?]"[63] The nihilist, who is also a disillusioned moralist, repeats this question in a varied form: "Why should it matter that there is something and not nothing?"

"Why," the nihilist might argue, "if there is only what is—why and for what purpose is there still something I 'ought' to do?[64] Why are the world and I not equally allowed to not exist as to exist? And why should I not exact my revenge for the fact that instead of being something that is meant to be, the world and I are simply 'here' and simply 'nature'? Why shouldn't I just take matters into my own hands and be the one who brings about this state of nonbeing? Wouldn't this deed—through which I would take a step that nature, in all its lawfulness, apparently has no intention of taking—be the only action I can still perform? Wouldn't this act be the sole means by which I could prove—even if just to myself—that I still exist and am something other than nature?"

Had the nihilist, in his panic, been able to find the words, his questions would have run along these lines. In brief, the Janus-faced quality of the formula *"everything is one thing," "it's all the same,"* is the key to nihilism—for nihilism was *monism in action*; or, rather, *nihilism was the revenge of the human against monism* (for more on this, see appendix II, "Portrait of the Nihilist," below).

What does this Janus formula have to do with our topic? With the bomb?

Although everyone has probably already found the answer themselves, it still needs to be spelled out: *The secret maxim of the bomb is identical to the maxim of monism, or rather, nihilism.* The bomb behaves just like a nihilist: whether it is human or device, bread or book, home or forest, plant or animal—to the bomb, it's all the same; it sees and treats everything as nature, which means, in this case, that everything is equally amenable to being contaminated by radium. And, apart from this, nothing exists for the bomb. If it were able to speak, its words would be those of the nihilist: "It's all the same to me, whether the world exists or not. Might it not just as well not exist?"

This is the bomb's maxim. And also that of those who have it; for as we know, everyone who has a thing also lives by its maxim. Regardless of whether they want to. Regardless of whether they know they are doing so. And that's why it is not just some random comment but the simple truth when we call the masters of the bomb "nihilists."

Let's bring the forefather of nihilism and his great-grandsons face-to-face, for only these two herostratic figures are each other's equals; only they deserve to be considered the great nihilists. In the time between the Russian nihilists and today's masters of the bomb, history witnessed many other nihilists, including those usually presented as its *archetypal* representatives. There were indolent nihilists, who didn't annihilate but were merely negative or even just described "nothingness"; there was the great one who called himself "the Antichrist"; and then there were also those bloodstained nihilists who really did behave in the manner of the Antichrist.— Yet, regardless of how great, melancholic, profound, or horrendously abject they all were, when compared to those two antipodean figures, they turn into mere transitional types. Only these two—only the first and the most recent—have any direct bearing on the *being and nonbeing of the world as a whole.*

While the forefather was a philosophical figure of the first rank, his grandson is a philosophically uninteresting *bonhomme.* The forefather's utterly ruthless questions—asking why purposeless entities ought to exist at all, and why *he* still ought to respect any "oughts"—his boundless melancholy, and his desperate wish to annihilate the world that was annihilating him, assure him his philosophical status. Yet, despite all this, unable to actually accomplish the total destruction he wished to perform, he appears to have had no direct impact on history.

His grandson, however, who is as incapable of cynicism as of melancholy, is a limited, privately harmless figure, for whom total annihilation is no different to any other new technology that falls into his lap. But precisely because of this, he has a historical significance that overshadows anything that has been called "historical" so far; he doesn't just have the power to annihilate the world; he may perhaps even be incapable of *not making use* of this power.

"The gods of the plague," as one says in Molussia, "are peaceful gentlemen who bear no sign of pestilence themselves."— The deities ruling over today's annihilation closely resemble these gentlemen. Going by their appearance, the last thing one would imagine is what they can unleash. Their smile is benevolently kind and even without deceit. Yet there is nothing more terrible than the well-meaning smile on the face of the gods of corruption and doom.

21. *The Bomb and Nihilism Form a Syndrome*

Apart from the millionfold smile of those they have doomed.

For it is not only those who threaten us with the bomb who turn into nihilists but also those who *are being* threatened by it.

That is, *all of us*.

Roughly ten years ago, nihilism entered an entirely new stage. For the first time in history, it stepped out of its esoteric existence and began to shape—or, rather, subvert—the popular imaginary and turn into a mass-philosophy or, at least, mass-mentality. And soon after, the various chronicles of nihilism written between 1940 and 1950 could already be bought in paperback—for a few short years, nothing "sold better" than nothingness.

Nihilism hence attained its new mass-appeal at the precise historical moment when the bomb was produced and deployed for the first time. The historical concurrence of *nihilism for the masses* and *mass-annihilation*—of a philosophy that negated the meaning of human existence and a device that annihilates humans—is highly conspicuous. Was this pure coincidence? Or was there a connection between the two?

It would, of course, be pointless to seek a direct causal link between these two historical phenomena. There is no direct bridge between Einstein's theory of equivalence (which underpins the construction of the bomb) and the formula "it's all the same" (which defines nihilism). But there are also other kinds of connections, and it is no mere coincidence that the two coappeared. We see two such connections:

1. Both share the same historical root: *National Socialism*.

   It is not necessary to demonstrate that National Socialism was itself a variant of nihilism. And it wasn't just nihilist in the vague sense in which this association is usually made, but in a strict sense, for National Socialism exhibits the attributes of naturalistic monism that were introduced as the quintessential aspect of nihilism above. It was the first political movement to negate the humanity *of* humans—of whole masses even—so as to then annihilate them as "nature," as raw material or discardable waste. In National Socialism—and the classic nihilist would have turned green with envy at this—the philosophy of nothingness and the act of annihilation fused to such an extent that one can even speak of "*annihilism.*"

Both the atomic bomb and the new (French) variant of nihilism were attempts to come to terms with this "annihilism." In the case of the atomic bomb, this was because its production originally tried to preempt the spread of this National Socialist "annihilism," while French nihilism, in its portrayal of "absurd existence," to some extent at least, also doubled as an attempt to portray life in the shadow of National Socialist terror. It sought to capture the feeling of human beings who are forced to experience themselves as a "nothing," as "there for nothing," and annihilable, as if it were nothing.— The bomb and French nihilism were thus responses to the same historical event, which makes them more than just two historically concurrent phenomena.—

2. But this commonality—the shared polemic origin of the annihilation device and philosophical nihilism—is perhaps not even the most significant link between the two. It is far outweighed by the fact that the bomb and nihilism *found each other* and formed into a "*syndrome.*" What does this mean?

It means that the bomb and nihilism have psychologically become one and the same. Over the last decade, they have formed into one single complex in the wider popular imaginary of our epoch (and even more so in public sentiment than in popular philosophy). The indissoluble nature of their bond is revealed in the way this relationship is framed in uninhibited speech. When the man of today spouts his unfiltered commentary on the current state of the world (and thereby divulges the items of faith the epoch holds onto), he couldn't care less whether the existence of the bomb points to the meaninglessness of existence or whether this meaninglessness is in turn used to legitimate the existence of the bomb. So much so that he isn't even sure himself which way he is arguing. Whenever a premise and an assertion become interchangeable in this way—whenever an argument is just as likely to start from the left as from the right, as if an image were being described—this is *the sure sign* of a "real" syndrome. "Real" in this context, of course, doesn't mean that the two elements have grown from the same root, but rather that they have grown together like two successfully grafted plant parts that have organically fused. Whenever a premise and an assertion become interchangeable, one is dealing with an indissoluble ideological aspect of the age that has become emotionally true—a complex that is immune to being debunked by means of argument, and which can only be destroyed as a whole.[65] And the bomb and nihilism have already been interchangeable in this way for many years.

A small conversation I was able to witness during a train journey across Germany in 1952 can reveal how, in the popular imaginary of our epoch, the bomb and nihilism really are one and the same. I actually witnessed but a small fragment of this conversation, but this snippet was already enough for me, and it will also suffice as an example here. And it was in this conversation that the nihilistic smile of those under threat showed itself—the smile I spoke about above.

Diary Entry (1952)

"Boo!" one of them cried, as I was jolted from my sleep just before we arrived in Frankfurt (it was the bony one, who didn't hold back last night when telling us about his "experiences" in Poland). "Boo-hoo! Why not? Why shouldn't they throw the thing?" And it was immediately clear that "they" meant the Americans and that the "thing" he was referring to was the atomic bomb.

The Swedish lady sitting next to me, opposite him, only waved her hand with her usual "whatever you say" response.

"Well, can *you* perhaps tell me," he said as he smiled in turn, "what the point of them is and why they exist?"

"Who do you mean, they?"

"Well, *they,* of course—the Russians," some other guy said (flicking his head back against the direction of travel).

"And *you*?" the lady responded with an angry frown. "*You* know exactly why *you* exist?"

"I didn't say that. Potato, potahto."

"Potahto?" she politely inquired.

"Well, what else could it mean!" he said. "*Them* or *us!* Tomato, tomatho—same difference!" And, after smiling into the round to ensure that every one of us passengers was with him, he added, "And what exactly is the point of *your* existence?"

Since the lady from Sweden was not traveling in metaphysics but in matters concerning an international nursery for kids, she couldn't provide an answer. Upon which the guy, with a triumphant grin, said, "There you have it." And then something incredible happened (which, apart from me, nobody noticed). For he now suddenly presented his argument the other way around. "Or do you believe that there is still a point to our existence now that this thing exists? Or to the kiddies you are swanning around for?" And with this he threw himself back into his seat, as if the final word had been spoken and the lady from Sweden had been shown up once and for all. Nobody in the compartment tried to contest his victory. He smiled.

In this crudely nihilistic conversation (which was probably just a random example of a conversation unfolding daily in a thousand other ways), the "pointlessness of the human" and its annihilability—nihilism and bomb—are one and the same.—

The bomb is OK because there is no point to us anyway, and because the bomb is here, we are no good for anything anymore anyway . . . and because we are no good for anything, the bomb cannot make matters any worse. Because, because, because . . . and so, the spurious argument nauseatingly whirls on, like a carousel spinning around the axis of nothingness. It spins without revealing its impuissance and false triumph, without revealing where it starts or where it ends, or what is based on understanding and what is mere assumption. And those who do not jump off, those who do not smash up the ride, will not escape alive.

VII. Concluding Remarks

"What a blessing," I read in the newspaper, "that the bomb isn't in the hands of nihilists."

This sigh of relief belongs to an unthinking person.

Whose hands the bomb is in might indeed be decisive, especially if the one who controls it really does make use of it. But if they decide not to use it, then nothing has been decided at all.

The bomb doesn't just loom over us—over those of us who are alive today. The threat will never be over. It is only ever deferred. What might be averted today can happen tomorrow. Tomorrow, it will loom over our children, and over theirs the day after. Nobody will be rid of it. No matter how far into the future coming generations may advance, wherever they will head to flee from it, the bomb will flee alongside them. Yes, it will even rush ahead, as if it were leading the way as a dark cloud above their path. Even if the ultimate outcome were never to occur, even if the bomb were only ever to tower above us without unleashing its power—we are, from this day onward, beings who are condemned to live in the shadow of this inescapable companion: We are henceforth condemned to live without hope, without a plan, as if we didn't matter.

Unless we muster the courage to make a decision.

The newspaper writer's sigh wasn't just thoughtless but also nonsensical. As nonsensical as if he'd written, "What a blessing that the dirt wasn't placed into dirty hands!" As if it mattered what state the hands were in before. They are dirty *now. Because* of the dirt. This is the only thing that counts. And dirty they shall remain.

Unless we muster the courage to make a decision.

Unless single individuals were to begin, like conscientious objectors, to publicly attest, under oath, that they, in full knowledge of the possible dangers, will:

never participate in any form in anything that could have even the slightest, most indirect bearing on the production, testing, or use of that thing—no matter what public pressure or physical duress they are placed under;

swear to speak of the thing as if it were a curse;

educate those who shruggingly reconcile themselves to its existence;

demonstratively distance themselves from anyone who seeks to defend the thing.

Unless such a start is made, and unless others follow their lead, and others theirs; until those who refuse to take this oath become notorious and known as scabs and strikebreakers in humanity's fight for its continued existence.

For the bomb has achieved one thing: It has turned the fight into one of all of humanity. The bomb has thus achieved what religions, philosophies, empires, and revolutions did not: It has turned us into *one* humanity. What has the power to affect everyone is everyone's concern.

The collapsing roof turns into *our* roof. For the first time, we as *morituri* are truly an "*us*."

There is little honor in the fact that the bomb is what was required to achieve this. But let's forget this, for we *are* now one. Let's prove that we can also be one as living beings; and let's hope for the day when the apocalyptic fears of today will have turned into the nightmares of yesteryear; let's hope for the day when we will have no other comment in response to the words spoken here than "what unnecessary pathos."—

VIII. Appendices

Appendix I. On Section 13

On the Plasticity of Feelings

The task of "grasping" [*fassen*] and "comprehending" [*auffassen*] things that are "too big"—in other words, the voluntary expansion of the capacity of our soul—is, as our everyday use of language reveals, not as unusual as it might at first have appeared. "*To brace oneself for something*" is one such phrase, and it points to a task we regularly need to perform in the course of our lives. When saying, "I am bracing myself for bad news," for instance, we are in effect implying that we are preparing ourselves for something that is "actually" too big for the usual capacity of our soul (and we use this phrase even though we may have no idea what it is that we are doing to prepare for it). We undoubtedly do something to prepare ourselves "internally." We do, in some way, "expand" ourselves to attain the capacity to comprehend this thing that is "too big," despite its "incomprehensibility," and to comprehend it in such a way that we are not seized, rendered "speechless," or even destroyed by it.[66]— This serves as just our first pointer.

But this is of course not nearly enough. To get to grips with the malleability of feelings, we must digress much further—into the very depths of "philosophical anthropology."

The "*indeterminacy of the human*"[67]—that is, the human's lack of a specific binding nature; or, expressed in positive terms, its ceaseless self-production and unabating historical transformation—makes it impossible to decide which aspects of the human should be considered "natural" and which should be deemed "unnatural." The very alternative between the two is wrong: "Artificiality is the nature of human beings."

The history of changing styles and morals is a never-broken chain of enterprises, by means of which humanity has attempted to compensate for its own indeterminacy through self-imposed restrictions. Socially and psychologically, it continually had to try to determine itself anew, always making something new out of itself—something that, "by its nature," it was not, but which it nevertheless had to be. For, insofar as it wanted *to be* at all, humanity could only ever function as a *specifically determined* society, however artificial this society may have been.

While every genus or species of animal appears to have a given, specific world- and social-schema (or has developed one that is now as definitive as the schematic of its bodily morphology), only the human is endowed with "*sociality as such*": Humans, so to speak, are issued with a blank check that must be filled out retroactively in some way if they want to function at all. In other words, they must themselves always produce the schematics of their own world and societies. This production, this human *praxis,* is thus a response to the empty "as such" ("sociality as such")—a response to the indefinite ontological endowment the human is invested with. And since human artifacts—when considered in terms of the empty "as such"—are "unnatural" and contingent, and because these contingent products must be made compulsory if they are to become even somewhat durable, every foundation and enforcement of a social schema entails stringency and violence. Especially violence against those whose class is overpowered during the act of foundation and who therefore lose out. And since the human is only ever adapted to "sociality as such" and never to "precisely this" societal schema, this violence is in a sense always also directed against human nature as such.

This is also why a given societal schema can only ever succeed if it manages to shape the human *as a whole.* Such total formation can only be achieved if human *feelings* are modeled, too—and with this we have arrived at our topic.

In most cases, it was not necessary to specifically model new feelings by force: Whenever the demands of a new world- or societal-schema started to be felt, humans usually tried to emotionally adapt and measure up to this new situation on their own initiative. They thus tried to turn the new artificial world, which they happened to be compelled to live in, into the world that was "natural" to them. This process of "fitting oneself to measure" [*Anmessen*], of adapting to the artificial as a "home" [*Wohnung*], is called "*habituation*" [*Gewöhnung*]. For those who succeed in this, the artificial world turns into what seems "natural": The *a posteriori* seems to become the *a priori,* the contingent, the necessary.

But because of the *phenomenon of the Promethean gradient,* this automatic process of habituation cannot always be relied on—and the wider connection to the topic now becomes clear. Our feelings change at a much slower pace than the world; the human, as a feeling being, lags behind itself. There is hence a recurrent need to lend feeling a helping hand or to explicitly create feelings. This need becomes especially acute whenever the "gradient" represents a political risk—whenever it threatens to delay or even prevent some specific change, that is, when there is a danger that humans might refuse the new "world" being forced on them because it is too novel or too violent. The year 1933 acts as a clear example of this. The Nazi propaganda we were witnesses and victims of was nothing other than a production of feelings on a most colossal scale. The party deemed this production necessary because it calculated that if its victims were equipped with these feelings, they would find it easier to accept (if not even enthusiastically welcome) the overwhelming system of terror.[68]

For the philosophy of history and the humanities as a whole, addressing the lack of a "*history of feeling*" (a term coined in analogy to the history of *res gestae* and ideas) probably represents the most desirable task. Such a history, if it exists at all, can currently only be encountered in unintentional versions—for instance, in the form of specific works in the history of art and religion. This gap is caused by the prejudicial assumption that emotional life is *the* constant, *the* nonhistorical element in the history of humankind. We cannot pursue the roots of this prejudice in detail here.[69]

But it is a prejudice. In the course of history, it was not just our practices and categories, not just our ways of perceiving things, that changed and gave way to others; feeling has had its history, too (although a much slower one). The core tenet of Rationalism and Deism—that the world religions only differ in their content but are otherwise identical, since all are "beliefs" and thus "feel the same way" (a view that can still be encountered in the United States today)—is utterly implausible. All this tenet proves is that Rationalism was much more Christian than it realized, for it mistook its own historically specific mode of feeling for "feeling as such." Every religion is, in fact, a system of feeling *sui generis;*

and every newly founded religion has represented a true revolution in the human history of emotion, a genuine *founding of feeling.*—

Just how nonsensical the idea that human feeling is constant is, becomes clear the moment one lets the thousands of types—no, singularities—of feeling invested in the artworks from different epochs pass before one's mind's eye. One cannot assume, surely, that Delacroix's way of feeling was no different from Parmigianino's, or Berlioz's from Palestrina's—that these artists all shared the same feelings while "merely" producing totally different artworks. Anyone who claims to have already been familiar with the specific feeling of the Giorgionesque without this feeling having been mediated by Giorgione's paintings cannot be taken seriously. The respective historical worlds of these artists—the worlds in which, for which, or in resistance against which, they created their work—also made each one of them feel in an entirely different way. With this abundance of historically distinct nuances of feeling before one's eyes, one will certainly no longer find our notion of "new feelings" that can measure up to our world today all that alienating. New feelings are nothing new. What is new is that we now acknowledge this reality and call it by its name.—

And the special metamorphosis suggested here—*the voluntary expansion of the capacity of our feeling*—is likewise nothing new. On the contrary, it would easily be possible to compile a long list of examples of such expansions. The history of magical and religious practices, as well as mysticism, provides ample material. But there is no need to venture into such remote territories. Such expansion technologies already are an integral part of our own lives; they are just never acknowledged as such or introduced by this name. The specific technology I am referring to here, the technology I will use as an example to illustrate the possibilities of voluntary expansion of feeling, is *music.* Anyone who wishes to understand this example would do well to temporarily forget the terminology with which music is usually discussed.—

Example: We are listening to a Bruckner symphony.

The "Bruckner-world" that builds up through the music is so expansive and voluminous that the expanse and voluminosity of our everyday world disappear. The moment we open ourselves to this expansive and voluminous happening, or *are opened to it* through its force, it begins to flow into us and fill us up—we "hold it," we "comprehend it" [*wir "fassen" es, wir "fassen es auf"*]. In other words, *our soul starts being expanded*; it takes on a capaciousness and capacity [*Fassungsweite*] that we couldn't have given it ourselves. That such experiences of "hyperextension" are often hard to bear—that we deem them too much—is thus nothing surprising.

But how should we understand the notion that "we" cannot give our own souls this voluminosity "ourselves"?

The symphony, after all, is a work by Bruckner—it is something man-made. In a certain sense, it is thus also something made "by us."

Once we view the symphony in this light, it attains an entirely new appearance: It now appears to be *an apparatus* that was made by us, *a device that helps us expand the capacity of our own souls.*

Of course, one is not allowed to imagine this elasticity and expansion in mechanical terms: It is not like the inflation of a rubber balloon that is filled with gas and now "holds" it inside. When the soul "holds" the "content" (the world of the symphony), this always also means that it "*comprehends it.*" The music can only ever be "held" as long as it is "comprehended"; only while the soul actually experiences the symphony can it also experience (in the sense of "undergo") changes in size. The relation here is thus not between two *res extensae*—not between an elastic container and its content—but between a *comprehending subject* and an *object of comprehension.*

Or perhaps this isn't the case either.

Do we really understand what the word "object"—*Gegenstand*—means in this context? Does the symphony really stand "*gegenüber*"—that is, "*opposite to*" or "*across from*"—the soul as any self-respecting "object" would?

Certainly not. The listener is *in* music, and the music is *in* them. So is one really allowed to use the word "object" to characterize such an ambiguous and peculiar interrelation?

Certainly not. In the sphere of acoustics, or at least in music, the notion of an oppositional subject–object relation becomes meaningless—which comes as a great shock to the epistemologist (who had conceptualized this opposition in response to the optical relation between human and world and deemed it universal). While making music, or in any competent listening to music, human beings become identical with their object, and the object becomes identical to them. Every act of making music is a piece of identity philosophy become real; in every such act, the human "takes in" [*einholt*] its product—that is, it "catches up with it" and "reels it in." The aim here is not to turn matters over to "subjectivistic music aesthetics" (although this misunderstanding seems inevitable), for, as we have seen, the alternative "objectivism/subjectivism" is inadequate. Nor do we intend to suggest that the purpose of music is to evoke some sort of inarticulate mood in the listener—music only has this effect on the unmusical. Quite the opposite. Human beings are, in fact, never more articulate than when they are "in music": for when they are in music, their mood [*Stimmung*] cannot be detached from the "voices" [*Stimmen*], harmony, and resonance [*Stimmigkeit*] of the music as a concrete object—that is, its processual movement. The highly articulated nature of the "object"—its size, tensions, movement, "space," tempo, et cetera—all these things are assumed by those who make music or listen to it. They "*co*"-function "*with*" the music; they *become* the object.— Yet, at the same time, the object also exhibits all the traits that are usually exclusively attributed to a subject. It has an "expression," a "mood," and so on. In other words, subject and object, state [*Zustand*] and thing [*Gegenstand*], form one single unit. Since both "subjectivistic" and "objectivistic" music aesthetics assume a subject–object distinction that is actually inapplicable to music, both are equally inadequate. And nothing we have said here should surprise us. Other art forms (e.g., dance) have already accustomed us to the shared identity of subject and object—the dancer and what they dance—or even of the shared identity of the musician, the instrument, and the music being played. In these instances, the separation of subject and object has not yet taken place.— Subjectivistic aesthetics becomes downright foolish, however, the moment it views artworks as nothing but a reflection or "expression" of something that is felt anyway. This version of aesthetics explains nothing, as it leaves us utterly in the dark about how we can possibly be so profoundly moved and unsettled by works of art if these works are merely expressing something that is "felt anyway." *While composing their piece of music, the composer feels something that they could only ever feel by means of it. This means that the artworks create these feelings—feelings sui generis—* that could not actualize themselves without these created objects. If these feelings were independent of the structure of these objects—that is to say, if they were mere sentiments—they would lack all substance. Even the emotional states that artworks [*Kunstwerke*] invoke in us are artificial [*künstlich*]—they are, if you will, "artworks."

As remote and far removed from our topic as the examples we've digressed to may seem, they nevertheless reveal something fundamental about its underlying problematic: *They show that humans do not need to resign themselves to a set of pregiven feelings that have been formed once and for all. Humans, on the contrary, are constantly inventing new feelings—feelings that exceed the normal, everyday capacity of their soul,* and which place demands on its holding power and elasticity that can certainly be called "excessive"—that is, "overwhelming."

That we needed to source our evidence from the realm of art—or rather, music—is no coincidence. It's only in the realm of imagination that we have been left with some freedom; in the realm of freedom, however—that is, in the realm of morality—our imagination has been curtailed.—

Appendix II. On Section 20

Portrait of the Nihilist

I. The Dialectic of the Herostratus.— The appropriate revenge against the formula "it's all the same" could only ever have consisted of *total herostratism,* a herostratism of cosmic dimensions. It was this, the total destruction of the universe, that the classical nihilist was genuinely holding out for. He believed this destruction to be the only vehicle for "truth," the sole means by which his desire for revenge could be satisfied and his melancholy brought to an end. This made him a *philosophical figure* in the truest sense, the dark equivalent of the metaphysician. The nihilist and the metaphysician are, in fact, closely connected, and especially amongst the mystics, there are also hybrid figures who could fit into either type: Hölderlin's Empedocles, for instance, who couldn't bear that he would only ever have to make do with singular things on account of his own singularity. Empedocles was an infatuated neighbor of the nihilist and, in fact, belonged to both types. In his passion for the whole, he would have preferred to see everything special destroyed *because it was* something special; and, in the end, he only saw one way out: self-annihilation.

This is not just some random example, for both the nihilist and metaphysician share an appetite for totality that is unquenchable—that is to say, their shared trait is *failure.* Because no technologies capable of facilitating his dream of revenge were available back then, it belonged to the nihilist's nature that he was *categorically unable* to make this dream come true. If a classical nihilist were transported into the present, the "opportunity" nuclear weapons represent would leave him breathless with joy, if not paralyzed by fear, as there would no longer be any excuse for his failure.—

The "frustration" the nihilist was made to feel, of course, didn't make him lose his destructive rage, just as frustration doesn't make the metaphysician lose their passion either. And so his rage ended up fragmented, and despite all his principles, this gave rise to those unmotivated, uncalculated, uncoordinated, panicked, playful, and downright childish tantrums we now view as the nihilist's distinctive traits. He always had to settle for partial terror; instead of destroying the universe as a whole, he was only ever able to scratch the surface of some specific province of being. *Faute du mieux, faute du pire,* he only ever settled for assassinations and programmatic vices, and despite his best efforts, he always found himself back in the same situation: Instead of the nothingness he longed for, he ended up with something meaningless staring back up at him from the pool of blood at his feet—whether it was the (always wrong) district governor, torn to shreds by one of his bombs, or some (always random) violated child. He was always condemned to the experience of these destroyed humans *still being there* (and still there, clearly, only to make a mockery of his hopes). He could only ever establish all over again that none of his extraordinary measures provided the compensatory satisfaction he craved. On the contrary, as all novels of the epoch attest, his actions only ever resulted in that most terrible *tristitia post,* which erupts, as every enthusiast for the "whole" knows, whenever it becomes newly manifest how pitifully small the feeling of satisfaction remains in comparison to the hunger. Faust's opening lament about the futility of his "labor keen"[70] can easily be translated into the nihilist's language.

Despair over the meaninglessness and nullity of the world [*die Nichtigkeit der Welt*] has thus always culminated in *a second despair—the despair over the indestructibility of meaninglessness.* This second form of despair ultimately builds into self-derision—that is, into the mockery of one's *own herostratic impotence.*

There is no doubt, however, that this self-derision would immediately cease and make way for a monumental arrogance if this classical figure were to be transported into the present. And there is indeed no more deeply disturbing thought than the idea of a generation in which the dark passion of this ancestral figure has merged with the absent-minded power of his grandchildren. Which has almost already come to pass.—

Yet it was not just technological impotence that gave rise to this "second melancholy," but also the "*dialectics of herostratic desire*" as such. What does this mean?

That every Herostratus suffers from an internal contradiction. If he is not to cheat himself out of the enjoyment of destruction, he must renounce his boundless desire and his very dream of total annihilation.

For regardless of whether he knows it, he (just like anyone else who is seeking revenge) is pursuing the secret goal of self-validation. Since he is already feeling "annihilated" within, he now needs an "*I annihilate therefore I am.*" This means, however, that his self-validation must be validated by others. *To be able to enjoy the satisfaction of annihilating the world, he needs a world he doesn't annihilate.* That's why, even in his dreams, his hunger must be, and always had to be, a hunger for *a scene of destruction*—for a theater of destruction in which he committed his acts of violence before eyewitnesses. Or else, it was a hunger for victims, for victims who continued to live even though they had died, and who remained there even though they were bleeding.[71] He has always needed to see horrified people and hear their screams, which means that he wasn't allowed to destroy these victims alongside all the others. Not even in his dreams was he allowed to, for he had to ensure that some were spared. He was thus always like a painter creating a picture of the apocalypse. Which, of course, also smothered any possibility that his desire could ever be fulfilled, for he was no longer allowed to take things all the way.

To sum up, this man could not be helped, for if he were to fully realize his dream of annihilation he would lack an audience. And without an audience, destruction would fail to generate a sense of affirmation; so *his herostratic desire was and could only ever remain melancholic.*

II. The Dialectic of the Monist.— Every monist bears within him the following contradiction:

Although, by recognizing himself as part of nature (which he treats as entirely deterministic), the monist dehumanizes himself, and should therefore believe himself to be totally unfree, he simultaneously believes himself to be free *because* he is part of nature: He believes himself to be free from all "supernatural" authorities, free from any "thou shalt."

This highly peculiar coupling of freedom and nature (which would have been incomprehensible to Kant) is confirmed by the history of the last century. Any movement that wrote "freedom" onto its flags (conversely) also used "nature" as its motto; theories of freedom and naturalistic theories consistently appeared "*junctim*." Up to the First World War, for instance, it was deemed good ideological form for progressive liberals to also be Darwinists; and if one was a Socialist without being a Darwinist, one was even deemed to have failed in one's basic duties. Only rarely was it considered irritating that the two theses—"the human is nothing but nature" (which actually ought to divest it of all rights to freedom) and "freedom is the human's highest good" (which thereby establishes an "ought")—weren't able to have a common denominator. Marxism exemplifies just how peacefully these mutually contradictory theses were able to coexist for many decades. Within Marxism, these theses were, in fact, even able to come to each other's aid as creatures originating from the same system-paradise; yet, to reach this paradisiacal state, they had to agree to form a united front against the common denominator they both opposed. And this negative common ground was found in religion. As a Socialist, anyone who was fighting against claims to entitlement based on the "supernatural" also believed themselves to be making a plea "for nature" and for the natural sciences. *With this, the "iron necessity" of nature paradoxically turned into the slogan of freedom.*[72]

Yet, within the nihilist's soul, the antagonism between freedom and nature (which systems thinking had neutralized) never quite abated, of course. His frantic behavior suggested that there was something fundamentally wrong with truth. His agitation, translated into words, would read:

"Because I am nothing but 'nature,' I am unfree."

"Because I am nature, I am free—of all prohibitions."

"Because I am free, there is nothing that could prevent me from rebelling against my unfreedom."

"And so, to prove that I am free, I now blindly behave like nature, the nature I am anyway."

It is perfectly clear that this is a "*circulus vitiosus.*" But this circle is no more absurd than the secret absurdity of the epoch itself; it's just that the nihilist was the *enfant terrible*, unable to keep the secret to himself.

III. Sublimation and Deferral of the Monism Shock.— Monism only had a genuinely shock-like effect in places where it appeared overnight, and where it annihilated the prevailing concept and sense of the world in one fell swoop. Central Europe, however, which had itself been responsible for the scientific transformation of the worldview, and where the link between naturalism and technological and industrial advancement could be daily witnessed in action, was much more capable of absorbing this shock.— Over many decades, monism was thus able to be domesticated and sublimated. The Monist League, the German "monistic movement" founded by Ernst Haeckel,[73] for instance, didn't just succeed in divesting the term of any shocking connotations but even managed to give it a celebratory feel. For young people who grew up before the First World War and devoured "Kosmos booklets"[74] in their hungry enthusiasm for the natural sciences—I am speaking from my own memory here—monism amounted to a downright "religion" (what was called a "religion" back then). For us, therefore, the very thing that had provoked the nihilist's desire for annihilation—a cosmos from which God had been removed—was something divinely astonishing [*überwältigend*]. This experience was most peculiar: Because one felt oneself to be "nature," and exclusively "nature" at that—or, at least, because one believed oneself to feel this way (for "ideologies" also exist in the realm of emotion)—every other part of nature was also felt to be something one was related to. One felt "related in descent," so to speak. European monism, therefore, in contrast to its Russian iteration, shimmered across into an enthusiastic form of pantheism, into a "*pan-atheism*," to be exact. What theology had always asserted—that pantheism is identical to atheism—was here a real event. On top of this, the movement automatically integrated the optimism of the age—that is, the idea of progress—into its conception of nature. As with the theory of descent, the progress that was "proven" to be happening in nature was now used to prove the universal validity of the category of "progress" itself. European monists thus perceived nature in much more optimistic, dynamic, biological, and natural-*historical* terms than Russian monists, whose focus was especially on mechanical laws (since their natural sciences and industrialism were still in their infancy).— Many decades later, this difference would still be reflected in the differences between the National Socialist and Communist conceptions of nature.—

It is, of course, no matter of dispute that there were earlier conceptions of a uniform world in which "everything was one"—that is, "nature" (e.g., eighteenth-century French materialism).

Yet there was no corresponding *world-feeling* [*Weltgefühl*]—let alone a feeling that could have taken hold of mass-humans. Such a feeling was only brought about by monism. If one were to write the "history of human feeling" that we have already held to be desirable on several occasions, monism would have to play a dominant role. No earlier movement was able to render the dualistic world-feeling of the Judeo-Christian tradition questionable with nearly as much success.

This became fully evident the moment the nihilistic component of monism could no longer be suppressed—that is, when National Socialism came into power. It was in this "new start," this "awakening," this "break" [*Aufbruch*] that monism likewise "broke open" (alongside many other things) to reveal the long encapsulated second meaning of its motto "everything is one." What was brought back to light, in other words, was the immoralist sense of the motto, and this in a most bloodthirsty way. And this time around, the nihilists didn't just consist of a few scattered individual desperados, but of a compact and compactly ruling power that became manifest as a mass. And its spokesman didn't just limit himself to being a spokesman for nihilism. As *Führer,* he now also put the equation "power equals right" into action, and, imitating the forces of nature, he really did liquidate anyone who was weaker. That he embellished the National Socialist movement almost exclusively with nature words (such as "blood," "race," and so forth) provides further evidence that this movement was a variant of monism. "Nature"—once the banner of freedom movements—had now turned into the favorite vocable of terror (however Orphic or blood-covered its embellishment may have been).—

IV. The Penultimate Phase of Nihilism.— If National Socialism is set aside as a distinct phase, then three stages of nihilism can clearly be distinguished, or three types of nihilists:

1. The original ancestor (already discussed in detail) who felt "annihilated" by the collapse of his world and wanted to exact his revenge by striking back, by becoming an "annihilist" and destroying the world for real—which he was unable to do.
2. Today's grandson, the lord of the bomb, who has the capability to destroy the world even though he has no desire to do so (or who at least doesn't desire this as fervently as his ancestor). Between these two extreme figures, we can now also add:
3. The indolent European nihilist, who limited himself to "breaking tablets"; that is, to symbolic or verbal acts of destruction. In most cases, what drove this type to despair—and we are now speaking of the type, by the way, who usually features as "the" nihilist in the history of ideas—wasn't even the collapse

of the religious conception of the world. No, since the religious world had already collapsed and naturalism had already become a *fait accompli,* his despair was partly the melancholy of an individual who exists "for nothing," and it was also fueled by his realization that the "moral" and "cultural" world erected over the course of a century in place of the religious (so as to screen off the monism shock) was but a mock-up and nonbinding. This, of course, might seem to be contradicted by the fact that the French nihilists of the 1940s (with whom the interim period currently being replaced by nuclear nihilism found its definite conclusion) explicitly made "the death of God" their central subject. This could give the impression that they were lagging behind history, for bourgeois middle-class morality had long since been secular, and, in the context of scientific–technological praxis, the question of God was no longer even debated: The "death of God" was itself already dead. But it would be wrong to suspect an anachronism here. When mobilizing this phrase, these nihilists didn't pretend to have discovered something new or to have set an actual precedent. They used it, rather, to point out that when even the "mock-up" begins to collapse in on itself, it suddenly becomes *apparent once again* that there is nothing to see; that the place behind the facade really is empty; that the thing it had covered over for such a long time really has vanished. And it was on the basis of this experience that, for the grandchildren of the monists and the atheists, "God" died a second time.[75]

V. The Irrefutability of Nihilism.— Nihilism's basic question, as we know, reads, "Why should there even be something we ought to do?" [*Warum sollen wir sollen?*] Or, "Within a frame that is itself unsanctioned and suspended in a moral vacuum, what basis could there still be for moral obligation?"— Or, rather, this is how its questions would read if the crisis that nihilism represents were to translate itself into the more harmless form of "questions."

Crises can only be abolished; they cannot be disproven. Any attempt to refute nihilism would be foolish; only the naive and the opportunistic set themselves this task.

If one at least tries to make the crisis that nihilism represents comprehensible, its fundamental affliction turns out to be that the "ought" is an "*endogenous phenomenon.*" That is, the question of why "the ought" is something "that ought to be done" can only ever be justified within the framework of a life that has already been affirmed—only if life finds life agreeable anyway (and finds it agreeable for extramoral reasons; no, for no reason at all). In other words, *the moral necessity of the world and the human is itself no longer morally justifiable.*

I know of no text that captures the crisis of nihilism more impressively in words than the following moral fable, which I lift from a Molussian *Introduction to the Basic Problems of Nihilism:*

"With its lights dimmed," the textbook states, "not guided or welcomed by any god, but also not rejected or obstructed by any other (let's just say, not known to any of them), a huge ship is navigating through the constellation of Orion. Where it is from (if anywhere) and where it is going (if anywhere) is also unknown. There are many reasons why even mentioning such a ship would be superfluous; sooner or later, it will have dissolved into the darkness like anything else of its kind. In other words, it will only have existed as if it had never existed at all. Despite all this (and this is the only thing we know for certain about this ship), its cabin walls are plastered with onboard rules and regulations—rules and regulations that must have been sanctioned by someone who was not sanctioned by anyone at all. There is no need to deny that these rules make the busy life on deck run smoothly.— *Question:* Are these rules binding? *Discuss.*"

Editors' Postscript

Scenes of Obsolescence

Günther Anders and the Method of Occasional Philosophy

> What method should the truth now adopt so it can compete with the lie and also be believed?
>
> —Günther Anders

Sometime in late 1929 or early 1930, Günther Anders stepped out of his front door with the radio still playing inside. A few meters down the hallway, the same song was playing from another flat, and then, out in the streets of the Neubabelsberg neighborhood of Berlin, it was met again, sounding from windows across the entire city block. Instead of shrugging off this "spooky" yet mundane encounter with the singer's inescapable voice, Anders turned it into one of his earliest accounts of the dynamic later discussed in *The Obsolescence of the Human:* "When humans let their products loose into the world, they reap their haunting in return."[1]

As one of the most radical, academically overlooked, but also popular works of twentieth-century continental thought, *The Obsolescence* lends itself to multiple, often contradictory framings. In the German-speaking world, its haunting vision of life amid increasingly capable machines has turned it into one of philosophy's bestsellers. Since its first publication in 1956, the German edition of *The Obsolescence* has sold over ninety thousand copies and has never been out of print, finding a wider readership than many more canonical works.[2] A significant portion of Anders's prolific body of work has also been translated into French and Italian, and partially into many other languages. Until recently, however, the name Anders mostly found its way into English in texts that were themselves translated or written by multilingual authors. It appears in the margins of discussions of the politics of consumerism and ongoing technological disruption, conceptualizations of the posthuman, the future-oriented politics of the climate apocalypse, and the Anthropocene,[3] and upon mention, Günther Anders is consistently introduced as "the first"

(Bernard Stiegler), "the most profound and most daring" (Jean-Pierre Dupuy), and the "too little mentioned" (Jean-Luc Nancy) thinker of the technological present and the planetary situation we find ourselves in.[4]

Over the last decade, the gradual "discovery" of Anders has led to multiple detailed engagements with his work in English. Rather than replicating this rich and diverse literature here,[5] we will focus on an aspect of *The Obsolescence* that is easily overshadowed by its very topicality: the mode of thought from which it emerged. The feeling of uncanny prescience the text can still generate today is not just the product of our ever-deteriorating world situation (preface), it also stems from the way the "Promethean gradient" is conceptualized and dramatized in the text. By challenging the notion that the word *technology* refers to a set of tools, devices, or solutions—a way of framing technology that is still ubiquitous today—Anders's approach seems to invert conventional ways of knowing. Rather than focusing on empirical data, stated intentions, or the tangible uses and sense of empowerment that technological assistance can promote, the text focuses on much less definite aspects of human–machine interaction: on phenomena that are usually too repressed, too ignored, too embarrassing, too obvious, or too strange to be considered anchor points for knowledge. And by giving substance to the lack of comprehension and control that mediates all that technology does on our behalf, *The Obsolescence* also presents itself as a call for new modes of discourse and new notions of expertise.

This dual focus—on the dynamic of the ongoing "technification" of our being and the question of how this technification might be adequately conceptualized and portrayed—is prefigured in the highly diverse materials that survive from Anders's long years of exile (1933–50), a time during which he wrote with "maximum intensity" yet "minimum prospects."[6] To situate the immediacy with which *The Obsolescence* speaks to the present, this postscript will relate its unique mode of address to the reflections on style and method that run through these earlier materials. As a point of departure, we can step back into the streets of 1930s Berlin and return to the uncanny sonic echo Anders discusses in "Spuk und Radio," the short essay quoted above. By latching onto the unsettling "presence" of the absent singer's multiplying voice, the essay documents the psychic impact of a small tweak in radio technology, one that was perhaps as transformative as the smartphone's much more recent reinvention of the radio antenna: the introduction of user-friendly radios with an integrated loudspeaker, which had made obsolete the cumbersome precursors solely equipped with earphones.[7] "Spuk und Radio" thus marks the moment radio went loud,[8] and as such, serves as an early example of what Anders would later call "occasional philosophy": It takes a contingent, clearly localized, and tangible phenomenon and develops it into an exaggerated portrayal of the monstrous potential contained by an object that had already disappeared from notice into the fabric of everyday life. And like later works, "Spuk und Radio" was cruelly prescient.

"When a mass-rally comes into our house through the radio," *The Obsolescence* notes, this is not just a representation; "we believe we hear the actual crowd" (chapter 2). It is the desire for precisely this kind of intrusion that can be read in the pages of Joseph Goebbels's diary, which documents the long Nazi struggle to get onto the airwaves. In Weimar Germany, radio started as a heavily regulated medium, with all political content prohibited until the late 1920s, when a change in the law allowed progovernment broadcasts (though extremist parties, like the Nazis, were strictly banned).[9] Yet, on January 30, 1933, when Hitler was declared chancellor in an ill-fated coalition agreement, this legislation was effectively voided. And so, the jubilant processions that marked Hitler's appointment also resulted in the very first Nazi radio broadcast, an occasion that Goebbels's diary triumphantly records: "At seven o'clock Berlin resembles a swarming beehive. And then the torchlight procession begins. Endlessly, endlessly, from seven o'clock in the evening until one o'clock in the morning crowds march by the Chancellery. . . . Germany has awakened! . . . For the first time the German people in demonstration is being broadcast. We speak for the first time over all German transmitters. I can say nothing, but that we are happy beyond words. . . . The German Revolution has begun!"[10]

The situation for Anders—a politically active journalist and antifascist Jewish intellectual—rapidly deteriorated after this. In the weeks following Hitler's appointment, the Nazi presence in the city and on the airwaves grew into an all-dominating force, and Anders left Berlin in early March 1933, just days after the Reichstag fire. His flight from Germany was preceded by a different kind of exclusion. In 1931, the growing "wave of anti-Semitism" at German universities had already required him to indefinitely suspend his academic career.[11] Apart from teaching two courses at the New School for Social Research in New York (1949–50), Anders never held an academic post and instead adopted the position of being "philosophy's outsider."[12]

The material that survives from these years of hardship is highly diverse, both in style and focus, and the many diary entries and anecdotes from the time of exile that are part of *The Obsolescence* are not the only way these exile writings feed into the text. Besides a few academic essays that appeared in French in *Recherches Philosophiques* (see below) and in English in the American journal *Philosophy and Phenomenological Research*, Anders primarily published poetry during this period. "At this murderous moment in time," the outline for an unrealized 1940 anthology reads, lyrical poets "can contribute little toward the resolution of the conflicts in the world." What poetry can provide, Anders adds, is a means of cutting through "the mechanized production of propaganda" and letting "the truth of feeling become language"; it can "create clarity" and bring "consolation in distress."[13] In keeping with this wider logic, Anders's remarkable poems were mainly published in German-language newspapers that served the émigré community, and they seek to comprehend the incomprehensibly horrendous accounts from overseas—the world émigrés had been forced to depart. Published amid statistics of numerical death tolls, reports of Nazi crimes and ruin, and next to small ads and adverts, these poems often reflect on the discrepancy between the horror conveyed in the paper and the ease with which it is possible to turn the page. Many motifs that reappear in *The Obsolescence*—reading the newspaper being a case in point (chapter 4)—are prefigured in these poems, which suggests that writing poetry served as one of Anders's principal vehicles of thought. Yet, alongside the materials that were published during exile or else incorporated directly into *The Obsolescence*, Anders also compiled hundreds of other texts: diaries, various methodological reflections, film ideas, and essays on literature, art, and philosophy (to name just a few genres). And one of the key methodological markers to emerge from these materials and Anders's continual process of revision, editing, and textual stitching is the notion of "philosophical exaggeration."

Philosophical Exaggeration: Imagination and the Disappearing World of Machines

The first mention of "philosophical exaggeration" can be found in "The Pathology of Freedom," one of two philosophical essays Anders published in French translation during his exile in Paris (1933–36). In contrast to the later works, these essays are written in a highly academic style that reflects Anders's training in phenomenology and his deep immersion in philosophical anthropology, especially in the work of Helmut Plessner and Max Scheler.[14] Philosophical anthropology, to cite Plessner's foundational text, originated from the "profound tensions between natural science and philosophy" and the desire not to "sacrifice the one to the other" when studying fundamental aspects of human existence.[15] This dual fascination—which results in phenomenological accounts of human emotion, sensory perception, language, tool use, and culture that are also situated in the natural sciences, evolutionary theory, and the study of matter—can also be detected in *The Obsolescence*.

Anders's own contributions to philosophical anthropology relate to what *The Obsolescence* calls "*negative intentionality*"—that is to say, the sphere of involuntary bodily phenomena, such as shame and hunger, that traditional phenomenology is said to have overlooked on account of a fixation on intentionality and the "cognitive, perceiving I" (chapter 1). Especially in its discussions of affect and emotion, *The Obsolescence* repeatedly refers to the complex exposition of this problematic in the Paris essays. It also cites the "paradoxical" definition of the human they advance. As human beings, Anders notes in these early texts, we are naturally artificial, unstable, and born unfinished. And to bridge the "strangeness to the world" that marks our existence, we need to learn to inhabit artifice (e.g., language, clothing—culture, in the widest sense). "Man," as the consequence of this formative relationship to technology is specified, "is in plural 'men' in an entirely different way than the animal is in plural animals [. . .]. He is not realized according to a single form foreseen in principle and valid in general, but, as daily life and history indicate, according to many different types."[16] It is in this context, while trying to reconcile the paradox that humans are authentically inauthentic—natural beings who lack "a specific" nature—that Anders first turns to "philosophical exaggeration." This paradoxical trait, as Anders notes, becomes manifest in a feeling of "non-identity with oneself" that programmatically compels us into self-articulation. This feeling affects us, as Anders elaborates, whenever we collide with the limits of our own freedom and are uncomfortably confronted with "precisely who we are," with facets of our being—our body, line of descent, et cetera—that we have not chosen, even though we can imagine them to be different. Since this feeling is always mediated by our own life and specific situation, since we have always already found strategies to contain and counteract it, this nonidentity can only very rarely be experienced in its pure and general form. Because of this, Anders concludes, its basic philosophical motif needs to be exaggerated, or else it cannot be studied at all.[17]

In terms of the reception of Anders's work, it is worth mentioning the impact of these early writings on twentieth-century French thought. "Une interpretation de l'aposteriori," the first essay Anders published in Paris, was translated into French by Emmanuel Levinas, who went on to develop Anders's discussions of shame into *On Escape*;[18] while "The Pathology of Freedom" went on to influence thinkers such as Jean-Paul Sartre, Gilles Deleuze, and Frantz Fanon. Yet, when compared to Anders's later writings, the highly academic style of these early essays is not the only thing that is conspicuous; the way in which philosophical exaggeration is used to define the human in opposition to animal life is also notable. By August 1942, when Anders revisits the notion that the human is "naturally artificial" in California during a seminar convened by Theodor Adorno and Max Horkheimer, the existentialist perspectives of the Paris essays have already shifted onto the realities of the material world. "In the course of history," the logic now reads, "the artificiality of the human increases; humans become the product of their own products." And although this phrase was already deployed almost verbatim in "Spuk und Radio," it is now developed into the logic *The Obsolescence* introduces as "the Promethean gradient": "In an economy that is tailored to the needs of the market and not the human," Anders continues, "a differential, a '*gradient*' [*Gefälle*] opens up between humans and products, because humans can no longer live up to the demands their own products place on them."[19]

During Anders's years in exile, while working "odd jobs" in Los Angeles factories, in the supply chain of the movie industry, and in other settings, his early focus on "the fact" that humans are naturally artificial thus shifts onto the dynamic of "the Promethean gradient" itself, irrespective of what the precise philosophical definition of the human may be.[20] This shift is elaborately dramatized in the way *The Obsolescence* rejects the human–animal opposition as an "anthropocentric delusion of grandeur" (chapter 1), yet it also finds expression in the way "philosophical exaggeration" is reframed. Rather than presenting it as a method that can, in the words of "The Pathology," push a phenomenon to a "bare truth" and generally valid state,[21] *The Obsolescence* deploys it more pragmatically. Exaggeration is likened to microscopy, telescopy, and caricature, introduced as a representational device for the study of things that are too big, complex, small, inconspicuous, phantomlike, and ontologically ambiguous to be visible in an "undistorted" state. No longer directed at the abstract realm of philosophical concepts, it now becomes a means to expose human attitudes and concrete constellations of power that take shape as soon as human self-articulation entails complex technological objects—products that are deemed too singular, too strange, and too surreal to be grasped without this representational aid.

And it is no coincidence that *The Obsolescence* speaks of "surrealistic configurations" when introducing the realm of human–machine interaction (introduction). Already in "Spuk und Radio," Anders singles out "surrealist literature" as the form of writing that is trying to "keep up" with "our own excessive *innovations*";[22] and a short, posthumously published exile text from 1934 gives a sense of what kinds of effects "keeping up" might specifically entail. Surrealist images, Anders notes, reveal "*the messiness*," "*the disorderliness*" of the world: "While the image *en général* hopes to make viewers step closer and identify with it, the surrealist image wants to make the viewer take a step back—it tries to startle and make its viewers feel shocked." The thing rendered shocking, the text concludes, "is the fantastical nature of the truth."[23]

In *The Obsolescence*, the truth of "the messiness" of the world is in a sense presented as "fantastical" in and of itself. This is conveyed by impressing on us how experientially distinct objects and realms of being are simultaneously entangled with each other in the most bizarre ways: The atomic bomb appears alongside our inability to feel shocked by the reality it represents; reality appears alongside its countless TV images; luxury and consumerism appear alongside a thousand forms of exploitation and towering mountains of waste; the small joys of everyday life alongside global catastrophe. If the world is surreal in itself, consisting of countless overlapping spheres of reality that are as immediate and concrete as they are mediated, artificial, virtual, and to various degrees remote, then the philosophical eye must likewise be furnished with surrealistic powers. Just like surrealists, who, according to Anders, do not shatter the world but merely the illusory realism of our perception, philosophy must artificially "rectify" the image of the world to bring it "up to date" with the help of disturbing, dazzling, and quasi-Dadaistic images.[24] This can almost be taken literally. In a world that transforms us into "postliterary" illiterates (introduction) with its image streams, the written word must learn to affect and address us in ways that can emulate the technological media we find ourselves immersed in. And in developing his style, Anders also seems to have transposed visual effects into writing to create the estrangement effects he ascribes to surrealism. The "double take" introduced by American slapstick films[25] is as much a source of inspiration as Brechtian theater[26] and the "x-ray like unmasking of seemingly worthy people"[27] that the "painting Marxist"[28] George Grosz was able to perform.

In light of the question implied in the epigraph—in a world shaped by artifice and media, is the truth required to lie as well?[29]—Anders's engagement with the visual artist John Heartfield (born Helmut Herzfeld) deserves special attention. In 1938, Anders delivered the opening address for a Heartfield exhibition in New York, in which he noted how Heartfield's photomontages are often accused of being "fakes." Yet, when Heartfield "distorts reality," as Anders redefines the meaning of montage, he doesn't falsify. His work simply adds a contextualizing layer to the "documentary photograph," an addition that tries "to set the world to rights." What this distortion shows, is "the motor that conceals itself behind the seemingly honest visibility of things."[30] To exemplify this process, Anders points to Heartfield's visual reinscription of the Hitler salute, which adds a hand with money to the boastful words "Millions are backing me!" And once seen, the truth-making distortion of Hitler's pronunciations of his own popular support cannot be unseen.

The images that Anders's own writings construct often appear to follow a similar logic. They seem intent on making the reader instantly "see" the unsettling reality his highly composed texts seek to reveal: "The fact that the Englishman says 'I see' when he understands an explanation," as Anders concludes in an unpublished typescript titled "Gelegenheitsphilosophie," "is a contribution to epistemology that replaces entire volumes."[31] Written in the United States between 1942 and 1950, the sixty-eight pages that make up the typescript represent Anders's most sustained reflection on his own method, with some material later reappearing in the introduction of *The Obsolescence* and the "Methodological Postscript" of *The Obsolescence, Volume 2*. Occasional philosophy, the typescript notes, doesn't aim at a philosophical system; it is not meant to lead to a specialization or a contribution to a specific discipline. In fact, it may not even be philosophy, but an attempt at social psychology—or even a "psychology of technology."[32] Yet, amid its probing reflections, the typescript is clear about one thing: The starting

John Heartfield, *Der Sinn des Hitlergrußes: Kleiner Mann bittet um große Gaben. Motto: Millionen stehen hinter mir!* (The meaning behind the Hitler salute: Little man asks for big donations. Motto: Millions stand behind me!), title page of the magazine *Arbeiter-Illustrierte-Zeitung* 11, no. 42 (October 16, 1932), rotogravure with retouching, 38 × 27 cm. (c) The Heartfield Community of Heirs / VG Bild-Kunst, Bonn 2024. Akademie der Künste, Berlin, Art Collection, inventory no.: JH 217.

point must consist of concrete and singular phenomena—"I only wish to speak of things that exist"[33]—and it must always relate to identifiable situations. These situations, as is specified, usually consist of an "embarrassment" [*Verlegenheit*] that occurs "in the course of life"; they are situations of "incomprehension."[34] "Occasional philosophy," we can read in the concise definition, "is philosophy about a specific life situation that emanates directly from this situation itself. It is not philosophy that generates a difficulty; the difficulties of life generate philosophy."[35]

The Obsolescence provides us with countless examples of what these situations and phenomena might be. They often consist of uncanny or otherwise noticeable collisions with technology—such as the spooky radio echo isolated above—or else of instances in which overheard remarks or specific words and images lead to some sort of agitation. They can also take on more anecdotal forms and relate to wider social attitudes that are translated into quasidramatic scenes—for instance, when the monstrous nature of nuclear weapons shows itself in the self-satisfied smile on the face of some guy who is mansplaining the apocalypse (chapter 4). Viewed through the lens of Anders's methodological reflections, these situations and phenomena seem to share one paradoxical trait: They are instances in which tangible phenomena are no longer legible in the traditional sense but instead show that they are concealing something. One of the many ways that Anders specifies this self-concealment is by reflecting on the inconspicuous appearance of technological devices. "If one understands the word 'phenomenon' to mean 'something that shows itself,'" as put in *The Obsolescence, Volume 2,* "then our devices are no longer 'phenomena.'" They look ordinary, "*like nothing,*" and remain "unrecognizable and unknowable," even though they can in some form be perceived—their look "no longer reveals anything about their nature."[36] If they are noticed at all, it is only in the fleeting and at times violently shocking ripples of perception that hint at something monstrous, precisely because this unsettling friction is so habitually ignored. "Make things up so as to discover them"—this is the short motto that Anders arrives at to characterize Heartfield's montages.[37] It also applies to his own work.

The great lengths to which *The Obsolescence* goes to dramatize the small collisions, glitches, and irritations that mediate our interaction with machines, and the world this interaction helps to configure—the text's turn to poetry, fables, diary entries, and (at times possibly fictitious) encounters—is thus no mere stylistic device. Creative imagination is introduced as a methodological necessity. In a world where our perception routinely transforms "abject monstrosity" into something "inconspicuously ordinary," as a 1944 diary notes in a variation on the logic of philosophical exaggeration, "imagination needs to become an empirical method, an organ of perception."[38] And the many iterations of this logic that can be found in Anders's work underline that the turn to exaggeration is as much directed at concrete objects and societal tendencies as it is concerned with our bodily and societal tendency to normalize the formative powers of technology to such an extent that they are no longer actively perceived. The principle of "inversion" at work in many of Anders's images at times also appears to inform the way he performs his own textual persona.[39] Its perspectives are grounded not in traditional truth claims but in the author's elaborately dramatized struggles to comprehend or find the right words. As readers of *The Obsolescence*, we are often directly included in this struggle and thus made to read our very inability to read the technological world.

This approach is especially pronounced in the context of the topic that is itself largely absent from the materials that survive from Anders's years of exile: the atomic bomb. On August 6, 1945, President Harry Truman released a statement via radio informing the world that an American airplane had "dropped a bomb on Hiroshima." This bomb, as the broadcast message gradually revealed, was no ordinary device; it was "a harnessing of the basic power of the universe," a device capable of unleashing "a rain of ruin from the air the like of which has never been seen on this earth."[40] Anders would later relate the absence of the bomb from his exile writings to this very broadcast and "the state of stupor" it effected in him. "The first thing I was able to write," as Anders portrayed this long silence late in life, "*was little more than an assertion of my—no, our—inability to imagine what 'we' had perpetrated and produced*"; it was but "*an admission of failure.*"[41] This comment not only provides a biographical note but also spells out one of the narrative techniques by which *The Obsolescence* seeks to render the disappearing world of devices visible, while also presenting it as an absence—a procedure that gives substance to the very thing that escapes. And in its call to pierce the neat demarcations, narratives, and sense of normalcy that our cognition habitually imposes on our world, the typescript "Gelegenheitsphilosophie" provides us with an earlier, much more playful image for the specific "look" occasional philosophy is meant to assume: "In Hollywood movies, drowning people who are being pulled from the water have perfect hair and they look perfectly neat. This, in a sense, is precisely how philosophy looks in Hegel when it tries to pull itself out of the swamp of immediacy."[42] For Anders, it is thus not enough to simply attest to the immediacy of the singular, arbitrary, "messy and murky" world; this reality also needs to be conveyed as "messy and murky" in the text. The "*choque*" needs to become intuitive, turned into something that can actually be felt.

The irritation we as readers are made to feel on account of seemingly obvious connections, repetitions, an overly moralizing tone, or a general lack of trim thus often seems to be an intentional device to make us "take a step back" by means of linguistic friction. "Human engineering," we can read in *The Obsolescence,* does not simply wish to establish the natural "limits" of our own bodily and cognitive capacities; it seeks to find the "soft parts" that might still be remodeled to serve the desires of industry and machines (chapter 1). In analogy to this image, Anders's mode of writing might be seen as a continual experimentation with the machineries of text, a way of probing for "soft parts" in feeling that might make us reconnect the realm of individual feeling with the technologies that are acting on our behalf. And to help us bear these often-involuntary self-examinations, *The Obsolescence* makes use of a device that is as complex as it is immediate in its effect: humor.

This brings us to the last source in this short and selective account of the materials from which *The Obsolescence* emerged: the 1942–43 outline for a new genre of animated film that Anders introduces as "Caricartoons" (one of the few texts he wrote in English). "Real laughter," Anders notes in this posthumously published text, always needs to be directed at some sort of superiority. Regardless of how hidden this superior power may be, it must be real and also feared, or else laughter does not set in. And the image Anders deploys to illustrate this logic perhaps also helps to contextualize *The Obsolescence*'s tendency to impersonate: "The pupil who bursts out laughing when his teacher is being imitated effectively by one of his schoolfriends laughs really since suddenly his fear ceases and the superiority of his superior collapses before his eyes."[43] The schematic outlined in this passage in effect equates laughter with an act of perception and thereby links this involuntary bodily motion back to the logic of philosophical exaggeration.[44] More than this, by laughing, the pupil does not just "see" the feared authority in a new and unexpected light; the one who laughs also becomes complicit with a challenge to authority that may otherwise perhaps not even have happened in thought. "Caricature," Anders adds, "was always a sort of unbloody defense against the suppressor," "a sort of counterpressure against real social or political pressure,"[45] and it is this pressure that the film idea seeks to harness and direct. Approached from this angle, exaggeration not only serves the analytic functions foregrounded in Anders's frequent allusions to microscopy and telescopy; it also seems intent on activating the bodily automatisms of laughter to automate a small internal revolt against the repressed and inadequately acknowledged part technology is playing in our lives.

It is this particular image of friction that also opens a perspective onto how easy it is to forget that *The Obsolescence* is responding to a technological environment that is over seventy years old and was not, therefore, written as a commentary on global warming and the world of touch screens, smart devices, social media, and artificial intelligence. The often humorously absurd ways in which translating Anders into English results in what seems like a direct commentary on the technological present can be contextualized with the help of what Cliff Kuang calls "one of the greatest cultural achievements of the last fifty years": the emergence of "user-friendly design." On account of this quiet and decidedly comfortable revolution, our ever-advancing technology is able to "disappear from notice," just as its formative powers are increasing at an unimaginable scale.[46] "Today," as Kuang spells out the implication, "we take it for granted that the most advanced technologies ever created should never need any explanation whatsoever."[47] Although Kuang's words here specifically refer to what Wendy Hui Kyong Chun calls "the profound ignorance"[48] that mediates our interaction with our devices—as users we do not need to know, and cannot know, how they actually work—they can also serve to illuminate the absurd lack of accountability and explanation with which new technologies are developed and released into the world at an accelerating rate. Whereas pharmaceutical innovations are subjected to extensive trials to study their risks and potentially harmful side effects, no such processes are in place to mitigate the possible consequences of the devices, services, and autonomous and automated technological capacities that are inserted into our lives. "It may be true," as Anders notes, that we are turning to technology "in the blind hope, or at least on the pretext, that this self-reification will reduce or even abolish all other forms of human frailty and degradation." What counts, however, are not intentions, but "what is actually done": "In order to combat the limitation of their liberty, humans limit themselves in a new way" (chapter 1).

In its attempts to render the escaping world of machine mediation visible, *The Obsolescence* provides multiple access points to confront the formative power of technology and the way this power is being instrumentalized. More significantly, perhaps, it can also help reimagine our very relationship to technology and the social constellations of power and material relations of production our devices embody but also powerfully deflect from. Technologies, Anders insists, are materialized modes of action. "Nothing is more misleading," as Anders notes in *The Obsolescence, Volume 2*, "than the assertion that machines are 'morally neutral.' [. . .] Once a machine exists, the fact that it works is already a kind of use, and its specific performance gives each machine a (socially, morally or politically) prejudicing role."[49] By conceptualizing the human as a "naturally artificial" being, *The Obsolescence,* despite its often pessimistic tone, introduces technology as a space of possibility that we inhabit, one that configures the way we feel, think, speak, and interact. It does not simply seek to make us "understand and feel" the power of technological mediation but it asks us to comprehend the fact that we inhabit a world in which its power to outsource categorically exceeds our physiologically limited capacities, one in which the image of the human as a "self-aware and thinking and feeling I" conveyed in Western thought is hopelessly obsolete. In its place, it gestures toward the need to develop ways to account for and contain the unbridgeable gap between the reality of our devices and that of our bodily perception and cognition, which tend to bracket out the hypercomplex relations each interaction with technology actively contributes to.[50] One of the things Anders's distortions of language thus tries to impress on us—in line with the wider posthumanities—is that our supposedly "human" capacities are necessarily distributed, artificial, and outsourced, blinkered by habit, custom, and physiological limits. What he adds to this picture is that it is precisely the effort, uncertainty, and at times insurmountable friction that we need to navigate as we seek to "fit into" and "measure up" to our artificial worlds (chapter 4) that makes it so attractive to outsource all this doubt and effort onto things.

To gesture toward the wide fields of questioning *The Obsolescence* continues to open up today, we will thus end by aligning Anders's notion of "philosophical exaggeration" with a counterimage with which it is uncannily compatible. In the late 2000s—at the dawn of the data-driven internet—the potential of machine vision, data analysis, and computational modes of perception and cognition was often introduced with the very analogy to microscopy and telescopy that Anders so frequently deploys: "Just as the telescope enabled us to comprehend the universe," a popular book on big data applications relates, just as "the microscope allowed us to understand germs, the new techniques for collecting and analyzing huge bodies of data will help us make sense of our world in ways we are just starting to appreciate."[51] Looking back at these words from today's perspective, it is clear that the capacity to detect patterns indetectable to humans has led not to an age of clarity and insight but to a world in which our flaws and blind spots can be coercively exploited and augmented out of all proportions. By directing the resources of humanistic thinking to consider the limits of the human from the other side of this equation—from the perspective of those who have to live in a world where our limited sight, cognition, and comprehension can be instrumentalized, mimicked, and exceeded in ever more ingenious ways—*The Obsolescence* doesn't lead back to some lost human essence or authenticity, but instead provides us with an interface for modeling the effects of technological augmentation, the "oligarchic tendencies"[52] of technology, and the very nature of human emotion as such.

—Christian Dries and Christopher John Müller

Translator's Acknowledgments

This translation has taken shape over the best part of a decade, and it would not have been completed without the generous collegial support, editorial input, and encouragement of a long list of friends and colleagues. Marija Grech's countless editorial suggestions on the complete draft have significantly enhanced the readability and helped fine-tune its overall feel, and I am truly grateful for this work. Shirley Müller's generous feedback on multiple iterations of evolving sections was invaluable in developing the wider strategy adopted and in shaping my overall approach. I would like to thank Chris Weedon and Rodney Livingstone, whose support and editorial suggestions were instrumental in the translation of "On Promethean Shame," which first appeared in an earlier version in my book *Prometheanism*. Over the years, this translation has benefited from the countless responses and comments of interlocutors who have heard me speak about this project and its wider framing. I would especially wish to thank the organizers and participants of the ZeM Spring-Lecture hosted by Brandenburgisches Zentrum für Medienwissenschaften (2022), and the conference "The Plasticity of Emotions: Günther Anders's Contribution to the History of Feelings" (2023). The final stages of this work were supported by the Faculty of Arts Outside Studies Program (2023), Macquarie University, Sydney, and I would like to thank my wonderful colleagues there for all the support received while working on this project.

The wider editorial approach adopted in this volume was developed in collaboration with Christian Dries, whom I would like to thank for his camaraderie and patience. This work has also immensely benefited from collaborations, conversations, and previous cotranslations with various colleagues, and I would especially like to thank Benjamin Nickl, Jason Dawsey, Manuela Koelke, Edgar Eckert, Kerstin Putz, and Bernd Bösel. Tom de Bruin and Etienne Morel provided valuable assistance with the translation of Greek, Latin, and French phrases, and Alina Kozlovski helped identify possible allusions to ancient texts. Ivan Callus, Tai Neilson, Lucy Menon, Christine Hentschel, Helen Wolfenden, Aaron

Aquilina, and John Martin provided feedback on drafts of individual sections, translations of poems, or the wider conceptualization and framing of this project. Existing English translations of Anders's work served as important sources of inspiration. I would especially like to foreground Norbert Guterman's powerful translations of the radically abridged versions of "The World as Phantom and Matrix" and "Reflections on the H-Bomb" for *Dissent* (1956), and the directness of Charlotte Zelka's (unacknowledged) 1957 translation of "Commandments in the Atomic Age," which were instrumental in starting to introduce Anders's voice to an English-speaking readership. Martin Esslin's 1965 translation of "Being without Time" and Benjamin Steege's translations of passages of Anders's work were also points of reference.

This project could not have been realized without the generous support of Wolfgang Beck and the C. H. Beck publishing house, as well as the friendly encouragement of Gerhard Oberschlick. As an editorial team, Christian Dries and I would also like to thank Cary Wolfe, the readers of the proposal, Douglas Armato, and the entire editorial and production team at the University of Minnesota Press for making this volume possible. Our special thanks go to Nicholas Taylor for his tremendous editorial work, which helped streamline and clarify this translation, and to Zenyse Miller and Rachel Moeller, whose kindness, work, and support were instrumental in preparing the final manuscript.

—Christopher J. Müller

Notes

Translator's Preface

1. Günther Anders, *Burning Conscience: The Case of the Hiroshima Pilot, Claude Eatherly* (1961; repr., New York: Monthly Review Press, 1962), 1.

2. "Herbert Marcuse an Günther Anders, Newton (Massachusetts), 20.9.1955," in *Gut, Dass Wir Einmal Die Hot Potatoes Ausgraben: Briefwechsel mit Theodor W. Adorno, Ernst Bloch, Max Horkheimer, Herbert Marcuse, und Helmut Plessner,* ed. Reinhard Ellensohn and Kerstin Putz (Munich: C. H. Beck, 2022), 103. I would like to thank Jason Dawsey for bringing Anders's correspondence with Herbert Marcuse to my attention.

3. The essays in question are "Reflections on the H-Bomb," and "The World as Phantom and Matrix," which both appeared in the journal *Dissent* in early 1956 (translated by Norbert Guterman). The latter essay was republished in 1957 under the title the "Phantom World of TV" as part of the pioneering volume *Mass Culture: The Popular Arts in America,* ed. Bernard Rosenberg and David Manning White (New York: The Free Press, 1957), which also included work by Marshall McLuhan, Siegfried Kracauer, and David Riesman.

4. Günther Anders in Elke Schubert, eds., *Günther Anders antwortet: Interviews und Erklärungen* (Berlin: Tiamat, 1987), 35. The "others" Anders is here specifically referring to are Hannah Arendt (his first wife) and Herbert Marcuse.

5. See "On Promethean Shame" (chapter 1). The method of philosophical exaggeration is further contextualized in the postscript.

6. Lawrence Venuti, *The Translator's Invisibility: A History of Translation* (London: Routledge, 1995), 5.

7. In German, the dash, as Joseph Vogl puts it, is "rather strangely" called *Gedankenstrich*—that is, a thought- or idea-line—and acts as a constant reminder of the textuality of the text. "Anything that is written," as Vogl elaborates on this wider point, "begins with a shorter or longer line [. . .]. The written differs from the spoken through the presence of the line. With the line [i.e., the dash] language begins to fall silent and make itself noticeable as text." See Joseph Vogl, "Der Gedankenstrich bei Stifter," in *Die Poesie der Zeichensetzung: Studien zur Stilistik der Interpunktion,* ed. Alexander Nebrig and Carlos Spoerhase (Bern: Peter Lang, 2012), 275 (my translation). I would like to thank Endre Malcolm Holéczy for alerting me to this text.

8. Christopher John Müller and Benjamin Nickl, "Real Is Not Real Enough: Script for an Audio Adaptation of the Californian Exile Diary by Günther Anders," *Anders Journal,* December 2021, 1, https://www.guenther-anders-gesellschaft.org/s/anders-journal-muller-nickl-2021.pdf.

9. For a detailed discussion of this aspect of Anders's mode of thought, see Christopher John Müller, "Hollywood, Exile, and New Types of Pictures: Günther

Anders's 1941 California Diary 'Washing the Corpses of History,'" *Modernism/Modernity Print+* 5, no. 4 (2021), https://doi.org/10.26597/mod.0185.

10. Günther Anders, "Caricartoons: A Suggestion for a New Type of Animated Pictures" [1942/3], *Schriften zu Kunst und Film,* ed. Reinhard Ellensohn and Kerstin Putz (Munich: C. H. Beck, 2020), 73–83.

11. A short discussion of the intersection of technology, shame, and laughter in Anders's work is provided in the postscript.

12. "Günther Anders an Herbert Marcuse, Wien, 2.9.1978," in Ellensohn and Putz, *Gut, Dass Wir Einmal Die Hot Potatoes Ausgraben,* 132.

13. A detailed discussion of the publication history and immediate reception of *Die Antiquiertheit des Menschen* can be found in Anna Pollmann, *Fragmente aus der Endzeit: Negatives Geschichtsdenken bei Günther Anders* (Göttingen: Vandenhoeck und Ruprecht, 2020), esp. 51–101.

14. Günther Anders, *Die Antiquiertheit des Menschen, Band 1: Über die Seele im Zeitalter der zweiten industriellen Revolution* (Munich: C. H. Beck, 2018).

15. Günther Anders, *Der Blick vom Mond: Reflexionen über Weltraumflüge* (Munich: C. H. Beck, 2024), 12.

Preface to the Fifth German Edition

1. [The reference is to the 1972 collection of essays *Endzeit und Zeitenende: Gedanken über die atomare Situation* (End time and the end of time: Reflections on the nuclear situation), which, from the sixth German edition onward (1981), was published under the title *Die atomare Drohung: Radikale Überlegungen zum atomaren Zeitalter* (The nuclear menace: Radical reflections on the nuclear age); the 1959 diary *Der Mann auf der Brücke: Tagebuch aus Hiroshima und Nagasaki* (The man on the bridge: Diary from a visit to Hiroshima and Nagasaki), republished in the 1982 collection *Hiroshima ist überall* (Hiroshima is everywhere); and the 1961 volume and exchange of letters *Burning Conscience: The Case of the Hiroshima Pilot, Claude Eatherly,* which, in its English edition, also features a preface by Bertrand Russell and a "Postscript for American Readers" by Anders.—Eds.]

2. [The reference is to *Die Antiquiertheit des Menschen, Band 2: Über die Zerstörung des Lebens im Zeitalter der dritten industriellen Revolution* (The obsolescence of the human, volume 2: On the destruction of life in the age of the third Industrial Revolution), first published in 1980.—Eds.]

Introduction

1. [English in the original.—Trans.]

2. [The German here reads "das Ganze ist das Wahre," which might either be a play on Theodor Adorno's "das Ganze ist das Unwahre" (the whole is the false) (*Minima Moralia,* aphorism 29) or a direct allusion to Georg Wilhelm Friedrich Hegel's premise "das Wahre ist das Ganze" (the true is the whole) (*Phenomenology of Spirit,* 13), which Adorno inverts, and to which Anders returns throughout.—Trans.]

3. [English in the original.—Trans.]

4. [Molussia is a fictitious land of Anders's invention. Anders's writings frequently evoke stories, sayings, and songs from Molussia as a stylistic device. The

1951 short story "Der Ahnenmord" (The murder of the ancestors) even contains an ethnological account of Molussia that refers to fictitious academic events and works with English titles such as "*Molussic Excavations, the Tablets A-D. Preliminary Report by Sir Allen J. Godefroy,* Calcutta University Press 1932," "*Molussic Studies,* a Symposium, Princeton 1952," and "*Molussic Proverbs,* London 1949" (see 241, 249). The substantial antifascist novel *Die molussische Katakombe* (The Molussian catacombs), posthumously published in 1992 and written between 1930 and 1932, seems to be the original gesture of finding recourse to Molussia as a vehicle to critique the present.—Eds.]

5. Because the movement presented itself as "a thousand-year" future—a future that, measured by human standards, might as well be called "eternal"—it could, in fact, present the critic as something much worse: It could in a sense denounce them as a "saboteur of eternity," and thus brand them a sacrilegious, subhuman subject.

6. [English in the original.—Trans.]

7. [English in the original.—Trans.] [This reference could not be identified. It may refer to the voices calling for the establishment of a world government amidst the formation of the United Nations in the immediate aftermath of the horrors of the Second World War.—Eds.]

8. [It is unclear whether the term *Okkasionalismus* is here meant to serve as a specific reference, but in terms of the way it is discussed, it is largely consistent with Carl Schmitt's detailed conceptualization of the term in the context of a discussion of the romantic attitude in *Political Romanticism* (1919).—Eds.]

9. [This distinction is made in Leibniz's *Monadology,* a text which Anders discusses in detail in other works, for instance in *Die Kirschenschlacht*.—Eds.]

10. [This passage cleverly plays on the multiple contextual meanings of the word *Grund:* "the ground," "reason," "basically," and "to get to the bottom of." The original reads, "Im Grunde ist es müßig, den Dingen auf den Grund zu gehen. Denn wäre der Grund ihr Grund, dann wären sie eben nicht diese Dinge hier, diese vom Grunde abgerissenen einzelnen Dinge. Es muß einen Grund dafür geben, daß sie nicht im Grund sind, sondern sie selbst; oder, wer weiß, soviele Gründe, als es Dinge gibt."—Trans.]

11. [The German attributes this foolish act to a *Schildbürger*, thereby locating it in the satirical and highly popular medieval folktale tradition about the residents of Schilda.—Eds.]

12. [English in the original.—Trans.]

13. [*Das prometheische Gefälle* is one of the key concepts of the text, yet there is no obvious English translation. The word *Gefälle* can describe the incline of a slope, and the text repeatedly plays on the image of an even steeper slope. Yet the word can also be used to describe a "gap," "discrepancy," or "disproportion" between two things—for instance, different social strata (e.g., the income gap between blue- and white-collar work). In a third meaning, it can refer to inclinations and biases one might hold—that is, to be inclined toward a certain point of view. To signal this multiplicity of meaning, I have often provided two possible translations in the text or varied its formulation. Further alternatives that can be encountered in publications on Anders include "Promethean slope," "Promethean differential," and "Promethean disproportion."—Trans.]

14. [The German here reads "Tempo-Gleichschaltung" and thus equates the

process of needing to adapt to the acceleration of technology to *Gleichschaltung*—that is, to the process of Nazification by which all aspects of societies under Nazi rule were "coordinated" and forced into conformity.—Trans.]

15. On this point, see the author's essay "Über die Nachhut der Geschichte" [On the rear guard of history] in *Neue Schweizer Rundschau,* December 1954.

16. [The German text here subtly invites the reader to think of these as yet undefined traits (*Züge*) as trains (*Züge*) that have not yet been placed on rails they could run along, thus creating a grotesque, caricatured image of a face *qua* railway network map still in the making.—Trans.]

17. This case of "nonidentity with oneself" shows very clearly that the existence of the "Promethean gradient" can lead to traumata or neurotic *idées fixes.* Today, it would not at all be far-fetched if technological factors were considered as possible causes for psychological disorders [*Gemütsstörungen*].

18. [English in the original.—Trans.]

19. [English in the original.—Trans.]

1. On Promethean Shame

1. "To the Cogs," in *Molussian Industrial Hymns,* translated into German by G. A.

2. [*natum esse:* having been born.—Trans.]

3. The rejection of being born [*natum esse*] is a motif that constantly reoccurs, especially in the context of religion. Founding figures are often stripped of the flaw of having been born (Moses).— The last echo of this religious rejection can be found in reactions to theories of natural descent—that is, in the indignation with which the proposition has been met that we humans have evolved (from other entities).— The latest variant of such discreditation of the fact of birth comes from the philosophy that accompanied the bourgeois revolution. *Fichte's* notion that the "I posits itself" is the speculative version of the self-made man, who does not want to be born or have become, but rather wishes to owe his existence to himself as the product of his own making. This particular rejection of the fact of birth has its origins in the resistance to the notion that a "noble" or "lowly birth" is the source of either privilege or the lack of rights. In Fichte's philosophy, the wish to "be made" (that is, to be self-made), rather than born, did not, of course, have any technological connotations but was exclusively moral and political: The self-made man is autonomous and the citizen of a state that is made by citizens. *Fichte's* famous omission of a philosophy of nature has its origin in this resistance to *natum esse:* As *Schelling* writes, "The arrogance of man rises up [*sträubt sich*] against this origin from the ground" [*Philosophical Investigations into the Nature of Human Freedom,* 29—Eds.].— *Heidegger* is a later variant of Fichte. His notion of the "*thrownness of Dasein*" is a protest not only against being created by God—that is, against supernatural origins—but also against having developed—that is, against natural origins. For further elaboration on this point, see the author's "On the Pseudo-Concreteness of Heidegger's Philosophy," *Philosophy and Phenomenological Research,* March 1948. "Thrownness" is an *evasive concept.* With Heidegger, however, the process of self-fashioning (the transformation of "existing" into "existence") has lost all its political connotations. The process remains a purely solipsistic enterprise. It is no coincidence that (at least at the time of *Being and Time*)

natural philosophy is missing, as in the case of Fichte: Had he not left it standing at the door, his denial of the fact of birth could not have been accomplished.—

4. [English in the original.—Trans.]

5. [The text is here playing on the German *Dasein* (existence), which literally means "to be here." This has added (implied) significance because *Dasein* is the term Martin Heidegger deploys in *Being and Time* to suggest that humans exist in the sway of an experience of what it means "to be."—Trans.]

6. "To the Cogs," in *Molussian Industrial Hymns,* translated into German by G. A.

7. On this point, see the more detailed analysis of shame in section 11 below.

8. The desertion of the one who feels threatened into the camp, or at least into the value system, of the one who extends the threat is a motif we are well acquainted with in the context of today's politics. And in the last two and a half decades [1930s to the mid-1950s—Eds.], we have often enough witnessed how those who are threatened take on not only the criteria of judgment of those who extend the threat but also their feelings. They are almost always brought to the point where they do this in a craze of voluntariness. In fact, this more or less gentle enforcement of voluntary conduct is the counterrevolutionary accomplishment of our times. Outside of the political arena, however, this process has not yet been documented.—

9. [English in the original.—Trans.]

10. [English in the original.—Trans.]

11. [English in the original.—Trans.]

12. [The reference is to the opening scene of Nietzsche's *Thus Spoke Zarathustra: "I teach you the overman.* Human being is something that must be overcome. What have you done to overcome him?" (5).—Eds.]

13. Nothing is more misleading than to subsume this "overcoming of the body" under the category "jewelry." It is nothing less than the inversion of this category. Whereas jewelry seeks to enhance the beauty of the living body, bodily modification aims to lend the body the beauty of fabricated things.—

14. [English in the original.—Trans.]

15. See chapter 2 ["Grasping at Heaven"—Eds.] of Robert Jungk's *Tomorrow Is Already Here.* This example is in keeping with an account provided by *Thomas Power,* the commander of the American Air Research and Development Command: In May 1956, an H-bomb had missed its target by six kilometers. When called on to account for a mistake of such gravity, Power, according to Reuter, offered the following explanation: "Whenever you have human beings you have this type of accident."* The pilot had forgotten to operate a certain switch.— One ought to listen to Power's words carefully. Anyone who only hears the old saying "To err is human" misunderstands their meaning. They imply much more: Because it is human to err—that is, because the human operates so unreliably—it is actually inappropriate to entrust humans with such a perfect machine. The human is here seen primarily as a *source of error.* [*These words are reproduced from "H-Bomb Switch Forgotten" (*Guardian Journal,* July 18, 1956), an English version of the Reuter report Anders quotes here. It provides an account of Thomas Power's testimony before Congress about this incident.—Eds.]

16. The failure is "shameful." As we will see below, the fact that a failure is not one's fault does not make a defect less shameful, but more so. See section 11.

17. In his 1930 essay "Die Weltfremdheit des Menschens" [The human's estrangement from the world], which appeared in French translation as "Pathologie de la liberté" in *Recherches Philosophiques* in 1936, the present author defined human beings as "indeterminate," as "*indéfini*" and "unfinished."— He defined humans as "*free and indefinable beings*" who are only defined and definable by what they make of themselves in each specific instance. (And soon after, Jean-Paul Sartre also formulated his credo along very similar lines.) By overemphasizing the philosophic-anthropological dimension of freedom, both Sartre's and my own early work represent belated attempts to obscure the already then valid circumstance that the "subjects of freedom and unfreedom" need to be substituted. Such definitions only seem plausible because animal existence as a whole is here used as a counter-image against which the human is measured (and this, by the way, is the case in almost all nontheological anthropologies). They seem plausible, therefore, only because "*the* animal" (an expression that already represents an ad hoc abstraction) is presumed to be unfree, a prisoner of the fate of its species. The trouble was not taken to verify this presupposition because it was considered self-evident (not least because of the theological tradition).— Today, I find this highly questionable, not least because I deem it philosophically audacious to provide a definition of the human that relies on a counter-image that doesn't match up to the actual backdrop of human existence. After all, we do not live against the backdrop of bees, crabs, or chimpanzees, but rather, lightbulb factories and radio sets. But even from the perspective of natural philosophy, the opposition between "man and animal" is unacceptable. The idea that the single species "human" can be opposed to the thousands upon thousands of immensely different animal species and types is simply an anthropocentric delusion of grandeur. As a warning against this cosmic immodesty, each textbook of "philosophical anthropology" should have "The Fable of the Ants" as a preface. This fable recounts how, at their universities, ants learn to distinguish between "plants, animals, and *ants*."— If one looks at what actually forms the backdrop to human existence, that is, the manmade world of products instead of choosing "the animal kingdom," then the image of "the human" changes immediately. The singular word "the human" immediately disintegrates—and with it, the notion of human freedom.— [During his exile in Paris (1933–36), Anders developed the lecture script "Die Weltfremdheit des Menschens" (1930) into a two-part publication, which appeared in French translation in *Recherches Philosophiques.* Anders here only mentions "The Pathology of Freedom," the second essay. The first part is titled "Une interprétation de l'a posteriori" (An interpretation of the *a posteriori*) and appeared in 1934 in a French translation by Emmanuel Levinas.—Eds.]

18. [English in the original.—Trans.]

19. By the way, a certain boomerang effect is part of every machine in so far as it is sent out to do its job and "bring in" a return.—

20. [*Ὁμολογουμένως ζῆν:* to live as one with the gods.—Trans.]

21. "To the Cogs," in *Molussian Industrial Hymns,* translated into German by G. A. [The German here reads "hieß uns 'Opfer unerhört,'" which may be a clipped version of the phrase *Menschenopfer unerhört* from Goethe's poem "The Bride of Corinth," which Anders reproduces in full in the next section. See below.—Eds.]

22. [*imitatio instrumentorum:* imitation of instruments.—Trans.]

23. [*nolens volens:* like it or not.—Trans.]

24. [This phrase is in English in the original throughout the text.—Trans.]

25. This term was created ad hoc following the template of Karl Jaspers's famous discussion [of "limit situations." The notion of "limit situations" is a central component of Jaspers's existentialism. They describe confrontations with limits that cannot be transcended, altered, or surpassed because they fundamentally demarcate finite human existence (examples of such limits are chance, death, suffering, and guilt). See Jaspers, chapter 7, "Limit Situations," in *Philosophy, Volume 2: The Illumination of Existence*.—Eds.].

26. The excessively strenuous demands that humans here place on their bodies so they can live up to the excessive tasks of their machines resemble, in a highly peculiar manner, the excessive demands that those who practice speculative metaphysics have placed on reason: both *ignore* the *fact that human power is limited* [*Faktum der Leistungsgrenze*]; here, too, the limit should be extended or destroyed. Only that the human in this instance does not pretend to be omniscient "*sicut deus*" [like God] but aims at becoming "*sicut gadget*" [like a gadget].

27. [In German, the word *fatal* also carries the meaning "really embarrassing."—Trans.]

28. [Psychotechnics was an immensely influential school of applied psychology dedicated to aptitude testing and behavior modification pioneered by Hugo Münsterberg and Anders's father William Stern. In Weimar Germany, it found wide use in educational contexts, the advertising industries and human resources departments. The intelligence quotient measured in IQ tests is one of the offshoots of this praxis which is still in use today.—Eds.]

29. Today, every industry produces these two kinds of products. The branch of production that makes the second product (the product that generates the need for the first product) is called the "Publicity Department." [English in the original.—Trans.]

30. [*déjà vu . . . déjà lu:* already seen . . . already read.—Trans.]

31. [English in the original.—Trans.]

32. I say "robots," and I say "conformity" [*Gleichschaltung*].— "Robots" because today's true robots aren't the pretend-humans [*Scheinmenschen*] assembled from dead components and actualized in "computing machines" but the machine-parts that living human beings function as. And I say "conformity" because the process in question here is a variant of the same behavior we are familiar with in the context of political reality as the "enforced conformity and coordination that totalitarian systems of government effect." Only that the variant treated here is more pronounced because humans are in this case actually attempting to "turn themselves into things," whereas in the political sphere enforced conformity through reification only ever turns people into a "thing" metaphorically (even if these people have been utterly robbed of their freedom).—

33. While employed as a worker in California, there was no phrase the present author had to listen to more often than "forget about it."*— In this impatient reminder not to know more than one is required to know, a resentment of the "highbrow" unfailingly resonates. It is directed at anyone who has openly provided proof in the form of too much knowledge or too many interests that they are or have pretensions to something "better" and that they do not belong to the people. *[English in the original.—Trans.]

34. [English in the original.—Trans.]

35. [English in the original.—Trans.]

36. [The German here reads "Menschenopfer unerhört," probably a reference to Goethe's poem "The Bride of Corinth," a classic of Gothic literature. The poem, which Goethe himself referred to as his "vampire poem," recounts how a young man is visited in his bedroom by a girl who he believes to be a nun transgressing her vows, only to later realize that she is an "undead spirit making love to a human." See Franklin, *Longman Anthology of Gothic Verse,* 33. The scene Anders quotes narrates the first encounter in which the girl warns him that she is no friend and mourns the departure of the "jocund gods" of Greece and rise of Christianity, with its "outrageous human sacrifice" of crucifixion. The stanza is reproduced here in Aytoun-Martin and Christopher Middleton's translation:

> They have left our household, left it lonely,
> The jocund gods of old, no more they reign;
> One, unseen, in heaven, is worshipped only,
> And a saviour crucified and slain.
> Sacrifices here-
> Neither lamb nor steer,
> But man himself in misery and pain.
>
> "The Bride of Corinth," 137—Eds.]

37. [*rejectio fati:* rejection of fate.—Trans.]

38. [The text here for the first time deploys a formula that is varied throughout *The Obsolescence.* One of its earliest articulations can be found in the poem "Der Mensch ist kleiner als er selbst" (Humans, you are smaller than you are), which was published in German on August 2, 1940, in the New York–based German Jewish newspaper *Aufbau* (provided in translation here): "Sure, you can kill a thousand. / But your tears, / can mourn no more than one. / How small sadness's compass is. / Poor, abjectly meager we wished and made the soul. / A fool, who called it godly, infinite. / See, the power given you to act is greater, / more tremendous / than the space freed in your heart to comprehend. / Your heart is outclassed by your hand's blind grip. / This instructs. / Humans, you are smaller than you are."—Eds.]

39. The expression "metaphysically conservative" here isn't meant to allude to people who insist on referring to existing metaphysical philosophies, but to those who (in parallel to people who are politically conservative) insist on the world's allegedly metaphysical status.—

40. [This distinction is made by Immanuel Kant in *The Critique of Pure Reason.*—Eds.]

41. The "autarky of the moral" as well "nihilism" are ultimately both just euphemisms (albeit with different emphases) that circumscribe the moral's quality of "being like a metaphysical bunch of cut-flowers."—

42. [As used by Anders, the German *hybrid* acts as an adjective that can denominate hubristic or hybrid qualities. I have used both words to underline this.—Trans.]

43. [Anders is playing on the words for measure and measuring—*Maß, messen*—that link the words *Anmessen* and *Anmaßung,* which otherwise do not share a direct link in meaning.—Trans.]

44. [The followers of the early Christian theologian Marcion held the belief that because Jesus is the embodiment of love, he could not possibly be the son of "the cruel and wrathful" (i.e., Marcionite) god described in the Bible.—Eds.]

45. [By stealing fire, Prometheus hubristically refused to honor Zeus.—Eds.]

46. [English in the original.—Trans.]

47. By the way, for many of these products, "immortality through replaceability" might already be redundant because they can be made to be practically indestructible. This is the case, for example, with certain razor blades. The patents for these blades have been acquired by companies that seek to prevent their production because the actual immortality of a particular razor blade would result in the death of razor blade production. Production lives from the death of the single product (which has to be purchased again and again). In other words, the "eternity" of production rests on the mortality of its products.—

48. [1 Corinthians 15:55—Eds.]

49. [English in the original.—Trans.]

50. [English in the original.—Trans.]

51. [English in the original.—Trans.]

52. [English in the original.—Trans.]

53. [All quoted phrases are in English in the original.—Trans.]

54. [English in the original.—Trans.]

55. [*aeternum . . . semperitum:* "something eternal . . . something that is always." Anders is here playing on the nuances of these two Latin words meaning "always." The former evokes the eternal realm of the "afterlife," whereas the latter has no such transcendental resonance.—Trans.]

56. [*numerus clausus:* the fixed number of places on specific courses, which dictates competition for entry to university.—Trans.]

57. [English in the original.—Trans.]

58. Only the logician finds it hard to bring this ideal of conformity into sync with the ideals of "freedom" and "individuality" (for these two ideals have not been thrown overboard). In our real historical situation, they do not interfere with each other, for the unfree are simply *declared to be* "free." Because conformists enjoy the liberty to pursue their happy extrovert career paths, to which they have a right, without coming up against the edges of their own individuality, *they* are now those who are "free." "Freedom" is hence equated with "smoothness" and "frictionless advance," whereas "individuality" becomes synonymous with being entangled in "barbed wire" and with "disablement."— Incidentally, in contrast to theoretical systems, there is no actual social order that cannot afford to sustain contradictory ideas. As long as each of these ideas has a function that is useful for a given order, nothing stands in the way of their coexistence (e.g., the coexistence of the first book of "Genesis" and the principles of "genetics"*).— On the contrary, only humans who point to these contradictions are deemed oppositional, and their impolite fanaticism for the truth makes them suspect of being intolerant. As a matter of fact, there is no contradiction or inconsistency that could not be justified through a false appeal to tolerance.— *[English in the original.—Trans.]

59. [English in the original.—Trans.]

60. It is therefore plausible that death and dying become more conspicuous when the exemplary and enviable status of products is widely acknowledged because a given country has high industrial potential and, in this country, the

frailty of being human plays itself out in front of the backdrop of deathless serial products. Death is here more likely to be regarded as an unnatural embarrassment than in industrially less advanced countries, in which the mortality of creatures hasn't yet been covered over. It is hence plausible that in industrialized countries, the attempt to hush up the fact of death is more widespread and conducted more vehemently than elsewhere. I am, of course, speaking about the United States, whose prevalent attitude toward death will probably be adopted by us all in the future. For it is well-known that in the States, it is a widespread custom to apply makeup to corpses for their funeral, in order to transform the dead body into a pretty, ready-made product *post festum* and undress the corpse from its deadness. Furthermore, cemeteries are advertised as "lovely sites"* with the aid of temptingly alluring images (in the style of travel publicity).* The living—that is, those who will die (the *morituros*)—are hence addressed as if they will become "posthumously immortal." The suggestion is that, once they have moved in and become "owners" of their graves, they will daily be able to enjoy the pretty setting, as if they were pensioners. It can hardly be denied that no matter how inarticulately, a *preservation of life after death* is implied here (one, of course, that has nothing to do with religious beliefs). The little book by Evelyn Waugh dedicated to this macabre business of immortality is also well-known here. *[English in the original.—Trans.] [The reference is to Evelyn Waugh's highly satirical novel *The Loved One*—first published in 1948—which revolves around Dennis Barlow, a British poet who begins working at the pet cemetery and funeral service Happier Hunting Ground after losing his job as a Hollywood scriptwriter. The story follows Barlow's cynical attempts to win over Aimée Thanatogenos, a cosmetician employed as a mortician's assistant at the (human) cemetery Whispering Glades.—Eds.]

61. [English in the original.—Trans.]

62. [The word *star* is in English in the original throughout.—Trans.]

63. [English in the original.—Trans.]

64. [English in the original.—Trans.]

65. [The German here is "über-antwortet," and it is again possible to hear a resonance with Heidegger's vocabulary. *Being and Time* presents human existence as unfolding in the space opened by the experience of "being delivered over and made answerable" (*Überantwortung*) to the inescapable fact that one "is and has to be" (see esp. the discussions of "thrownness" and "facticity"; section 29ff.).—Eds.]

66. [English in the original.—Trans.]

67. [Moira: the goddess of fate; Tyche: the goddess of chance and fortune.—Eds.]

68. [The text is here playing on divergent strands of meaning the word *rechnen* can carry: (1) "to calculate, count, and account"; (2) to "anticipate," to "weigh up," to "process" and "work something out"; but also (3) to "rely on" or to "count on." Someone who is *unzurechnungsfähig* is of "unsound mind" and therefore declared "not responsible" and "not liable" for their actions. There is an added irony here because today one of the German words for a computer is *der Rechner*.—Trans.]

69. The fact that the engineers were incapable of associating anything other than "to feed"* when responding to the word "brain"* is to me the icing on the cake. *[English in the original.—Trans.]

70. [English in the original.—Trans.]

71. This type of shame is, of course, not a variant of "Promethean shame." On

the contrary, for anyone who is ashamed in this new sense, the "disgrace" is the very existence of "Promethean shame"—the fact that there is such a thing. This is thus "shame about being ashamed"—that is, "iterated shame." As such, this variant may be the first corrective remedy for "Promethean shame."

72. [In 1952, Douglas MacArthur joined Remington Rand, the company then producing and developing the UNIVAC computer.—Eds.]

73. [English in the original.—Trans.]

74. The philosophical novelist may even be inclined to weave this dialectical thread further into a text that is now entirely fictitious. He would show how MacArthur's act of revenge was always destined to fail—that he had to "find death in triumph" because he had to learn from experience that the word of machine oracles is as indispensable to the management of companies as it is to the waging of war. That is to say, that in the interest of revenge, MacArthur had to make use of the verdicts issued by the very creatures he wanted to revenge himself upon. These machines thus had ample opportunity to take pleasure in ridiculing him and spoil the satisfaction he sought to obtain.— This is not the place to follow the "dialectic of Promethean shame" all the way to this point.

75. [The Latin *coram* means "to be in the presence of," "to be before"—for example, the eyes of the public, the law, or an authority.—Trans.]

76. This originally "negatively intentional" shyness is transformed into positive religious awe the very moment it in some way turns to the authority instead of fleeing it. Now, it anxiously faces and subjects itself to this authority from a distance. When this happens, shyness turns into "awe," in the shape of quivering worship (of that which is "*awe-inspiring*"). Because the worshipper before the authority is a "nobody," the "interference in identity" sets in, which gives shyness the character of "shame."

77. Because the "authority" in front of which one feels ashamed is often "oneself"—and this is especially the case in historically recent varieties of shame—shame can be reflexive three times over: *one* is ashamed; one is ashamed *of oneself;* one is ashamed of oneself *before oneself.*

78. I do not believe that there can be a philosophical system based on reflection [*Reflexionsphilosophie*], departing from the fact of the self's recognition of itself [*Faktum des sich selbst vorfindenden Ich*], that can avoid being either a philosophy of freedom or unfreedom, or a philosophy of pride or shame. In philosophies of shame, identification fails; in philosophies of pride, the "I" occupies or "constitutes" the "me," or it simply denies its existence.

79. [In the aid of stylistic clarity, I have used "the self" to translate *das Ich.* "The I" or "the ego" would be more literal alternatives.—Trans.]

80. On this point, see my "Pathology of Freedom," 283–84.

81. When "*Dasein,*" at the beginning of *Being and Time,* asks itself "Who?" it is, and finds—although it only crystallizes into a "self" by asking this question—that it is already constituted by the "impersonal public" Heidegger calls "*das Man,*" it discovers itself as an "it." In fact, the action *Being and Time* portrays (in the form of a theoretical ontology) is nothing other than a systematic fight against shame. *Being and Time* is about the attempt of the "I" (which is ashamed of itself) to overcome the disgrace of being an "it" and become "itself." [See *Being and Time,* section 9, for Heidegger's introductory discussion of the question "Who?"—Eds.]

82. The conventional assumption that shame occurs especially or even only

in the case of something "that was one's fault" clearly inverts the true state of affairs. This assumption is a symptom of the immoderate pretensions we have about freedom, which we just mentioned, whereby humans attempt to appropriate the pain of unfreedom (which shame represents) in order to present it as a "pain that acts as punishment" and employ it as such.— The truth, rather, is that one is ashamed of an action because, in the eyes of the authority, one is someone that one shouldn't be.— How little it is actual guilt and responsibility that generates shame is evident in the fact that one is also ashamed when wrongly accused—yes, especially then. In this instance, one is not ashamed because one is hurtfully believed to be capable of this or that misdeed—this would be far too subtle to be true. No, one is ashamed because in the eyes of others—that is, socially—one actually is guilty. In brief, one is not ashamed of guilt; on the contrary, that of which one is ashamed often turns into guilt. *"I am ashamed—therefore I am guilty" is more valid than "I am guilty—therefore I am ashamed."* In dictatorships, for example, it has happened hundreds of times that actions or failures to act became shameful the very moment when they had to be justified in front of a jury, even if these acts or omissions were motivated by moral concerns. And it cannot be denied that the onset of shame convinced not only the jury that the accused was guilty but also the accused.

83. See the Second Objection as outlined in section 2.

84. [The text is here playing on the shift in meaning the verb *herausnehmen* (to take out of) undergoes when its reflexive form is deployed: *sich etwas herausnehmen* means "to take the liberty to (do something)."—Trans.]

85. Philosophically speaking, Sigmund Freud's discovery of the "Trauma of Birth" cannot be rated highly enough: for what more incisive event could happen to life than being torn from the "ground"?— The feelings Freud brought into view ("Oceanic Feeling"; "Death Drive") are metaphysical in every respect, even if Freud's language was masked by the scientific vocabulary of his century. This is equally true of the "Trauma of Birth" with which he described the *shock of individuation,* no matter how well he disguised it.— In an analogy with our question "who is ashamed?," the question to ask in relation to this trauma is the following: "*Who* is actually shocked here?" Is it the individual itself? Or, since the shock precedes individuation, is it not rather "life that has not yet been individuated"? Is not the process of individuation shocking *in itself?* Is the individual not merely the *heir* to this shock, which it now drags along with it throughout its entire life as the pain of being an individual, a pain that is never overcome?— It seems to me that this was Freud's view; for the "Death Drive" is ultimately nothing other than the longing of the individual to be rid of the agony of being an individual.

86. This theory of shame was first formulated in "The Pathology of Freedom," where I employed the expression "leap from the origin" [*Sprung aus dem Ursprung*]. [See "Pathology of Freedom," 288ff.—Eds.]

87. This type of shame includes what is probably the most basic form of shame—metaphysical shame: the shame the individual feels at having been taken out of the "ground" and existing *as* an individual, *as* a singular being; the shame of having "fallen off" the ground [*abgefallen*] instead of still belonging to it in the cosmically proper manner; the shame of now being forced to exist as a "*cosmic exception.*" The equation of this form of shame is: nonbelonging = improperness =

insubordination. Anaximander and (Hölderlin's) Empedocles are the principal witnesses of this form of shame.—

The following can be read in the Molussian metaphysics of individuation, which did not yet know our speculative inhibitions: "The ground-of-the-world [*Weltgrund*] is ashamed of its individuations, of the fact that it has distanced itself from itself in the form of individuals." Further we can read (in a similar vein to a famous formula by Spinoza): "The shame the individual suffers from is only a part of the shame with which the 'ground' is ashamed of its individuations."*— The "*it*." *[The exact formula being varied here could not be identified, but it is possibly a reference to part 3 of Spinoza's *Ethics*, "Of the Origin and Nature of the Emotions."—Eds.]

88. [*μετάβασις εἰς ἄλλο γένος:* a step across into a different area of meaning.—Trans.]

89. [The word *stumpf* does not translate smoothly because English makes a distinction between "dull" and "blunt" that German does not. The paragraph is bringing together the sensation of handling "blunt objects" and feeling "dull pain."—Trans.]

90. It is hard to conceive a philosophically more ridiculous preoccupation than the one concerned with the variants of what is called "synesthesia" (the allegedly retroactive confluence of characteristic impressions belonging to different sensual realms). Those who are astounded by these phenomena should also be astonished that tree branches link up with a tree trunk. The philosophical problem is not the question of how these divergent characteristics "converge," but rather how these characteristics branch out from their originary, pre-specific state in the first place.

91. [This seems to be an allusion to the Fourth Movement of Arnold Schönberg's String Quartet No. 2 in F♯ Minor (1908), which is titled *Entrückung* (Transcendence). Unusually, Schönberg's quartet includes song (which uses poetry by Stefan George). The singing voice enters with the lines "I feel air from another planet. / The faces that once turned to me in friendship / Pale in the darkness before me." (here reproduced in Richard Stokes's translation).—Eds.]

92. After completing my book, I have found that an emphatic account of this mutuality of visibility is part of Walter Benjamin's discussion of his notion of "aura" in his admirable Baudelaire essay. Benjamin's account anticipates what is asserted here. In the same essay, there is also a treatment of Valéry's characterization of perception in dreams: "In dreams there *is* an equation. The things I look at see me just as much as I see them." [See Benjamin, "Some Motifs in Baudelaire," 339.—Eds.]

93. [The term *das Man* is central to Heidegger's early thought, in which it is used to name the "faceless public" in view of which "one does what one does." See, for instance, *Being and Time*, section 27.—Eds.]

94. ["Jazz" here probably serves as a blanket description for popular music, and the text here replicates similar cultural prejudices known from a number of émigré German thinkers.—Eds.]

95. [*das Geschlecht* can refer to sex and gender as well as the human species as a whole.—Trans.]

96. To achieve this end, the music-machine, with special glee, especially likes to incorporate monotonous folkloric elements.

97. [*Chthonian forces* are the those that bring life and end it (fertility and death). They are embodied by a number of ancient Greek deities and mythological beings.—Trans.]

98. [English in the original.—Trans.]

99. Art history, unfortunately, has not produced a specific study of the significance of the face in different historical epochs. Such a study would show that these "degradations of the face" also existed in earlier times. In Mannerism, for example, a period in which all faces not only tend to look the same but are also depicted in far too small a size (this is notorious, for example, in Bassano and El Greco). Unless the painting is explicitly a portrait, the same degradation of the face is also at work in the Baroque style, where the transformation of the human into a vividly fluttering being embellished in elaborate garments is a much more important motif than the face. This is not the place to investigate whether the face was at that time also being devalued because of two deindividualizing forces that sought to consume the human. This suspicion seems reasonable, however, because the epoch in question coincided with the burgeoning natural sciences and the Counter-Reformation, and because in the art of this time, a most uncompromising realism was constantly colliding with supernatural exuberance.—

100. [In ancient Greek, the word *ethos* could be used to describe the formative power of music and its ability to structure thoughts, feelings, and attitudes.—Eds.]

101. [English in the original.—Trans.]

102. The sense of time of a person working in conformity with the machine no longer has anything to do with the subject's sense of time, which (as all modern philosophies of time deem self-evident) irreversibly moves forward into the future. The machine-worker's sense of time is cyclical. Time is felt as consisting of small durations, which are coextensive with the continually recommencing, identical action of the machine. Hence, this is time as measured by the second hand of a watch, although its full revolutions are in this instance not registered as minutes and hours as they usually are. Only when starting work at the machine do the repetitions "swim along" in the broad advancing stream of time; and they are only perceived as being boring during this initial phase. That is, they are boring as long as the irreversibly advancing flow of time has not yet lost its hold, which happens after a short while.— Once work has become attuned to the machine, the repetitions no longer appear to follow each other in succession. They no longer form a chronological sequence, just as one breath of air is not registered as following another by the one who is breathing (because they are always breathing but never know the number of the breath they are taking). It is at this point that the repetitions stop being boring. With this, I naturally do not want to suggest that they become interesting, only that they are now devoid of time. In fact, these repetitions are so devoid of time that workers are often surprised by the end of the shift: Having been robbed of a sense of time, they have the feeling of having just started their working day. This is similar to someone who has just woken from a dreamless sleep and is convinced that they only just went to bed.— If temporality is not just understood as one trait of subjectivity among others, as it has been since Kant, but rather as its decisive attribute (as is apparent in the title of Heidegger's major work), the fact that time "ceases" [*eingehen*] while working at a machine can be seen as a symptom for the "decease" of subjectivity itself.

103. [The term *Identitätsstörung*, translated as "disturbance of identity" throughout, here takes on an additional nuance. The text begins to foreground the word *Störung*, which has several specifically machine-related meanings, such as interference (in a radio or phone signal), a malfunction, or a defect. I have used "interference" here because Anders's writings frequently evoke the phenomenon of being uncomfortably confronted with oneself on account of a power cut or a missing signal or, as in this instance, when a human factor interferes with the working of a machine.—Trans.]

104. [*Nachlässigkeit* also means negligence.—Trans.]

105. [The word *Gerätbedienung*, translated here as "operation of machines," could also have been translated literally as "serving the machines." Anders elaborates on the word *Bedienen* in the second volume of *The Obsolescence:* "If we were a bit more keen-eared, we would probably daily be astonished to realize that the word *Bedienung*—once employed to describe what servants 'do' when serving 'ladies and gentlemen'—now describes what we do when operating machines . . ." (69, my translation).—Trans.]

106. [The text is here playing on variations of the German word *werfen* (to throw) in a manner that suggests an engagement with Heidegger's concept of *Geworfenheit* (thrownness), which Anders has already evoked earlier in this chapter.—Trans.]

2. The World as Phantom and Matrix

1. [This chapter makes frequent use of the verb *prägen*, which has two interrelated layers of meaning. When used in relation to workable material, it means to "mold, coin, emboss, stamp," or otherwise impress a shape on something. This meaning is retained in the figurative use of the verb, which conveys how things, texts, lives, and so forth have assumed a certain form because they were "shaped" or "influenced" by something that has "left some sort of impression." As note 125 below suggests, Anders seems to have at least partially modeled his use of this term on the gramophone industry and the stamping of records from copies made from an original matrix—that is, a master copy. To retain this link, *prägen* has mostly been translated as "to mold," but at times I also use "to influence," "to shape," or "to mold and influence" to soften the mechanistic feel that only partially captures the German.—Trans.]

2. [The German here reads, "Wo der Gang zum Brunnen / so gut ist wie der Trunk."—Trans.]

3. [English in the original.—Trans.]

4. [English in the original.—Trans.]

5. [English in the original.—Trans.]

6. [The saying *Eigner Herd ist Goldes wert* is often translated as "there is no place like home," but I have here reproduced Norbert Guterman's elegant literal translation from the radically abridged English version of "The World as Phantom and Matrix" that appeared in 1956 and then reappeared in a slightly modified form in 1957. See Anders, "Phantom World of Television," 360.—Trans.]

7. [The reference is to "Télévision—macht die Familie familiärer," a *Presse* article by Alfred Lang.—Eds.]

8. [The reference is to "Der Fernsehapparat war wichtiger," a short *Weltpresse* article without a named author.—Eds.]

9. [These French expressions—to converse, to entertain, to provide support (especially financial support)—spell out the three meanings of the verb *unterhalten* this section plays on.—Trans.]

10. [English in the original.—Trans.]

11. [The phrase "small earthen lamps" is not in the German and has been added here to clarify the text's implicit reference to a widely used motif in ancient Roman and Greek poetry and drama. See, for instance, Martial's epigram "A Night-Lamp": "I am a night-lamp, privy to the pleasures of the couch; do whatever you please, I shall be silent" (14.73).—Trans.]

12. [English in the original.—Trans.]

13. [English in the original.—Trans.]

14. [*ζῶα λόγον ἔχοντα:* living beings with *logos* (i.e., the capacity to speak). —Trans.]

15. [The word *infant* is derived from the Latin *infari,* "to be unable to speak." —Trans.]

16. [*ἄνευ λόγου εἶναι:* being *logos*less.—Trans.]

17. [The text here repeats the nouns *Unmündige* and *Hörige,* which are used in the title of this section. These are words that evoke sensory organs and bodily capacities. *Unmündig* means to not be *mündig* ("to not be of age, legally responsible, to not have the right to decide"). It is a close homophone of the German word for mouth (*Mund*). *Hörige* means "to be in bondage, to be under someone's spell or fully obedient." It is derived from *hören,* meaning "to hear, listen, or obey"—that is, "to only be allowed to listen." I tried to retain this direct reference to speaking and listening in the title of this section through paraphrase.—Trans.]

18. The prelude to this now universal impoverishment of language is already over. I am speaking of the dead art of letter-writing, a demise that the fifty years spent on the telephone have brought about. This decline is so profound that even the letters written by averagely educated people a century ago seem like masterpieces of clarity and affection. Because the human is an expression of the way humans express themselves, it is not just the subtlety of communication that is in decline but the human too.

19. Nothing is more misguided than the irrationalist's lofty or whiny complaints about the inability of language to keep up with the richness and depth of our lived experience. The past greats, who had a depth and linguistic wealth we scarcely match, were perfectly able to linguistically process their lived experiences. Their speech was able to reach into the most peripheral spheres of experience; when they did report linguistic incompetence and failing speech, they did so very late, only in closest proximity to the furthest limit. Yet the less one has to say, the more readily one turns destitution into infatuation and poverty into wealth—the more lavishly one begins to show-off the failure of language as evidence for the abundance of one's own lived experience: youth has no time for the unsayable. No, today's true plight and cause for embarrassment is not that we might "exhaust" our apparent depth and richness "by bringing it to word." For if we possess this wealth at all, it is in danger of trickling away and silting up because the constant supply of language is beginning to make us unlearn speech.

20. [The reference is to the "Song of Lynceus" (*Faust, Part Two*), which opens with the canonical lines:

I am born for seeing,
Employed to watch,
Sworn to the tower,
I delight in the world.

I see what is far,
I see what is near,
The moon and the stars,
The wood and the deer.

In all these I see
Eternal beauty,
And as it has pleased me,
I'm pleasing to myself.

O happy eyes,
Whatever you have seen,
Let it be as it may,
How fair it has been!

See Goethe, "Lynceus, the Watchman, Singing on Faust's Observatory."—Eds.]

21. The notion that people erect "ivory towers" and retreat into them so they don't have to face reality is utterly outdated. For some time already, reality (figuratively speaking) has been erecting such towers itself and now also acts as their landlord and investor: Today, we don't sit in these towers as refugees looking to escape from reality but as tenants who were forced to move in. And once we're inside, we're not meant to become immersed in fantastical images that transport us to fictitious alien worlds; reality accommodates us, rather, to ensure that we live within its picture. Not in a true picture of course, but in the false representation presented to us as "reality itself"; a false picture selected for us on behalf of very real interests. We're only included in this putatively "real" image to give reality an opportunity to distract us from reality itself. This distraction is again motivated by very real aims: The aim of genuinely influencing and working on us until our human reality has become useful to the interests that reality fronts. Those of us who are compliant victims are called "extroverts," those who resist "introverts."

22. [The text here plays on Goethe's small poem "Erinnerung":

Willst Du immer weiter schweifen?
Sieh, das Gute liegt so nah.
Lerne nur das Glück ergreifen,
Denn das Glück ist immer da.

See Goethe, "Erinnerung."—Eds.]

23. The notion that "the world is coming to us" has become so intuitive that everything crossing our earthly paths is now taken to be a visitor: Yesterday, we

were visited by frightful flying saucers; today's visitors are superhumans from Sirius.

24. [This section plays on the title of Arthur Schopenhauer's central work *The World as Will and Representation*.—Eds.]

25. [This appears to be a reference to the addition of paragraph 44 of Hegel's *Elements of the Philosophy of Right*.—Eds.]

26. The classic image of the world as a "gift" can be found in the book of *Genesis*, which depicts the world as something that was created *for* the human.— It is thus no coincidence that these modern forms of idealism postdate Copernicus: In a sense, each represents an attempt to rescue the biblical "for us," which was compatible with the pre-Copernican worldview but is no longer congruent with a post-Copernican one. These idealisms thus amount to covert attempts to preserve geo- or rather anthropocentrism within a decentralized universe.

27. [The text here plays on the verb *fahren*—"to go, fare, travel, drive, venture out"—which also acts as the root of the German noun *Erfahrung* ("experience") and the verb *erfahren* ("to experience"). In its literal meaning, the German word for experience thus denotes what we acquire as we are "underway" in our lives and the world. The verb *fahren* also provides the root for the German word for "danger" (*Gefahr*) and thus conceptually links experience to exposure to danger or uncertainty, a connection this entire section plays on.—Trans.]

28. On this point, see the author's essay "Une interprétation de l'a posteriori" in *Recherches Philosophiques*, 1934.

29. It is hardly a coincidence that this form of "coming into the world" is waning at the very historical and cultural moment in which the trauma of physical birth is also in the process of being abolished with the help of technology.

30. [See Marcel, *Homo Viator*.—Eds.]

31. Television sets are in fact already being installed in cars. Since December 1954, for instance, in Cadillacs made by General Motors.

32. On the expression "*entfremden*" (to alienate): The prefix *ent-* signals removal and privation (as in "*enthüllen*," to *un*cover; or in "*entflecken*," to *remove* stains). *Entfremden* should thus not actually mean "to alienate" but "*to rob* something of its strangeness, otherness." It would be wasted effort to attempt to revert to this original meaning, a meaning that contradicts the established use of the term. From now on, we will thus replace the ambiguous expression *entfremden* with the unmistakable "*verfremden*" (to defamiliarize), which Brecht introduced to refer to certain theatrical stage effects. [The approach outlined in this note was not consistently implemented in the German text, which at times continues to use the term *entfremden* while also adopting *verfremden*. As the distinction Anders makes here doesn't apply in English, I have generally translated both *entfremdung* and *verfremdung* as "alienation."—Trans.]

33. The once revolutionary word *alienation* has now been claimed by journalism and literature. It is used with such gusto and fluency that the very process that the word had once attempted to condemn has come to seem highly familiar. The word *alienation* has itself been stripped of its alienating character. One can no longer detect that the phenomenon was introduced in the context of work, commodity, freedom, and property—that is, that it was first used in a revolutionary sense one hundred years ago. The word hasn't just become fit for polite society; it has pretty much turned into *the* ID card of avant-gardism. No self-respecting

commentator of modern art would ever dare leave their house without carrying it along in their pocket.— Whether intentionally or not, the ease with which this term is now used has robbed it of its *morally scandalous* edge. The expression has (in the literal sense of the word) itself been alienated: it has been "robbed of" its "strangeness" and "otherness." *To disown and disarm your enemies, make everything you acquire from them your own.*— The process by which alienation was rendered harmless has the following roots: (1) The branch of late 1920s German sociology (Karl Mannheim) whose sole contribution consisted of the forcible extraction of isolated Marxist expressions to insert them into other contexts or into everyday speech, thereby neutralizing them. This branch of sociology migrated to France in the early 1930s and then to the United States in the late 1930s.— (2) Surrealism, which, in the brief time it was allied to Communism, loved to drape itself in Hegelizing word scraps. Those who use the term "alienation" today suspect nothing untoward, of course, for they are already copying the copycats from the 1930s. Some of them would, in fact, be utterly amazed to learn to whom they originally owe their favorite expression, the one they so like to savor.— Even this brief account of today's use of the term "alienation" has thus revealed a counterforce: a process of pseudofamiliarization and means by which to make things seem more intimate. Because it doesn't just attack terms and isn't just intent on giving words an air of familiarity, this process is not equivalent to the caricature and stereotyping of words. This procedure, rather, views the world and everything in it as its prey, and its aspirations are just as universal: Just as alienation encroaches on everything with its Midas touch to turn what is familiar and intimate into something cold, thing-like, public, and strange, pseudofamiliarization takes possession of everything that is strange and different to transform it into something seemingly homely.

34. [English in the original. The official name for the US weapons test conducted on March 7, 1955, was "Turk," which was part of Operation Teapot.—Eds.]

35. Before we move on to examples, we will prophylactically note that "cozification" is not the same as "popularization," although the boundary between the two may at times be blurred. Whereas cozification displays a fundamental lack of respect for its object and seeks to profit from the damage and disadvantage it inflicts on consumers, true popularization does not just convey content in an accessible format: like all genuine instruction, it also conveys that this content is to be respected.

36. [English in the original.—Trans.]

37. [English in the original.—Trans.]

38. [English in the original. This reference could not be attributed, but the quoted line and wider logic of the passage is highly reminiscent of the language that film posters used in the early 1950s. The 1953 poster for *House of Wax*, for instance, introduces the experience of 3D film with the tag line "Right at you! The hand is at *your* throat . . . Right at you! The kiss is on *your* lips . . . Right at you! The horror that chills the spine!"—Eds.]

39. [The quote plays on the following line from Goethe's *Zahme Xenien* III: "Wär nicht das Auge sonnenhaft, die Sonne könnt es nie erblicken." Literal translation: "If the eye were not sunlike, it could never catch sight of the sun."—Eds.]

40. [English in the original.—Trans.]

41. [English in the original.—Trans.]

42. [English in the original. This appears to be a reference to article 1 of *Universal Declaration of Human Rights,* which opens with the sentence "All human beings are born free and equal in dignity and rights."—Eds.]

43. [English in the original.—Trans.]

44. [All quoted phrases are in English in the original.—Trans.]

45. [English in the original.—Trans.]

46. [English in the original.—Trans.]

47. [English in the original.—Trans.]

48. [At the time of Anders's writing, it is worth noting here, the technology required to create prerecorded television programs was still in its infancy, meaning that television was an inherently "live" medium, comprising live broadcasts and telecasts of studio productions.—Eds.]

49. I visited a television expo, where I enjoyed the questionable privilege of seeing and hearing seven simultaneous TV projections of a sketch being performed by an actor in an adjoining room. It was notable that (1) to the eyes, the actor split himself into seven identical brothers, but that he also only had one unsplit voice, which sounded across two rooms; (2) the images felt more naturalistic than the original, for his actual movements needed to be staged in order to ensure that his reproductions would attain an unstaged feel; and (3) another thing that was notable—no, more than notable, downright shocking—was that the encounter with a seven-bodied actor didn't even feel shocking: Such is the extent to which we have already come to only ever expect serial-products.

50. Events that are only significant in a subliminal manner, even when they are happening inside our own bodies, aren't something we are "present to." Such events are merely happening at the same time as us. This isn't because they aren't in some way consciously "given"; rather, they fail to register because they are insignificant.

51. [Eleatic philosophy was a school of pre-Socratic philosophy (esp. Parmenides), which centered on the notion that there is one unified realm of being.—Eds.]

52. [English in the original.—Trans.]

53. [*à concevoir la moitié d'aucune âme:* "to conceive a half of any soul." A reference to the synopsis of the Second Meditation: "We can understand no body except as divisible, but on the other hand no mind (*âme*) except as indivisible: for we cannot conceive a half of any mind, as we can with any body, however small it is." See Descartes, *Meditations on First Philosophy,* 10–11.—Eds.]

54. [English in the original.—Trans.]

55. If it is legitimate to consider cancer a disease *sui generis,* then the loss of control over single functions is the psychological analogue of a tumor. It is the psychic equivalent of the physiological condition in which the central power of the organism has lost its ability to maintain dominion over all its cells, which allows some of them to grow autonomously and in an uncontrolled manner.

56. On this point, see Beckett's *En attendant Godot* [*Waiting for Godot*]. Its author continually lets his figures either put on boots or take them off so as to ensure that their hands always have something to do.

57. This "passive and simultaneous performance of many different parts" is also motivated by the economic principle and by the (of course totally distorted) ideal of peak efficiency and performance at work. Transposed onto leisure, this

means that one seeks to be as invested in leisure as possible. Everything that is "fun," by the sweat of one's brow, is done at once: crosswords, and gum, and music from the radio, and so forth. Otherwise, one would let leisure go to waste.

58. [English in the original.—Trans.]

59. [English in the original.—Trans.]

60. This fraudulent provocation of excitement, and the arrangements that are made to ensure that a surrogate satisfaction flows into the void, are reminiscent of a practice that has become the norm in a very different sector of today's life: cattle-breeding. It is common knowledge that one lets breeding animals mount mock-ups instead of cows; they are made to mount so called "dummies." *Attrappe*, the German word for "mock-up," is derived from the French "*attrapper*," to catch; to "trap" means to trick, ensnare. Bulls are thus made to jump into a trap—that is, "fall for it." This is done because their sex drive was deemed to be operating in a highly inefficient and wasteful manner in its natural state. That is to say, its workings are most unprofitable; the reproductive process by which animals procreate falls embarrassingly short of the industrial ideal of reproduction. This is no longer the case now that it has been trapped and intercepted—that is, "*attrappé*." *The "mock-up" is thus a make-believe reality in the service of commodity production.* In this particular instance, it consists of make-believe flesh in the service of industrial meat production. In places where feelings have withered in the absence of partners, industry now creates new feelings with the help of mock-up partners. This is done because industry knows that such feelings will in turn always demand the creation of further mock-ups and thus sustain the industry all by itself.

61. [English in the original.—Trans.]

62. [English in the original.—Trans.]

63. [This section seems to be channeling a widely reported (anonymous) death threat to the TV announcer and actor Donald Gray. The threat, here cited from *The Sunderland Echo* (August 14, 1954), read: "You have wrecked my life and I no longer have my wife's affections, but I refuse to give up my TV set because of my small daughter. I have come to hate your face, voice, and everything about you. It has come to the point where I give you seven days to resign your job as TV announcer. I shall use all my energy and resources and even if it takes six months I will get you. Before I kill myself I will kill you." This particular story and the wider topic of death threats to TV personalities were widely covered in the Austrian press in September and October 1954, including in the newspapers Anders cites earlier in this chapter.—Eds.]

64. At times, the temptation to improve a product technologically is so irresistible that one gives in to it even when this development hinders rather than enhances the social function that is desired.— But major industries have also often acquired proposed technological improvements to slow down their development. The history of technology may leave the impression of unrestrained development; but it is, of course, also a history of technologies that were "suppressed" [*Verdrängungen*].

65. In contrast to the great in size, which has been a problem for philosophy since Longinus, the "small" has almost consistently been neglected. The young *Burke* is a rare exception: Practically emulating Rococo motives, he equated the small with the beautiful. This equation was based on the following experience: small = unthreatening = available = defenseless = moving = pretty = not infringing

on our freedom. Because this definition of "beauty" is informed by "freedom" (at least *modo negativo*), it has a certain affinity with Kant's definition. All the more so because Kant presents the "sublime," which surpasses all human proportions, as the counterpoint to the "beautiful."— Even though smallness in the form of the pretty, cute, or harmless has come to represent the only aesthetic category a great portion of humanity knows, it barely even exists in contemporary aesthetics.

66. [The text here returns to the famous line from the *Phenomenology of Spirit*, "the true is the whole." See note 2 in the introduction.—Eds.]

67. [This entire section plays on the word *Nachricht*—"the notification, the message, the news." It is derived from the highly versatile verb *richten* (to orient something or someone, to judge, to address) and thus intrinsically links the realms of notification and judgment that this section goes on to discuss. I have usually opted to translate this with either "notification" or "message," but I have at times also used "notification and message" to signal the range of meanings the word retains in German.—Trans.]

68. [*dico—δείκνυμι:* I say—I show.—Trans.]

69. [The text is here playing on the Latin verb *facere,* "to make, to do"—*factum* is the past participle; it denotes something that has been made or been done.—Trans.]

70. The thesis, which enjoys such strong backing today, that turning truth into a question of judgment distorts the very concept of truth, will probably need to be qualified and put into perspective the moment "judgment" is understood to mean what it originally means—that is, message. Through its messenger function, because it enables an absent person to adjust themself to something absent—that is, to treat something absent as present—the judgment performs a decisive "act of discovery" that makes something hidden present [*Entbergung*]. It is alone the exchange of news and messages [*Nachrichten*]—that is, speaking—that opens the world; this exchange constitutes the only truth of the human as a society; it alone establishes the "claim to universality" that is an attribute of anything that seems obvious and logical.

71. [The terminology and wider perspective adopted in this section are consistent with the work of Heinrich Rickert and Edmund Husserl, and the translation has tried to reflect this.—Trans.]

72. [*λέγειν:* speaking.—Trans.]

73. [English in the original.—Trans.]

74. [English in the original.—Trans.]

75. [English in the original.—Trans.]

76. In those programs that openly present themselves as adverts for soap powder or petrol this is so obviously the case that these programs do not need to be discussed here.

77. [This is again a play on Hegel's line "the true is the whole" from the *Phenomenology of Spirit*. See note 2 in the introduction and note 66 above.—Eds.]

78. Those peculiar creations the late nineteenth and early twentieth centuries brought forth and called "worldviews," were only the harmless and rejected prefigurations of today's "stimulus-response models." Of these "worldviews," none that was but a world*view* was able to survive; only those that were actually able to establish themselves as stimulus-response models prevailed. And today's stimulus models—although they still indulge those who, as a vain luxury, seek to

distil some kind of academic worldview from them—no longer even maintain the illusion of being worldviews.

79. [English in the original.—Trans.]

80. [English in the original.—Trans.]

81. [*Der Stürmer* was an anti-Semitic newspaper founded in 1923 and published until 1945; it was known for its violently anti-Semitic caricatures and cartoons.—Eds.]

82. [English in the original.—Trans.]

83. Because our imagination is already unable to hold the thought of the millions who were murdered, our horror will now scarcely increase by the added realization that this annihilation happened because of images or in their place.—To us who were raised with the assumption that the world is becoming more and more humane, there was one basic belief that seemed entirely obvious: that humanity started the moment human sacrifice was only offered figuratively [*bildlich*]—that is, at the historical moment when Isaac was replaced by the ram. This belief couldn't have been disillusioned in a more terrible way than by the millions of humans who were meaninglessly slaughtered in the place of their images.

84. [Julius Streicher (1885–1946), the founder and editor of *Der Stürmer*, was sentenced to death as an accessory to genocide at the Nuremberg trials for his role in inciting the extermination of Jews. For context, the following words are lifted from the verdict: "The crime of Streicher is that he made these crimes possible, which they would never have been had it not been for him and for those like him. Without Streicher and his propaganda, the Kaltenbrunners, the Himmlers, the General Stroops would have had nobody to do their orders.

In its extent Streicher's crime is probably greater and more far-reaching than that of any of the other defendants. The misery which they caused ceased with their capture. The effects of this man's crime, of the poison that he has put into the minds of millions of young boys and girls goes on, for he concentrated upon the youth and childhood of Germany. He leaves behind him a legacy of almost a whole people poisoned with hate, sadism, and murder, and perverted by him. That people remain a problem and perhaps a menace to the rest of civilization for generations to come" (Office of United States Chief of Council for Prosecution of Axis Criminality, *Nazi Conspiracy and Aggression*, 709).—Eds.]

85. This also has something to do with the international decline of caricature and satirical magazines. The act of ridiculing power (which is what true caricature always does) has simply become too risky.— The *Stürmer* drawings didn't really constitute caricatures or satire, for they categorically only ever selected the victims of power as their victims and targets of attack.

86. This lie follows a pattern that structurally corresponds to all of today's counterrevolutions, which categorically seek success by calling on the help of those they are directed against.

87. [The text here refers to the work of James George Frazer (e.g., *The Golden Bough*), Lucien Lévy-Bruhl (e.g., *Primitive Mentality*), and Ernst Cassirer (e.g., *Philosophy of Symbolic Forms*), which are steeped in racialized imaginaries that seek to distinguish between "primitive" and "higher" cultural forms.—Eds.]

88. [English in the original.—Trans.]

89. That this is the case is widely accepted today, but of course not as the philosophically relevant fact Marx introduced it as. For Marx, the origin of the

things he called "ideological" lies in a peculiar cross between "idealism" and Hegel's "master–slave dialectic." In Marx, "ideology" names the master's world-view, which servants without class consciousness take to be "the world" itself; it names that image of the world, therefore, that didn't actually belong to the servant but to the property owner. This is what Marx had in mind when he advanced his thesis that the philosophy of his epoch was but the philosophy of the ruling class.— This Marxian schematic, of course, cannot be applied *verbatim* to the conditions of today's mass society. Today, every type of commodity (be it a cigarette, film, or worldview) is from the outset produced to the specs that promise the highest sales figures; in other words, every commodity is designed to meet consumer desire or alleged desire. And because the producers are also consumers of the commodities they produce—cigarettes, films, worldviews—this has the dialectical consequence that the "ruling class" is now also shaped by the mass-products it did not produce for itself, but for the masses. Hegel's sentence "The master turns into the servant of the servant" should thus be replaced by the formula "The master turns into a servant *among* servants."

90. The human duty to let oneself be molded to templates is often justified via recourse to existing moral standards. Those who mount resistance against being molded, for instance, are often branded as "unchristian" or "undemocratic." The "logic" behind this is as follows: People who refuse to participate display a lack of humility—that is, a lack of Christianity; or else they are deemed to be demanding special treatment—that is to say, special privileges.—In Henry Link's famous book *Return to Religion,* for instance, anyone who entertains their own moral scruples instead of solely relying on home-delivered pangs of conscience is described as an "introvert," someone who suffers from a social pathology. Link's book, first published in 1936, appeared with one of the biggest publishing houses and reached eighteen editions within nine months, so it's not just some small booklet of manifesto-literature. For my discussion of this book, which presents Jesus Christ as the model "extrovert," see *Zeitschrift für Sozialforschung,* 1938, nos. 1–2. [The reference is to Anders, "Link, Henry C., *The Return to Religion.*"—Eds.]

91. [English in the original.—Trans.]

92. [English in the original.—Trans.]

93. [English in the original.—Trans.]

94. [English in the original.—Trans.]

95. [English in the original.—Trans.]

96. [English in the original.—Trans.]

97. [English in the original.—Trans.]

98. [English in the original.—Trans.]

99. [English in the original.—Trans.]

100. [English in the original.—Trans.]

101. [English in the original.—Trans.]

102. [English in the original.—Trans.]

103. [English in the original.—Trans.]

104. [English in the original.—Trans.]

105. Once this payment method is adopted, then the renunciation of freedom that begins as compliance with the command issued by the goods on sale becomes total: Any buyer who still owes installments suffers from a constant feeling of guilt—and not just guilt toward the supplier but also toward the commodity

supplied. They regard this ownership as something they haven't yet deserved, and by already using it, they now enter into an unfree relationship with the commodity. Because the commodity already allows them to live it up like a lord, they must now spend their lives trying to make up for this high standard of living by taking on additional work. With this, any chance of ever arriving at themselves is squandered once and for all.

106. [English in the original.—Trans.]

107. [The text here develops a logic that is encapsulated in the poem "A Summer Evening on the Hudson" from an unpublished collection of poetry from the years 1939–46, reproduced here in Kerstin Putz's translation:

> Now sinks the light. The air turns mild.
> And gently fades into the blue
> Pepsi Cola.
>
> From the opposite shore, the night's command
> Blinks red and urgent:
> Drink Pepsi Cola.
>
> And from the bushes one last time
> The night chant rises to the stars:
> Time for Pepsi Cola.
>
> Get up! Don't fail to heed the call!
> Who needs the night that God has made?
> We're talking Pepsi Cola!
>
> Somewhere in a drugstore waits
> That which God never did create
> but ours regardless: Pepsi Cola.
>
> Reproduced from Putz, "Improvised Lives," 236.—Eds.]

108. [English in the original.—Trans.]

109. This axiom was, in a sense, already introduced by the exacting natural sciences: For them, only phenomena that consistently recur under identical test conditions—only what adheres unfailingly to natural laws—"truly exists." Its maxim could thus read: "I do not regard as existent what I cannot repeat." The conceit with which natural scientists view history, and with which nineteenth-century scholars viewed religion (which was identified with "miracles"), is based on this very equation of being with the plural—that is to say, its equation to lawfulness.—

110. There is, of course, value in owning an item that is unique—an artwork for instance—for it represents a monopoly that demonstrates wealth and creates an aristocratic consciousness. Today, incunables and mass-products live in an intimate union: The more mechanized his products become, the more the person who produces them likes to adorn himself with items that are made by archaic methods.

111. What makes photography so peculiar—and so peculiarly fascinating—is that it combines two of today's main activities: reproduction and acquisition. There is the added facet that those who take photographs and acquire things by this method only need to pay for the gear required for the acquisition (just as is

the case for hunters and anglers); the thing they acquire (the sight) is available to them for free.— This represents a fairy-tale-like exception in today's shopping culture. It is, of course, no coincidence that taking pictures is, in English, called "to shoot," as if the subject were some hunted prey.— After all, photography is fascinating because it's fun while at the same time being a gainful activity: It's a form of leisure that makes itself attractive to illiterates of leisure by presenting itself as an occupation that seems to be productive or even a form of work. In short, it presents itself as a hobby, a form of relaxation that pretends to be work (if it isn't even a form of work done to relax from work). This hobby again belongs to the wider problematic of phantomlike things. But this connection really doesn't need to be spelled out: It reveals itself to anyone who takes a closer look.

112. [English in the original.—Trans.]

113. [English in the original.—Trans.]

114. [*κατά συμβεβηκός:* happy accident.—Trans.]

115. Too much, too many. Too many waters flow
superfluously out of the night.
Yes, much-too-much world was created,
too many coasts that all seek a mention,
too many winds that blow in vain—
Who could possibly list or exalt this many things?
What cartographer could chart the nameless
reefs on the ocean floor,
the veins of gold that are yet unseen,
the constellations of stars still in need of a name—
in vain, only to mock us and suddenly appear,
is this overflowing abundance here.
And even if there were a man, who did not miss a thing,
a man who listed everything,
who would count him in the end,
who would add him to a list?
Whose gratitude could he solicit? Or even just a listening ear?
Even he is too much! His praise but noise!
Too much! Drop the sails,
pull the curtains closed in front of me!

From "The Delirious Columbus" [by Günther Anders—Eds.]

116. The word "*einräumen*" here first and foremost means "to concede," "to cede"—that is, to give ground and yield to a factually acknowledged superior power. "*Ich räume mich ihm ein*"—I give way to it. In a second phase, however, this unfreedom is at least partially overcome: Cults and philosophical systems are means by which a designated place for this power is created. This localization may still acknowledge the power, but it is now also confined: The deity becomes the captive of its temple; the *κακὸν* turns into the prisoner of its place in the system.— Today, "*einräumen*" has exclusively come to mean: to position something in such a way that it is always available for me or doesn't bother me. "*Ich räume mir etwas ein*"—I arrange things for myself.

117. This astonishing sentence, a passage from a letter by young Rilke (1904), does indeed sound like a description of an eschatological state in which all matter

has been elevated into something finished. Because Rilke omits mentioning the production process by which he believes this condition to have been established, his description is overly decorative and highly pretentious; or, rather, it directs our imagination in a falsely positive way: It has us think of aging wines and jewelry-making, of production processes that are as delightful as they are untypical. His dream is nevertheless nothing other than a dream of the total rape of world-matter [*Weltstoff*]. And even if the image seems to present the alchemistic fantasies of a gilded world as virulent once again, this was only possible because they are reminiscent of the eschatological ideas of economic ontology. Sentences like this and especially ideas such as Nietzsche's notion of the "*Apollonian*" need to be newly assessed against the background of economic ontology (Erich Heller was right to point out that this astonishing sentence by Rilke cites Nietzsche almost verbatim). Nietzsche, by introducing the two mythologizing forces of the "Dionysian" and "Apollonian," created an entirely new version of the conceptual coupling of "matter and form." This, however, is something that cannot be recognized until one considers how in the age of industrialism "matter" (= the world of raw material) and "form" (= product) began to attain a pragmatic significance within the world that earlier metaphysicians couldn't have dreamed of.— The Rilke passage is reproduced from Erich Heller, *The Disinherited Mind* (1952). [Anders here seems to conflate Heller's commentary with the Rilke passage in question. My translation mirrors Anders's presentation and reproduces the relevant passage from Heller's book without the quotation marks that Heller uses to identify Rilke's words in both the original English edition and the German version of his book from which Anders quotes. See Heller, *Disinherited Mind,* 110.—Trans.]

118. "Language is the house of being": This assumption is not just entirely unfounded but also anthropomorphic in a way that only gets more curious (but not better) by appearing in disguise. For is it not curious how the "unhoused" human foists its own need for shelter and housing onto being as such? Is it not curious that the human convinces itself that it needs to be the shepherd of being or its host because it can no longer be its guest?— No, it is categorically only ever an individual entity that wants to have a "house"—be it a snail, a human, or a family—only ever an entity that is detached and individual. And it only feels this desire precisely because it is detached, lost, unprotected, and hence too little at home in the wide expanse of the world. The world itself, not to mention being, is never in need of a house. It has other "concerns" than looking for a home (that is, if it even has concerns at all). [See Heidegger, "Letter on Humanism," 239.—Eds.]

119. [See Scheler, *Human Place in the Cosmos.*—Eds.]

120. [The German here reads "Was sie aber, ehe sie gemein geworden, gewesen, das verdämmert in wesenlosem Scheine," and possibly channels a passage from Goethe's "Epilog to Schiller's 'Song of the Bell,'" published on occasion of Schiller's death. The passage speaks of the progression of the poet's spirit into the eternal realm of truth, goodness, and beauty, and is reproduced here both in German and in Edwin H. Zeydel's translation, which informed my rendition:

> Indessen schritt sein Geist gewaltig fort
> Ins Ewige des Wahren, Guten, Schönen,
> Und hinter ihm, in wesenlosem Scheine,
> Lag, was uns alle bändigt, das Gemeine.

Meanwhile his titan soul did forward stride,
In Goodness, Truth, and Beauty found duration.
Behind him lay, to empty shadows faded,
The mean and base, which keeps us all degraded.

See Goethe, "Epilog zu Schillers Glocke," 151–52.—Trans.]

121. A substantial portion of today's abstract painting is preoccupied with inventing such *angels of the Industrial Age*—that is to say, creations devoid of a body. The high level of popularity of these drawings, which consist only of wiry outlines that leave most of what they depict entirely without substance, would remain utterly incomprehensible if this style were a mere product of artistic play.

122. [This is a direct quote from *The New York Times*, and the original in-text reference provided in the German has been amended accordingly. Anders provides a German translation of this exact quote, but it is attributed to a *New York Herald* article dated September 13, 1954, which could not be identified. The words cited here with a minor grammatical modification are taken from "Broadcast Trials Scored by Medina," *New York Times*, September 3, 1954.—Trans.]

123. ["Im Anfang war die Presse / und dann erschien die Welt." See Kraus "Couplet des Schwarz Druckers." The English translation is taken from Reitter, *Anti-Journalist*, 23.—Eds.]

124. See the first essay in this collection for more on this reference to Plato.— In the United States, ushers can already tell a thing or two about the many hundreds of thousands who don't really care to be present at boxing matches or football games. Because they were designed to be transmitted, these original events already seem tinged by the unreal, as their realization seeks to emulate a preexisting idea; that is, the idea of the most perfect broadcast. There will, of course, always be aficionados who, full of contempt for copies, will continue to only ever enjoy original bloody noses, just as there are aficionados who, out of scorn for reproduction, only ever go to Padua to see their Giottos. But these snobs will soon only survive as exceptions who prove the rule.

125. This expression originates in the gramophone industry, where the hair-raising confusions that now shape the relationship between "original" and "copy" are more clearly visible than anywhere else. In this industry, the term "mother"—that is, "mother-matrix"—refers to the master copy. It is a copy that is itself the reproduction of a voice, which reproduces a musical composition. But, as the term "mother-matrix" (i.e., "mother-mother") indicates, this reproduction of a reproduction is now treated as the "original" and distinguished from the "son"—that is, the "stamper"—that is created from it. But each such son, each such reproduction of a reproduction of a reproduction, now itself serves as the primary matrix for the stamping of discs. Put otherwise, the son now becomes the mother of all the vinyl discs that are mass-produced and appear on the market to act as templates meant to shape our taste.

126. [The reference is to Wilhelm Dilthey's "The Origin of Our Belief in the Reality of the External World and Its Justification" (1890).—Eds.]

127. [It is unclear whether this is meant to be a reference to a specific text, but the ensuing imagery could be a reference to Hans Sachs's highly influential poem "Schlaraffenland" ("Sluggard's Land") (1530).—Eds.]

128. [Possibly a reference to the following passage from Sachs's "Sluggard's Land" (here in William Leighton's translation):

In every lake.
The finest fishes, one can take;
Boiled, roasted, salted, stuffed, or fried,
Along the shore they slowly glide.
So one, with ease, can take them up.
And on their daintiness may sup.
Roast fowl and geese and pigeons fly
So low that any passer-by
May catch them; if too idle he,
Into his mouth, most pleasantly
They come. Pigs, roasted, run about.
Each with green parsley in his snout.
And, in his back, a knife with which
To cut a slice quite fresh and rich. (36)—Eds.]

129. This is what we were alluding to at the outset of our investigation when we called today's consumer an "*idealist*" (this happened in the context of a discussion of "the idealist" as a person who only knows "their own world" but not the "world" as such).

130. [The German here invokes a famous passage from Nietzsche's *Thus Spoke Zarathustra* and has been translated accordingly: "The stillest words are those that bring the storm. Thoughts that come on the feet of doves steer the world" (117).—Trans.]

131. [This phrase is in English in the original throughout.—Trans.]

132. [English in the original.—Trans.]

133. [English in the original.—Trans.]

134. [English in the original.—Trans.]

135. [English in the original.—Trans.]

136. The term "creative" shouldn't be a cause for irritation here: Against the backdrop of an everyday routine of consuming ready-made products even the most modest attempt at doing something oneself already feels like an act of creation, or at least Michelangelesque.

137. [English in the original. This may be a reference to Elmer Ferris's book *Who Says Old!* (1933), which was reviewed and advertised with the tagline "Life Begins at Seventy."—Eds.]

138. [English in the original.—Trans.]

139. [English in the original.—Trans.]

140. [English in the original.—Trans.]

141. [The German here reads "die Jungfrauen der Weddas," and it is unclear what cultural group or indeed ritual this refers to. While the Vedda people of Sri Lanka were part of the wider imaginary of "the primitive" in early twentieth-century German thought—for instance, in the work of the pioneer of Gestalt psychology Max Wertheimer—no accounts of the kind of rituals suggested here could be identified.—Eds.]

142. Because the number of those who have this gap surgically removed just like V. far exceeds the tiny number of matrixes that are needed in film, there are now thousands of phantoms living in California who in no way reveal who they were in their previous life and how they had once looked back then. Never given the opportunity to become matrixes, they continue to look like phantoms for a

while and "temporarily" fill jobs in drugstores or as carhop girls—holding on to the illusion that it might not be too late to get a chance to serve as an illusion—until, that is, they renaturalize under their everyday pressures and their old nature reappears from beneath the phantom-glamour.

143. [The German for the section heading reads, "Nicht der Betrachter, der Betrachtete ist beträchtlich."—Trans.]

144. [English in the original.—Trans.]

145. On this point, see "On Promethean Shame," the first essay in this volume.

146. *Wilhelm Meister's Apprenticeship* provides a classic representation of the connection between "social hierarchy" [*Sozialgefälle*] and "image." As someone born into a bourgeois household and thus excluded from living a "great" life (the only kind that counts), Wilhelm Meister obtains this for himself by means of a "formative education" [*Bildung*], by means of representing and imagining a "great" life—that is, by performing plays.—

147. [The text here cites the opening line of "The Song of Mignon" from Goethe's *Wilhelm Meister's Apprenticeship* ("So laßt mich scheinen, bis ich werde"), which is also reproduced (and inverted) in the epigraph to this section. See Goethe, *Wilhelm Meister's Apprenticeship*, 316.—Eds.]

148. [This passage echoes the language of *Being and Time* and thereby adds a further layer to Anders's recurrent discussion of Heidegger's insistence that human existence—*Dasein*—can only be interrogated with the help of the personal question "who" and never with the impersonal "what is" of empiricism. —Trans.]

149. [English in the original.—Trans.]

150. [*ἐσθλός*: noble. This (folk) etymology may have been lifted from Nietzsche, who makes this point in *On the Genealogy of Morality: ἐσθλός* means, "according to its root, one who *is*, who has reality, who really exists and is true; then, with a subjective transformation, it becomes the slogan and catch-phrase of the aristocracy and is completely assimilated with the sense of 'aristocratic'" (14).—Eds.]

151. Yet we must also consider that every mass-product is also a reproduction of the model or prototype that gave rise to it. Every model is thus only a model for reproductions; and the better the model, the higher the number of reproductions and the more successful its mass-reproduction becomes.—

152. [English in the original.—Trans.]

153. [English in the original.—Trans.]

154. [English in the original.—Trans.]

155. [English in the original.—Trans.]

3. Being without Time

1. [The German text makes use of the French title *En attendant Godot* throughout.—Eds].

2. [The text here plays on the literal meaning of the Latin *abstraho*—"to remove forcibly, detach, split"—which Anders also plays on in earlier texts. See, for instance, "Pathology of Freedom," 279.—Eds.]

3. Viewed from this perspective, the low suicide rate in concentration camps that is so often deemed astonishing is no longer surprising.

4. [The text is here playing on Heidegger's demarcation of human existence (*Dasein*) as "being-in-the-world" from the mode in which physical objects exist. This is famously discussed in section 12 of *Being and Time*, with the example of the chair's inability to touch the wall (since "touching," according to Heidegger, is a mode of relation that is exclusive to *Dasein*). See *Being and Time*, 51ff.—Eds.]

5. [This reference could not be identified. It may be a playful variation on Molière's play *Le Cocu imaginaire* (*The Imaginary Cuckold*).—Eds.]

6. [See Beckett, *Waiting for Godot*, 6, 41.—Eds.]

7. [The reference is to Kafka's short story "An Imperial Message."—Eds.]

8. [During the 1940s, Heidegger mobilizes this line from Hölderlin's poem "Patmos" in a number of texts. It is unclear, therefore, what specific deployment(s), if any, Anders has in mind here, although "What Are Poets For?" (1946) seems a likely source. The translation of Hölderlin's poetry is here reproduced from the effective translation of this line in Heidegger's later "The Question Concerning Technology" (35), a text that postdates Anders's essay on Beckett.—Eds.]

9. On this point, see chapter 2 of the author's *Franz Kafka*. [The reference is to Anders's *Kafka, Pro und Contra* (1951), which appeared in English translation in 1960 under the title *Franz Kafka*.—Eds.]

10. The "transient" passing moment is here not a mere episode in "time"; it becomes synonymous with time itself, so much so that it is now time that turns into a passing incident.— Our daily existence doesn't just amount to a continual "being in time" either; it is succeeded over and over again by the intermezzo that nontemporal sleep represents. Or, from the perspective of sleeping, our existence is again and again interrupted by "passing" stretches of time called "being awake."

11. [The German here appears to quote from Elmar Tophoven's 1953 German translation of *Waiting for Godot*, yet Anders rearranges both the logic and sequence of the dialogue in a way that attributes words spoken by Vladimir to Estragon. My translation draws on the corresponding passage in the English version of *Waiting for Godot*, but it alters the pattern and wording to Anders's pattern. See Beckett, *Waiting for Godot*, 55. Anders's original German reads, "'Wie wär's,' schlägt Estragon vor, 'wenn wir uns freuten? Es ist doch schon etwas?' Und als Wladimir zurückfragt, was dieses 'etwas' bedeuten solle, antwortet Estragon: 'Etwas weniger.'"—Trans.]

12. [Beckett, *Waiting for Godot*, 27.—Eds.]

4. On the Bomb and the Roots of Our Apocalypse Blindness

1. [In its development of the concept of "apocalypse blindness," the German text heavily relies on tropes that associate an unwillingness or inability to perceive, comprehend, or consider the apocalypse with being deaf and blind. I tried to soften or avoid these associations wherever possible. The analogies to vision the German relies on, however, made alternative translations of "Apokalypseblindheit," such as "apocalypse imperception" or "apocalypse avoidance," awkward and difficult to integrate, and so the phrase "apocalypse blindness" was retained.—Trans.]

2. [The German here reads "zum 'man'-Sein verurteilt," a phrase that references Heidegger's discussion of "das Man," which is usually translated as "the They" (see, for instance, sections 25–27 of *Being and Time*). Rather than inserting

cumbersome Heideggerian vocabulary, the translation here adopts language Anders uses in a 1948 critique of Heidegger, which he published in English. For Anders, "das Man" denotes the mode of existence of the "average man, living in the *doxa* and feeding on average and pre-chewed vocabulary and emotion," and his essay is highly critical of Heidegger's portrayal of this mode of being and attitude toward it. See "On the Pseudo-Concreteness of Heidegger's Philosophy," p. 343.—Trans.]

3. [It is unclear whether "faustischer Mensch" is meant to be a specific reference, but my translation here mirrors translations of Oswald Spengler's work, which makes wide use of this phrase. See, for instance, *The Decline of the West* (1922).—Trans.]

4. The antithesis between the "Apollonian" and the "Dionysian," for instance. The Apollonian bliss of turning the finite entity one was into something finished and complete had its counterpart in the intoxicating Dionysian frenzy of exploding the boundary of finitude. Yet, since we are no longer finite, and since we have lived through this "explosion," this contrast has become unreal, an illusion.

5. Time and again, we can find our apocalyptic situation portrayed as a "*being toward the end*," and this is usually followed by some remark about how prophetic this Heideggerian phrase was when it was coined thirty years ago. Such remarks are totally vacuous. The very last thing Heidegger had in mind was a shared eschatological situation, let alone the prospect of collective suicide. And in the decade after the First World War, when *Being and Time* was written, no one would have had any cause to think such things. As hellish as the First World War was, it didn't yet pose a threat to the globe. No, the experience that led to Heidegger's philosophy was the experience of the loneliness of existence, the experience of the soldier who ceaselessly found himself confronted by death.— Heidegger also didn't mean to say that every one of us must die (rumor has it this was already known before him). What he attempted, rather, was to turn the confrontation with death into a part of life itself, even into its sole and absolute characteristic. That is to say, he sought to transform the "halt!"—the call that might at any moment confront *Dasein* with the fact that its time is up—into something *Dasein* could actively "hold on" to; no, he even tried to turn it into its sole support. In short, for him, *Dasein* was meant to become a conscious "*moriturus*" and thereby enlist in an "existential project."*— But as indicated already, all this was only ever meant for individual home use—that is, for individual mortals as individuals. Which just goes to underscore again that Heidegger's philosophy is but a final version of individualism—the macabre version—in which the individual is exclusively constituted by the fact that "*no one can take their death away from them.*"**

Yet it is easy to understand how the idea of applying this "being toward the end" to our collective situation today was able to arise. Heidegger's concept simply transposes the anticipation of apocalypse into the language of the solitary individual. And because the apocalypse was merely substituted for the notion of one's own death, it was inevitable that, the moment the pressure of the bomb's total danger was felt, individual death was translated back into the "language of all of humanity." But just because something is obvious doesn't mean that it is true. There is no evidence whatsoever that humanity, taken in its entirety, is mortal in the same way that individuals are; and neither is there any evidence to suggest that humanity's death is just as inescapable. At any rate, those who apply this Heideggerian formula to today's situation thereby also automatically turn the catastrophe into something positive—an opportunity for humanity to "become authentic." *[My translation of *existentiell werden* as "existential project" here echoes the title of section 53 of *Being and Time,* "Existential Project of an Authentic Being-toward-Death."—Trans.] **[To aid understanding, I have here simply cited Heidegger's famous sentence—"No one can take the other's dying away from him" (*Being and Time,* 223)—rather than translating Anders's circumscription.—Trans.]

6. [The translation here echoes Ecclesiastes 3:15, which appears to be the closest approximation of Anders's version ("Alles wird gewesen sein"). Guterman opts for the more accessible "There is nothing new under the sun" (Ecclesiastes 1:9), which would be a possible alternative.—Trans.]

7. [Both "responsibility" and "response" are in English in the original.—Trans.]

8. [This line is lifted from Norbert Guterman's translation.—Trans.]

9. Even the *laissez-faire* principle assumed that free, unguided competition was the best—if not only—means by which a goal could be achieved. At times, it was even deemed unnecessary to pursue a final (economic) goal, for this goal would pop up all the more reliably if matters were left to a kind of preestablished harmony between all various activities. It is not impossible that the deflection away from the goal, a deflection that characterizes our age, has its deepest root in this linking of freedom (of initiative) and trust (in a preestablished mechanics of means).

10. [The reference is to James Thurber and E. B. White's *Is Sex Necessary? Or: Why You Feel the Way You Do,* a bestselling spoof of psychoanalysis first published in 1929.—Eds.]

11. [These qualifications are not in the German original but have been added here because the English word "means" doesn't hold the wide range of meanings the German word *Mittel* does. To aid the reading experience, I have occasionally reverted to these alternatives or else used them in addition to "means" (e.g., "means and devices").—Trans.]

12. The idea of an ontologically unclassifiable object was anticipated by Kafka in his description of Odradek.— Many examples of "abstract sculpture" likewise hint at this new type of "monstrosity," and they do so to a much higher degree than abstract painting. Rather than merely portraying something indefinite, these sculptures are solid pieces of reality that at the same time positively exist as something indefinite.

13. The great efforts invested in constructing the bomb during the war were largely a product of the need to do so before Hitler, who had made mass-liquidation his core principle. But it's terrible that this preventative measure has since "caught the enemy's disease"—that the mass-murders of Nagasaki and Hiroshima (but also the "tests") have turned into twin events of Hitler's organized mass-liquidations.

14. [English in the original. This is possibly a reference to Operation Alert, which was conducted across fifty-seven cities on June 15, 1955, as part of a nationwide rehearsal for a nuclear attack against the United States.—Eds.]

15. The inscription on the memorial stone in Batavia, commemorating this first victim of the hydrogen bomb, reads:

You little fisherman,
we don't know whether you had merits.
(Where would we be if everyone had merits?)
But you had worries like us,
like us, somewhere, the graves of your parents,
somewhere, on the shore, a woman who waited for you,
and at home, the children who ran to meet you.
Despite your worries
you thought it was good to be there.
Just like us. And you were right, Aikichi Kuboyama
You little fisherman, Aikichi Kuboyama
even if your foreign name does not tell of merit,
let us learn it by heart for our brief term
Aikichi Kuboyama.
As a word for our disgrace
Aikichi Kuboyama.
As our warning call
Aikichi Kuboyama.
But also,
Aikichi Kuboyama,
as the name of our hope: for whether you
preceded us in your dying or only
departed in our stead—
that depends only on us, even today,
only on us, your brothers,
Aikichi Kuboyama.

(*Sydney Chronicle,* March 3, 1955)

[We here reproduce Benjamin Steege's translation of this poem, which is listed as ["Epitaph for Aikichi Kuboyama"] among Anders's publications in the bibliography. See Steege, "This Is Not a Test," 336. Anders presents this poem as taken from *The Sydney Chronicle,* yet no such paper existed in the 1950s. It is worth mentioning, however, that during his years of exile, Anders regularly contributed poems in response to world events to newspapers. One example of such a poem is provided in note 38 in chapter 1.—Eds.]

16. *Science,* July 1956. [Four issues of *Science* were published in July 1956. They feature multiple contributions on the effects of radiation along with public summary reports produced by bodies such as the U.S. National Academy of Sciences. Anders here seems to collate information from across multiple sources, including "Pathologic Effects of Atomic Radiation," which touches on effects of radiation on the thyroid glands of children, and "Meteorological Aspects of Atomic Radiation," which discusses the concentration of iodine-131 in the atmosphere.—Eds.]

17. [See Henrik Ibsen's *Ghosts* (1881).—Eds.]

18. [My translation. This reference could not be identified.—Trans.]

19. [English in the original.—Trans.]

20. [English in the original.—Trans.]

21. [English in the original.—Trans.]

22. [English in the original.—Trans.]

23. [English in the original. The reference is to Franklin Roosevelt's State of the Union Address on January 6, 1941, known as "The Four Freedoms Speech": "The fourth [freedom] is freedom from fear, which, translated into world terms, means a world-wide reduction of armaments to such a point and in such a thorough fashion that no nation will be in a position to commit an act of physical aggression against any neighbor—anywhere in the world." See Roosevelt, "President Franklin Roosevelt's Annual Message (Four Freedoms) to Congress (1941)."—Eds.]

24. [See Grimm and Grimm, "Story of a Boy Who Went Forth to Learn Fear."—Eds.]

25. [English in the original.—Trans.]

26. [English in the original.—Trans.]

27. [English in the original.—Trans.]

28. Because we only ever go along with things but never act ourselves, we modern conformists no longer know any *courage:* we're at best reckless or brave. *Reckless*: that is, ready to risk our lives for something arbitrary—something that we are either persuaded into believing is important, or else, leave for others to judge. This proves a total lack of moral independence.— Or else we are *brave*: capable of enduring a terrible situation—one that others have brought us into, but which we might not even hold them responsible for. Even though this situation may demand superhuman efforts, it again coincides with a lack of independence. If even cowards have no choice but to persevere with heroic grit, it may in fact be more difficult and take much more courage *to not be* a hero.— Courage would amount to a willingness to take the risk of assuming personal responsibility, even under pressure and even under the weight of public opinion (which, for the most part, is nearly impossible in situations of dictatorial rule). But where it is attempted nevertheless, this act *eo ipso* attains the stigma of treason (as was demonstrated, for instance, by the fate of those who attempted to assassinate Hitler). In the age of conformity, courage almost becomes synonymous with treason—or at least renders one suspect in the court of public opinion [*Mitwelt*].

29. That which is too big is not just not imaginable, by the way; it is also nearly impossible to remember. Perhaps it cannot even be "experienced": The *sight* of one thousand murdered people is no more horrific than the sight of hundreds. Counting stops on the other side of shock.—

30. And even if it is, what matters here is not to discover new psychological facts but to find the roots of our failure. And these roots, moreover, are not simply to be "found."

31. The expression "*vorstellen*"—"to imagine, to picture, to come up with ideas"—actually loses its validity in this context because its prefix "*vor*" signals an act of anticipation (usually the kind of planning that precedes the actualization of something in the course of daily production). But we are dealing with a perfectly counter-Platonic situation here, one in which already actualized objects and situations precede their *eidos*—a situation, therefore, in which they are here even before they are anticipated—no, without ever being anticipated by "an idea that *pre*conceives" their size and consequences. So much so that the act of coming up with ideas no longer amounts to "*vorstellen*" ("to mentally place something in front of oneself") but to "*nachstellen*" ("to be chasing *after* something"). It now

represents an attempt to keep up with one's own products, with the incalculable power in one's possession—the power one has become through one's own acts, a power that has become too much. It is an attempt to keep up, to somehow stay a match to oneself.

32. [English in the original—Trans.]

33. It is highly conspicuous that the language with which humans once signaled that they are struggling with themselves is becoming unfashionable (or is already out of use). The expression "to wrestle with oneself," for instance, which was readily employed even one generation ago (even by the youth), now already sounds old-fashioned, pathetic, and unbelievable.

34. [The reference appears to be to a passage from "The First Elegy," reproduced here in William Gass's translation:

> But should you long like this, sing of love's ultimate lovers:
> the fame of their feeling is not yet immortal enough.
>
> Those—the forsaken—you envied them almost, they so
> outstripped
> all love-appeased lovers in loving. Begin
> continually to accomplish their unachievable praise. (190)

This reference has added significance here as Anders published a coauthored essay on Rilke's *Duino Elegies* together with Hannah Arendt in 1930, which includes a detailed mediation on the nature of love (and also discusses the above extract). See Arendt and Anders, "Rilke's *Duino Elegies,*" 14.—Eds.]

35. See chapter 1, section 5.

36. The claim that belief in progress is "over" is far too categorical. It is still flourishing in the United States and Russia, and its victory march is only just beginning among the populations of developing countries. Yet it is undeniable that it is undergoing a profound change: the term "progress" is slowly morphing into the expression "project"; it is losing the element of "iron necessity." The prognosis "the end of progress" was characteristic for the Europe of the collapse in 1946, but viewed from a global perspective, it is incorrect.—

37. By the time our ancestors deistically diluted their idea of God, they had already transformed the devil into an allegorical figure. When they declared God "dead," they had already killed the devil, and when they entrusted heaven to music, hell had long since become a setting for theater. In our parents' intellectualized version of religion, all fear of hell had already been lost (even among those who steered clear of explicit atheism)—it was lost so completely, in fact, that it was no longer even noticed to be missing.

38. In places where the twentieth-century bourgeoisie stopped believing in progress, the idea of "permanent improvement" was not countered by that of a "bad ending," but was replaced by the (to varying degrees metaphorical) notion that the present is "*hell.*" This was already true for Strindberg, and since then, nothing has fundamentally changed (in Céline, Kafka, early Sartre). Yet although these authors had already given up hope in progress, they still relied on it polemically. Unable to imagine a future that didn't amount to "progress," they stopped believing in the future altogether as they lost their faith in progress. When discarding the notion of a "better future," they thus discarded the concept

of the future as such. This transformation is typical of a generation with roots in a progressive bourgeoisie, while having also lost all faith in the advancement of its class. The futurelessness that can be found in Beckett's Godot-world, for instance, has no equivalent in any Christian or political conceptualization of the future. What is decisive here, however, is that this futurelessness didn't flip into an anticipation of the apocalypse, but instead construed the present as the *nunc stans* of hell: *Nihilists, too, are blind to the apocalypse.*

39. When someone still struggles with the fact of dying despite all this and now begins to beleaguer the philosophy of progress with their questions, this philosophy emphatically fails (and this applies to both its socialist and bourgeois variants). Yet this doesn't happen all too often: By the time the children of progress arrive at the idea of asking questions, the questioning part of their mouths has already been stuffed with optimistic answers. The power of a worldview doesn't just show itself in the answers it provides, but also in the questions it knows to shut down.

40. [*partir c'est mourir un peu . . . mourir c'est partir un peu:* to leave is to die a little . . . to die is to leave a little.—Trans.]

41. [English in the original.—Trans.]

42. [This section plays on the word *Zukunft* (the future), which is a compound of the verb *kommen* (to come) and preposition *zu* (to, towards). In German, the word for future thus literally means "that which comes toward us," "that which arrives."—Trans.]

43. In planned economies—that is, in places where the word "project" has replaced "progress"—this has fundamentally changed. The future has here become a kind of "space" in which a target output is materialized. As a consequence, time has transformed from a "*mode of perception*" into a "*mode of production.*" Where this kind of future prevails, the classic questions (like whether it is possible to "foresee" the future) have become baseless, as the future is already nothing but a schematic of foresight.— *Cum grano salis,* one might thus say that the idea of progress has now reverted back into meaning "providence," the term from which it originates. This time around, however, it signals *human providence.*

44. [The reference is unclear but may be to "the seven generation principle" of the Iroquois, which is introduced here in words by Oren Lyons: "We are looking ahead, as is one of the first mandates given to us as chiefs, to make sure and to make every decision that we make relate to the welfare and well-being of the seventh generation to come, and this is the basis by which we make decisions in council. We consider: will this be to the benefit of the seventh generation?" ("Iroquois Perspective," 173).—Eds.]

45. [The epigraph translates as "an imagination broad and expansive enough to embrace the universe as if it were a city," and it echoes the following passage from Montaigne's "On Educating Children": "We are all cramped and confined inside ourselves: we can see no further than the end of our noses. When they asked Socrates where he came from he did not say 'From Athens', but 'From the world.' He, whose thoughts [*imagination*] were fuller and wider, embraced the universal world [*univers*] as his City, scattered his acquaintances, his fellowship and his affections throughout the whole human race, not as we do who only look at what lies right in front of us" (200).—Eds.]

46. [The use of the words *angehen* and *eingehen* here accentuates a bodily

boundary that I have tried to capture with circumscriptions. Whereas *an-gehen* signals an external factor that becomes one's concern, something that touches one (i.e., is one's affair), the verb *ein-gehen* signals a process of crossing the divide between the outside and inside, denoting something that sinks in, gets to us, is comprehended, or properly grasped.—Trans.]

47. [The German here cites a colloquial saying "Was ich nicht weiß, macht mich nicht heiß"—often translated as "Ignorance is bliss"—and then coins its own version in response: "Was ich nicht kann, geht mich nichts an."—Trans.]

48. [The text here begins to use the term *Betrieb*, which can be translated as "company," "factory," "plant," or "organization." It can also be used in a more general, abstract, and process-oriented sense as "operation." I have sometimes opted for "operation," not least because this also gestures toward the way the term has since been adopted in the context of computation—for instance, in the German word for a computer operating system (*Betriebssystem*), a connection that is anticipated by Anders's own use of machine interaction as an example.—Trans.]

49. [The German adjective *medial* (medial) carries the meaning "to correspond to the power and ability of the medium." I have incorporated this into the translation here to spell out the inference.—Trans.]

50. [Here and throughout this chapter, the German plays on the link between *wissen / Wissen* (to know, knowledge) and *Gewissen* (conscience).—Trans.]

51. The employee who feels deeply troubled in their soul because of their contribution at work is an entirely novel phenomenon. This figure didn't exist prior to the production of the atomic bomb. And this type of person will not emerge from one day to the next. The case of Oppenheimer demonstrates that in times of extreme conformity, moral scruples can themselves turn into a cause for scruples because such noncoordinated feelings seem morally suspect.— But even this form of scruple is still an exception today. The basic axiom is and remains *non olet*—the axiom that no form of work can be morally discredited by what it accomplishes. What makes this axiom so fatal is not that it lends even the most horrendous crimes the air of something innocent, but that it leads to downright nihilism. For when the greatest part of human activity (i.e., work) is excluded from the verdict of morality from the outset, nihilism is effectively ruling already.

52. [English in the original.—Trans.]

53. On the contrary, automatization represents the definitive replacement of conscience with the conscientiousness (reliability) of a mechanical function.

54. [The text here echoes Luke 23:34–38.—Eds.]

55. [The German here is *Bewandtnis-Charakter*, which evokes Heidegger's seminal discussions of "being-in-the-world." See, for instance, *Being and Time*, section 18: "Relevance and Significance: The Worldliness of the World."—Trans.]

56. [English in the original.—Trans.]

57. [These references could not be identified, but they may allude to projects such as Alfred Weber's *The Conquest of Nihilism* (1946), which was translated into English, or more likely perhaps, Rudolf Brunngraber's *Überwindung des Nihilismus*, which was published in Vienna in 1949. (Brunngraber, best known for his 1936 novel *Radium*, was a prominent scriptwriter and science fiction author whose work is infused by technocratic optimism.)—Eds.]

58. ["Die Anstrengung des Begriffs" is a famous phrase employed by Hegel

in the preface to the *Phenomenology of Spirit:* "What, therefore, is important in the *study* of *Science,* is that one should take on oneself the strenuous effort of the Notion" (35).—Eds.]

59. [The text here plays on the wording of Immanuel Kant's categorical imperative: "Act only in accordance with that maxim through which you can at the same time will that it become a universal law." See *Groundwork of the Metaphysics of Morals,* 31.—Trans.]

60. When we Europeans today speak of the outcomes of this collision as "a purely Asian phenomenon," this merely underscores the extent to which we are unable to recognize ourselves in the effects we are responsible for.

61. The sudden upheaval that affected Russian intellectuals in the nineteenth century is likely just the first precursor of upheavals yet to come. The trauma resulting from this sudden collision, which manifested itself as "nihilism," is just a precedent. It is far too rarely considered what it means that millions of non-Europeans, whose cultural history doesn't feature any traces of a naturalistic conception of the human and the world, are now suddenly confronted with this concept; just as one far too rarely reminds oneself that, for the majority of today's humanity, the epoch of natural science only began with the detonation of the atomic bomb.

62. It is highly characteristic that the nihilist's basic reference point, even when announcing the overcoming of nihilism, remains monism. When Nietzsche, for instance, believes he is accomplishing this through his yes to "the eternal return," this is nothing other than the affirmation of the glory of being part of nature's meaningless existence, which harbors no "ought." This is all the more plausible because the concept of the "eternal return" was quite evidently modeled on *the repeatability* of physical experiments and influenced by the concept of being that is operative in physics. The violent solution Nietzsche suggests is to involve oneself in this being, to leap into "the great noontide"* of that which meaninglessly circles on and on. Apart from its hymnal tone, this solution is in a sense no different from the desperate situation that usually causes the nihilist to despair (since he cannot deny, or believes himself unable to deny, that he, too, belongs to this world, devoid of meaning and orienting purpose). Nietzsche's impulsive solution ultimately amounted to turning the fact that nature is "beyond good and evil" into something that "ought" to be. *[The German here reads "hineinzuspringen in 'den großen Mittag'" and evokes a phrase that reappears across several of Nietzsche's works. My translation here adopts R. J. Hollingdale's rendering of the term in *Ecce Homo,* 66.—Trans.]

63. Lotze, "On the Being of Things," 36.

64. Kropotkin's attempt to solve this dilemma with the help of the phenomenon of "beneficial cooperation in nature" was philosophically weak, and it is no coincidence that it was undertaken in the country—yes, milieu—of nihilism. It was weak because even the fact that actions resembling morality can be found in nature is not enough to prove that morality has a binding character. [See Kropotkin, *Mutual Aid.*—Eds.]

65. I compiled a catalog of such "reversible associations"—the ones belonging to National Socialism—in 1940 in *Studies in Philosophy and Social Science.* [The reference is to Anders's review "Christoph Steding's *Das Reich und die Krankheit der europäischen Kultur.*"—Eds.]

66. [The German here makes use of the word *fassen* (to grasp, comprehend, hold, contain) in a way that gradually activates this multiplicity of meaning. —Trans.]

67. For a detailed discussion of these formulas, see the author's "Une interprétation de l'a posteriori" in *Recherches Philosophiques* (1934).

68. That the organizers of feeling especially liked to reach for nature categories such as "blood," "race," "organic," and so forth was, of course, no coincidence. This was done to sell the imposed system to the public, despite its horrendously artificial nature. The small word "nature" is the darling of terror.—

69. Besides the Promethean gradient, this prejudice is probably mainly rooted in *the self-deluding nature of habit and familiarity.*— What I mean to say is that a person who has become emotionally accustomed to a specific world-schema will no longer be able to even imagine an alternative one. That is, they will no longer be able to imagine that another world-schema would be possible; they might not even be able to acknowledge that it was once possible in the past. In brief, *habituation retroactively eternalizes whatever it has become used to.*

70. [The reference is to the opening monologue of Goethe's *Faust:*

> I've studied now Philosophy
> And Jurisprudence, Medicine,—
> And even, alas! Theology,—
> From end to end, with labor keen;
> And here, poor fool! with all my lore
> I stand, no wiser than before:
> I'm Magister—yea, Doctor—hight,
> And straight or cross-wise, wrong or right,
> These ten years long, with many woes,
> I've led my scholars by the nose,—
> And see, that nothing can be known! (41–42)—Eds.]

71. This also explains why cruelty holds such incredible significance for the nihilist: The one who inflicts torture is not just able to prolong his act of annihilation as long as he pleases; he also secures the victim as a witness.

72. This paradoxical "*junctim*" of a naturalistic worldview and the freedom movement, this coupling of "iron necessity" and "revolution," isn't over yet. Without this alliance, which has its origins in the European history of ideas, the doctrines of Soviet Russia, for instance, would even today remain incomprehensible.

Yet, over the course of the last century, as naturalism has triumphed everywhere in the guise of technology, this alliance is no longer as characteristic as it once was. Naturalism has become the universal worldview of our age; today, there is no country that isn't, in effect, naturalistic, none whose treatment of the world and of humans doesn't represent "naturalism *in actu.*" This, of course, doesn't mean to imply that naturalism is now everywhere propagated as an official philosophy. On the contrary, in some of the most powerful nations, it remains entirely unexpressed. But not because philosophy has "not yet" caught up with reality (this simple "not yet" is not where the ideological problem of our age is located). No, it is the other way around: Naturalism remains silent because it no longer even needs to be explicitly proclaimed as a philosophy in these highly industrialized countries. If naturalism doesn't reign in these places, it is only because its principles have already unambiguously prevailed as *modi operandi*. They rule without challenge and to everyone's satisfaction, and as such, they can afford to not be articulated in a programmatic philosophical manner. In these countries, protected by a defensive wall of laboratories and financed by naturalism itself, one can already once again allow oneself to propagate more or less bona fide humanistic conceptions of the world and the human ("*studium generale*"). And from this perspective, the naturalistic philosophies that are officially adopted by those countries that are still industrially underdeveloped, philosophies that might even be celebrated as doctrines of salvation, appear as "barbarism." Yet what is rejected with this word (or other words like it) is ultimately just the philosophical version of the program "science for the purpose of industrialization"—a program that no longer needs to be actively propagated because it is nothing other than the loud proclamation of the thing one is practicing oneself. If this philosophy is now attacked with such vehemence, it is not least because it represents the threat of industrial competition.

It seems certain to me that the official naturalistic philosophy that is either *still* reigning *today* (in Russia) or is *only just* beginning its career (in the "underdeveloped countries" that have been condemned to repeat the European nineteenth century) will automatically wither the moment a certain level of industrial saturation is reached. I also believe it certain that Marx's expectations with regards to metaphysics—that it will perish on account of being superfluous—will become true for its successor—that is, naturalism.

But once again, naturalism would here only vanish as an explicit or even official philosophy, and where it has already vanished, it has only done so as a philosophy. It reigns everywhere. For actual philosophy doesn't consist of words, but of the applied principles behind the treatment of the world and humans; it doesn't consist of what one believes the world or the human to be, but of the reasons why one makes oneself their keeper.—

73. [The reference is to the Deutscher Monistenbund, which was founded by Ernst Haeckel in 1906. Haeckel's highly influential work, which include *Die Welträtsel: Gemeinverständliche Studien über Monistische Philosophie* (1899) (published in English as *The Riddle of the Universe at the Close of the Nineteenth Century*) and a renowned collection of naturalist illustrations, *Kunstformen der Natur* (1899–1904) (published in English as *Art Forms in Nature*), played a significant role in shaping political and social movements and the wider perception of nature in Europe and beyond.—Eds.]

74. [Kosmos books were popular science booklets published by the Gesellschaft der Naturfreunde (Friends of nature club) from 1903 onward. Besides small booklets, the society also published the monthly magazine *Kosmos*, and substantial volumes called *Handweiser für Naturfreunde* (Guides for friends of nature).—Eds.]

75. The reason why it was only now (i.e., in 1940s France) that nihilism became a broad movement is that it was for the first time a perfect fit for the historical situation, and this for very specific parts of it. We should remind ourselves of the many totally disparate occurrences that could suddenly fit into its frame, and how many diverse phenomena it was able to simultaneously paraphrase:

1. The fact of Nazi terror, which made a mockery of any "ought."
2. The situation of the *Résistance* fighter, who sought to free himself from the "ought" in the fight against this terror. (This is why there is such a close relationship between this variant of nihilism and the philosophy of freedom.)
3. The hypocrisy of the bourgeoisie, which betrayed its own "oughts."
4. The situation of the once petit bourgeois intellectual who no longer knew what he "ought" to do and if he "ought" to do anything at all, since he had become a social nonentity.

It is thus perfectly understandable how a philosophy that simultaneously does justice to such a wide range of events, situations, and attitudes could turn into a dominant philosophy: Its arrival was overdetermined.

Editors' Postscript

All citations from German-language sources were translated by C. J. Müller.

1. See Günther Stern [Anders], "Spuk und Radio," *Anbruch: Monatsheft für moderne Musik* 12, no. 2 (1930): 66. Reprinted in Günther Anders, *Musikphilosophische Schriften: Texte und Dokumente*, ed. Reinhard Ellensohn (Munich: C. H. Beck, 2017), 248–51.
2. This figure was provided by the C. H. Beck publishing house.
3. See, for instance, among many other possible texts, Zygmunt Bauman, *Consuming Life* (Cambridge: Polity, 2007); Michael Hardt and Antonio Negri, *Multitude: War and Democracy in the Age of Empire* (New York: Penguin, 2004); and Bruno Latour, *Facing Gaia: Eight Lectures on the New Climatic Regime*, trans. Catherine Porter (Cambridge: Polity, 2017).
4. Bernard Stiegler, *The Age of Disruption: Technology and Madness in Computational Capitalism*, trans. Daniel Ross (Cambridge: Polity, 2019), 305; Jean-Pierre Dupuy, *A Short Treatise on the Metaphysics of Tsunamis*, trans. M. B. DeBevoise (Lansing: Michigan State University Press, 2015), 2; and Jean-Luc Nancy, *After Fukushima: The Equivalence of Catastrophes*, trans. Charlotte Mandell (New York: Fordham University Press, 2015), 19.
5. Some coordinates include Paul Van Dijk, *Anthropology in the Age of Technology: The Philosophical Contribution of Günther Anders*, trans. Frans Kooymans (Amsterdam: Rodopi, 2000); Veit Erlmann, *Reason and Resonance: A History of Modern Aurality* (New York: Zone Books, 2010); Günther Bischof, Jason Dawsey, and Bernhard Fetz, eds., *The Life and Work of Günther Anders* (Innsbruck: Studienverlag, 2015); Christopher John Müller, *Prometheanism: Technology, Digital Culture and Human Obsolescence* (London: Rowman and Littlefield, 2016); Rens van Munster and Casper Sylvest, *Nuclear Realism: Global Political Thought during the Thermonuclear Revolution* (Abingdon: Routledge, 2016); Christopher John Müller and David Mellor, eds., "Utopia Inverted: Günther Anders, Technology and the Social," special issue, *Thesis Eleven* 153, no. 1 (2019); Srećko Horvat, *After the Apocalypse* (Cambridge: Polity, 2021); and Babette Babich, *Günther Anders' Philosophy of Technology: From Phenomenology to Critical Theory* (London: Bloomsbury, 2022). A detailed, continually updated bibliography of individual articles and book

chapters on Anders's work is maintained by the International Günther Anders Society at https://www.guenther-anders-gesellschaft.org/bibliographie.

6. "Günther Anders to Max Rychner, 8 February 1947," quoted in Kerstin Putz, "Improvised Lives: Günther Anders's American Exile," in *Quiet Invaders Revisited: Biographies of Twentieth Century Immigrants to the United States,* ed. Günther Bischof (Innsbruck: Studienverlag, 2017), 235.

7. In the early years of broadcasting, radio sets "had very little in common with modern radios," resulting in an entirely different listening experience from today's. As Karl Christian Führer elaborates, "Up to 1926–27, the quickly expanding radio industry offered only sets that looked exactly like what they were: technical devices to receive and transform radio-waves. No effort was made to hide the technical character of the apparatus. Integrated loudspeakers were unknown; most listeners contented themselves with earphones." See Führer, "A Medium of Modernity? Broadcasting in Weimar Germany, 1923–32," *Journal of Modern History* 69 (1997): 733. For a discussion of the smartphone's various radio antennae, see Adam Greenfield, "Smartphone: The Networking of the Self," in *Radical Technologies: The Design of Everyday Life* (London: Verso, 2017), 16ff.

8. On this point, see Dominik Schrage, *Psychotechnik und Radiophonie: Subjektkonstruktionen in artifiziellen Wirklichkeiten, 1918–1932* (Munich: Wilhelm Fink, 2001), 221–22.

9. For a detailed discussion of the role radio played in the Nazi rise to power, see Maja Adena et al., "Radio and the Rise of the Nazis in Prewar Germany," *Quarterly Journal of Economics* 130, no. 4 (2015): 1885–1939.

10. Joseph Goebbels, "January 30, 1933," in *My Part in Germany's Fight,* trans. Kurt Fiedler (New York: Howard Fertig, 1979), 207–9. Nazi radio broadcasts are a frequent motif in Anders's writings, including *The Obsolescence.* The most sustained treatment can be found in "The Obsolescence of Privacy" in *The Obsolescence, Volume 2,* a text that links up directly to "The World as Phantom and Matrix": "Wherever totalitarianism emerges," as Anders condenses its central thesis, "the single individual is always its first 'occupied territory.' *Expansionism begins at home.*" See Anders, "The Obsolescence of Privacy," trans. Christopher John Müller, *CounterText* 3, no. 1 (2017): 27.

11. The precise circumstances that led Anders to abort his habilitation project at the University of Frankfurt are complex and difficult to reconstruct. We here cite a 1958 letter in which Anders asks Hannah Arendt, to whom he was married during his time at Frankfurt, to confirm his account of events to enable him to claim compensation from the German state for his loss of earnings due to his thwarted academic career. See Hannah Arendt and Günther Anders, "Günther Anders an Hannah Arendt, Wien, 22.9.1958," in *Schreib doch mal hard facts über Dich: Briefe 1939 bis 1975,* ed. Kerstin Putz (Munich: C. H. Beck, 2015), 77. Anders's habilitation project—a phenomenological investigation of the musical situation—was posthumously published in 2017, and it directly informs the conceptualization of "the plasticity of feeling" that is part of the bomb chapter in *The Obsolescence* (chapter 4). For a detailed account of how Anders's philosophy of music intersects with *The Obsolescence,* see Benjamin Steege, "Techniques of Feeling," in *An Unnatural Attitude: Phenomenology in Weimar Musical Thought* (Chicago: University of Chicago Press, 2021), 143–61. Anders's relationship to music and his time at Frankfurt is discussed

in detail in Reinhard Ellensohn's postscript to the posthumously published philosophy of music. See Anders, *Musikphilosophische Schriften*, 336–80.

12. Konrad Paul Liessmann, "Between the Chairs: Günther Anders—Philosophy's Outsider," in Bischof et al., *Life and Work of Günther Anders*, 73–82.

13. Günther Anders, "At This Instant" (unpublished book outline), quoted in Kerstin Putz, "Activist Poetry versus Lyrical Action: Günther Anders on Poetry and Politics," in "Utopia Inverted: Günther Anders, Technology and the Social," ed. Christopher John Müller and David Mellor, special issue, *Thesis Eleven* 153, no. 1 (2019): 28. For a detailed account of Anders's exile poetry, see also Jason Dawsey, "Fragile Apprehension: Günther Anders and the Poetics of Destruction," in Bischof et al., *Life and Work of Günther Anders*, 21–34.

14. Anders completed a doctorate under Edmund Husserl in 1924, attended courses taught by Martin Heidegger, and briefly worked as an assistant to Max Scheler.

15. Helmut Plessner, *Levels of Organic Life and the Human: An Introduction to Philosophical Anthropology*, trans. Millay Hyatt (New York: Fordham University Press, 2019), xv.

16. Günther Stern [Anders], "The Pathology of Freedom: An Essay on Non-Identification," trans. Katharine Wolfe, *Deleuze Studies* 3, no. 2 (2009): 286.

17. For a detailed account of the evolution of Anders's "negative anthropology" and the way these early writings informed *The Obsolescence*, see Christian Dries's postscript to Anders's collected writings on philosophical anthropology, "Nachwort: Von der Weltfremdheit zur Antiquiertheit des Menschen," in Günther Anders, *Die Weltfremdheit des Menschen: Schriften zur philosophischen Anthropologie*, ed. Christian Dries with Henrike Gätjens (Munich: C. H. Beck, 2018), 437–500.

18. On this point, see Emmanuel Levine, "Les traductions du jeune Levinas," *Bulletin D'Analyse Phénoménologique* 19, no. 2 (2023): 1–30, esp. 18–23.

19. Max Horkheimer, "Zu Günther Anders' Thesen über 'Bedürfnis,' 'Kultur,' 'Kulturbedürfnis,' 'Kulturwerte,' 'Werte'" (discussion minutes, August 25, 1942), in *Gesammelte Schriften, Band 12: Nachgelassene Schriften 1931–1949* (Frankfurt: S. Fischer, 1985), 579. See also appendix I of the bomb chapter (chapter 4), which sets out the connection between the position presented in the early work and the logic of the Promethean gradient in some detail.

20. Anders would later note that *The Obsolescence* could not have been written without the firsthand experience of mechanized production. See, for instance, Günther Anders in Elke Schubert, ed., *Günther Anders antwortet: Interviews und Erklärungen* (Berlin: Tiamat, 1987), 38. A particularly vivid account of one of the many "odd jobs" Anders worked in exile is provided by the 1941 Californian diary "Washing the Corpses of History: The Hollywood Costume Palace," which is available in an English translation by Christopher John Müller; see *Modernism/Modernity Print+* 5, no. 4 (2021), https://doi.org/10.26597/mod.0185.

21. Anders, "Pathology of Freedom," 284ff.

22. Anders, "Spuk und Radio," 66.

23. Günther Anders, "Was ist Surréalisme?," in *Schriften zu Kunst und Film*, ed. Reinhard Ellensohn and Kerstin Putz (Munich: C. H. Beck, 2020), 152.

24. Anders, "Was ist Surréalisme?," 154.

25. For Anders's detailed stylistic reflection on how a double take might manifest in words, see, for instance, some (untitled) 1948 notes "on poetics,"

posthumously published as "Über Rilke und die deutsche Ideologie," ed. Gerhard Oberschlick, in *Sans Phrase* 7 (2015): 127–28.

26. See, for instance, Günther Anders, "Über Brecht," in *Mensch ohne Welt: Schriften zu Kunst und Literatur* (Munich: C. H. Beck, 1984), 133–72.

27. Günther Anders, "Caricartoons: A Suggestion for a New Type of Animated Pictures," in *Schriften zu Kunst und Film,* 81.

28. Günther Anders, "George Grosz," in *Mensch ohne Welt,* 211.

29. This epigraph is adapted from a passage in "The World as Phantom and Matrix" (chapter 2).

30. Günther Anders, "Über Photomontage: Ansprache zur Eröffnung der Heartfield-Ausstellung 1938 in New York," in *Mensch ohne Welt,* 176ff.

31. Günther Anders, "Gelegenheitsphilosophie" (unpublished typescript), Nachlass Günther Anders, Literary Archives of the Austrian National Library, Vienna, LIT 237/W116, 45.

32. Günther Anders, *Die Antiquiertheit des Menschen, Band 2: Über die Zerstörung des Lebens im Zeitalter der dritten industriellen Revolution* (Munich: C. H. Beck, 2018), 463.

33. Anders, "Gelegenheitsphilosophie," 3.

34. Anders, "Gelegenheitsphilosophie," 54, 6.

35. Anders, "Gelegenheitsphilosophie," 66.

36. Anders, *Die Antiquiertheit des Menschen, Band 2,* 38–39.

37. Anders, "Über Photomontage," 179.

38. Günther Anders, *Besuch im Hades* (Munich: C. H. Beck, 1996), 39–40.

39. We are here following the lead of Henri Paucker, who presents the principle of inversion as the exemplary "corrective" Anders deploys. See Paucker, "Günther Anders," in *Deutsche Exilliteratur seit 1933,* ed. John M. Spalek and Joseph Strelka (Bern: A. Francke 1976), 1:228.

40. Harry S. Truman, "Statement by the President Announcing the Use of the A-Bomb at Hiroshima," August 6, 1945, National Archives, Harry S. Truman Library, https://www.trumanlibrary.gov/library/public-papers/93/statement-president-announcing-use-bomb-hiroshima.

41. Günther Anders, *Hiroshima ist überall* (Munich: C. H. Beck, 1982), xi.

42. Anders, "Gelegenheitsphilosophie," 41.

43. Anders, "Caricartoons," 74.

44. For a detailed exposition of the logic glossed here, see Müller, "Hollywood, Exile, and New Types of Pictures." Henri Bergson's *Laughter,* one of Anders's likely sources of inspiration, entails a lengthy discussion of caricature and the way it "exaggerates" to make a hidden cause "visible." See Bergson, *Laughter,* trans. Drew Burk (New York: Atropos, 2009), 27–33.

45. Anders, "Caricartoons," 75.

46. Cliff Kuang, *User Friendly: How the Hidden Rules of Design Are Changing the Way We Live, Work, and Play* (London: Penguin, 2019), 10.

47. Kuang, *User Friendly,* 45.

48. Wendy Hui Kyong Chun, *Programmed Visions: Software and Memory* (Cambridge, Mass.: MIT Press, 2011), 18.

49. See Günther Anders, "The Obsolescence of Privacy," trans. Christopher John Müller, *CounterText* 3, no. 1 (2017): 24.

50. This is exemplified, for instance, by the ease with which we misrepresent

our relationship to our technological devices in language. Even a simple sentence such as "I will call you later" profoundly obfuscates how little this technologically augmented "I" actually contributes to and comprehends this exercise of agency. It is the phone—encompassing hundreds of components, a telecommunications infrastructure, software, and service providers—that enacts the call on our behalf. And these hidden processes need not be as friendly to others or the environment as the user experience makes them seem. And even though we might "know" that these hidden processes are not as inconsequential as the user experience makes them seem, the idea that this knowledge could cause us to hesitate or feel scruples when interacting with the screen almost seems absurd. On this point, see Christopher John Müller, "From Radioactivity to Datamining: Günther Anders in the Anthropocene," in "Utopia Inverted: Günther Anders, Technology and the Social," ed. Christopher John Müller and David Mellor, special issue, *Thesis Eleven* 153, no. 1 (2019): 19. For detailed discussions of Anders and digital technology, see Christian Fuchs, "Günther Anders's Undiscovered Critical Theory of Technology in the Age of Big Data Capitalism," *tripleC* 15, no. 2 (2017): 582–611; Christopher John Müller, *Prometheanism: Technology, Digital Culture and Human Obsolescence* (London: Rowman and Littlefield, 2016), esp. 121–63; and Elke Schwarz, "Günther Anders in Silicon Valley: Artificial Intelligence and Moral Atrophy," in "Utopia Inverted: Günther Anders, Technology and the Social," ed. Christopher John Müller and David Mellor, special issue, *Thesis Eleven* 153, no. 1 (2019): 94–112.

51. See Kenneth Cukier and Viktor Mayer-Schönberger, *Big Data: A Revolution That Will Transform How We Live, Work and Think* (London: John Murray, 2009), 7.

52. See Günther Anders, *Die atomare Drohung: Radikale Überlegungen zum atomaren Zeitalter* (Munich: C. H. Beck, 2003), 194–203.

Bibliography

This bibliographical list is not part of the original German edition. It collects sources either directly quoted by Anders or used in the supplementary notes to identify possible references.

Adorno, Theodor. *Minima Moralia: Reflections from Damaged Life*. Translated by Edmund F. N. Jephcott. London: Verso, 2020.

Anders, Günther. "Der Ahnenmord." In *Kosmologische Humoreske und andere Erzählungen*, 241–95. Frankfurt: Suhrkamp, 1978.

Anders, Günther. *Die Antiquiertheit des Menschen, Band 2: Über die Zerstörung des Lebens im Zeitalter der dritten industriellen Revolution*. Munich: C. H. Beck, 2018.

Anders, Günther. *Die atomare Drohung: Radikale Überlegungen*. Munich: C. H. Beck, 2003.

Anders, Günther. *Burning Conscience: The Case of the Hiroshima Pilot, Claude Eatherly*. New York: Monthly Review Press, 1961.

Anders, Günther [Günther Stern]. "Christoph Steding, *Das Reich und die Krankheit der europäischen Kultur:* Hanseatische Verlagsanstalt: Hamburg 1939." *Zeitschrift für Sozialforschung* 3 (1939): 464–69.

Anders, Günther. *Endzeit und Zeitenende: Gedanken über die atomare Situation*. Munich: C. H. Beck, 1972.

Anders, Günther. ["Epitaph for Aikichi Kuboyama."] Translated by Benjamin Steege. In Steege, "This Is Not a Test," 336.

Anders, Günther. *Franz Kafka*. Translated by A. Steer and A. K. Thorlby. London: Bowes and Bowes, 1960.

Anders, Günther. *Hiroshima ist überall*. Munich: C. H. Beck, 1995.

Anders, Günther [Günther Stern]. "Une interprétation de l'a posteriori." Translated by Emmanuel Levinas. *Recherches Philosophiques* 4 (1934): 65–80.

Anders, Günther. *Kafka, Pro und Contra: Die Prozeß-Unterlagen*. Munich: C. H. Beck, 1951.

Anders, Günther. *Die Kirschenschlacht: Dialoge mit Hannah Arendt*. Edited by Gerhard Oberschlick. Munich: C. H. Beck, 2011.

Anders, Günther [Günther Stern]. "Link, Henry C., *The Return to Religion:* Macmillan: New York und London 1936." *Zeitschrift für Sozialforschung* 7, nos. 1–2 (1938): 249–50.

Anders, Günther. *Der Mann auf der Brücke: Tagebuch aus Hiroshima und Nagasaki*. Munich: C. H. Beck, 1959.

Anders, Günther. "Der Mensch ist kleiner als er selbst." *Aufbau* 6, no. 31 (August 2, 1940): 4.

Anders, Günther. *Die molussische Katakombe*. Edited by Gerhard Oberschlick. Munich: C. H. Beck, 2012.

Anders, Günther [Guenther Stern]. "On the Pseudo-Concreteness of Heidegger's Philosophy." *Philosophy and Phenomenological Research* 8, no. 3 (March 1948): 337–71.

Anders, Günther [Günther Stern]. "Pathologie de la liberté: Essai sur la non-identification." Translated by P.-A. Stéphanopoli. *Recherches Philosophiques* 6 (1936): 22–54.

Anders, Günther. "The Pathology of Freedom: An Essay on Non-Identification." Translated by Katharine Wolfe. *Deleuze Studies* 3 (2009): 278–310.

Anders, Günther. "The Phantom World of Television." Translated by Norbert Guterman. In *Mass Culture: The Popular Arts in America,* edited by Bernard Rosenberg and David Manning White, 358–67. New York: The Free Press, 1957.

Anders, Günther. "Reflections on the H-Bomb." Translated by Norbert Guterman. *Dissent* 3, no. 2 (Spring 1956): 146–55.

Anders, Günther. "Sein ohne Zeit: Bemerkungen zu Samuel Becketts Stück *En attendant Godot.*" *Neue Schweizer Rundschau* 21 (1953–54): 526–40.

Anders, Günther. "A Summer Evening on the Hudson." Translated by Kerstin Putz. In Putz, "Improvised Lives: Günther Anders's American Exile," 236.

Anders, Günther. "Über die Nachhut der Geschichte." *Neue Schweizer Rundschau* 22, no. 8 (December 1954): 496–500.

Anders, Günther [Günther Stern]. "Die Weltfremdheit des Menschen." In *Die Weltfremdheit des Menschen: Schriften zur philosophischen Anthropologie,* edited by Christian Dries with Henrike Gätjens, 11–47. Munich: C. H. Beck, 2018.

Anders, Günther. "The World as Phantom and Matrix." Translated by Norbert Guterman. *Dissent* 3, no.1 (1956): 14–24.

Arendt, Hannah, and Günther Anders [Günther Stern]. "Rilke's *Duino Elegies.*" Translated by Colin Benert. In Hannah Arendt, *Reflections on Literature and Culture,* edited by Susannah Young-Ah Gottlieb, 1–23. Stanford, Calif.: Stanford University Press, 2007.

Beckett, Samuel. *Waiting for Godot.* London: Faber and Faber, 2000.

Benjamin, Walter. "Some Motifs in Baudelaire." Translated by Harry Zohn. In *Walter Benjamin: Selected Writings, Volume 4: 1938–1940,* edited by Howard Eiland and Michael W. Jennings, 313–55. Cambridge, Mass.: Belknap Press of Harvard University Press, 2003.

"Broadcast Trials Scored by Medina." *New York Times,* September 3, 1954.

Brunngraber, Rudolf. *Überwindung des Nihilismus: Betrachtungen eines Aktivisten.* Vienna: Wiener Volksbuchverlag, 1949.

Cassirer, Ernst. *Philosophy of Symbolic Forms, Volume 2: Mythical Thinking.* Translated by Steve G. Lofts. New York: Routledge, 2021.

Committee on Meteorological Aspects of the Effects of Atomic Radiation. "Meteorological Aspects of Atomic Radiation." *Science* 124, no. 3212 (July 20, 1956): 105–12.

Committee on Pathologic Effects of Atomic Radiation. "Pathologic Effects of Atomic Radiation." *Science* 124, no. 3211 (July 13, 1956): 60–63.

Descartes, René. *Meditations on First Philosophy.* Translated by Michael Moriarty. Oxford: Oxford University Press, 2008.

Dilthey, Wilhelm. "The Origin of Our Belief in the Reality of the External World and Its Justification." Translated by Maximilian Aue. In *Wilhelm Dilthey:*

Selected Works, Volume 2: Understanding the Human World, edited by Rudolf A. Makkreel and Frithjof Rodi, 8–57. Princeton, N.J.: Princeton University Press.

"Der Fernsehapparat war wichtiger." *Weltpresse,* October 6, 1954.

Ferris, Elmer. *Who Says Old!* New York: Sears Publishing Company, 1933.

Franklin, Caroline, ed. *The Longman Anthology of Gothic Verse.* Harlow: Pearson Longman, 2011.

Frazer, James George. *The Golden Bough: A Study in Magic and Religion.* Oxford: Oxford University Press, 2009.

George, Stefan. "Transcendence." Translated by Richard Stokes. https://oxfordsong.org/song/entrückung.

Goethe, Johann Wolfgang von. "The Bride of Corinth." Translated by Aytoun-Martin and Christopher Middleton. In *Selected Poems,* ed. Christopher Middleton, 133–43. Princeton, N.J.: Princeton University Press, 1994.

Goethe, Johann Wolfgang von. "Epilog zu Schillers Glocke (Epilog to Schiller's 'Bell')." In *Goethe, the Lyrist: 100 Poems in New Translations Facing the Originals with a Biographical Introduction,* trans. Edwin H. Zeydel, 150–54. Chapel Hill: University of North Carolina Press, 1955.

Goethe, Johann Wolfgang von. "Erinnerung." In *Goethes Werke: Vollständige Ausgabe letzter Hand, Erster Band,* 67. Stuttgart: J. G. Cotta'sche Buchhandlung, 1827.

Goethe, Johann Wolfgang von. *Faust.* Translated by Taylor Bayard. 1871. New York: Grosset and Dunlap, 1935.

Goethe, Johann Wolfgang von. "Lynceus, the Watchman, Singing on Faust's Observatory." Translated by Richard Stokes. In *The Book of Lieder.* London: Faber and Faber, 2011.

Goethe, Johann Wolfgang von. *Wilhelm Meister's Apprenticeship.* Translated by Eric Blackall. Princeton, N.J.: Princeton University Press, 1995.

Goethe, Johann Wolfgang von. *Zahme Xenien.* Munich: C. H. Beck, 2014.

Grimm, Jacob, and Wilhelm Grimm. "The Story of a Boy Who Went Forth to Learn Fear." Translated by D. L. Ashliman. *Folklore and Mythology Electronic Texts.* University of Pittsburgh. Last modified December 10, 2018. https://sites.pitt.edu/~dash/grimm004.html.

Hegel, Georg Wilhelm Friedrich. *Elements of the Philosophy of Right.* Translated by H. B. Nisbet. Cambridge: Cambridge University Press, 1991.

Hegel, Georg Wilhelm Friedrich. *Phenomenology of Spirit.* Translated by Terry Pinkard. Cambridge: Cambridge University Press, 2018.

Heller, Erich. *The Disinherited Mind: Essays in Modern German Literature and Thought.* London: Bowes and Bowes, 1952.

Heidegger, Martin. *Being and Time.* Translated by Joan Stambaugh. Albany: State University of New York Press, 1996.

Heidegger, Martin. "Letter on Humanism." Translated by Frank A. Capuzzi. In *Pathmarks,* 239–76. Cambridge: Cambridge University Press, 1998.

Heidegger, Martin. "The Question Concerning Technology." Translated by William Lovitt. In *"The Question Concerning Technology" and Other Essays,* 3–35. New York: Garland, 1977.

Ibsen, Henrik. *Ghosts, or Those Who Return.* Translated by Rebecca Lenkiewicz. London: Faber and Faber, 2009.

"'I Hate Your Face, Voice': Husband's Death Threat to TV Man." *Sunderland Echo,* August 14, 1954.

Jaspers, Karl. *Philosophy, Volume 2: The Illumination of Existence.* Translated by E. B. Ashton. Chicago: University of Chicago Press, 1969.

Jungk, Robert. *Tomorrow Is Already Here.* Translated by Marguerite Waldman. New York: Simon and Schuster, 1954.

Kafka, Franz. "An Imperial Message." In *Selected Stories,* trans. Mark Harman, 182–83. Cambridge, Mass.: Harvard University Press, 2024.

Kant, Immanuel. *Critique of Pure Reason.* Translated by Paul Guyer. Cambridge: Cambridge University Press, 1999.

Kant, Immanuel. *Groundwork of the Metaphysics of Morals.* Translated by Mary Gregor. Cambridge: Cambridge University Press, 1998.

Kraus, Karl. "Couplet des Schwarz Druckers." In *Worte in Versen,* 362. Munich: Kösel, 1959.

Kropotkin, Peter. *Mutual Aid: A Factor of Evolution.* New York: McClure Phillips, 1902.

Lang, Alfred. "Télévision—macht die Familie familiärer." *Presse,* December 25, 1954.

Leibniz, Gottfried Wilhelm. *The Monadology and Other Philosophical Writings.* Translated by Robert Latta. London: Oxford University Press, 1925.

Lévy-Bruhl, Lucien. *Primitive Mentality.* Translated by Lilian A. Clare. London: George Allen & Unwin, 1923.

Lotze, Hermann. "On the Being of Things." In *Metaphysics in Three Books: Ontology, Cosmology, Psychology, Volume 1,* 31–52. Translated by T. H. Green. Oxford: Clarendon Press, 1887.

Lyons, Oren R. "An Iroquois Perspective." In *American Indian Environments: Ecological Issues in Native American History,* edited by C. Vecsey and R. W. Venables, 171–74. Syracuse, N.Y.: Syracuse University Press, 1980.

Marcel, Gabriel. *Homo Viator: Introduction to the Metaphysics of Hope.* Translated by Emma Craufurd. London: Victor Gollancz, 1951.

Martial. "A Night-Lamp." In *Epigrams of Martial,* 609. London: George Bell and Sons, 1897.

Montaigne, Michel de. "On Educating Children." In *The Complete Essays,* 189–219. Translated by M. A. Screech. London: Penguin, 2003.

Nietzsche, Friedrich. *Ecce Homo.* Translated by R. J. Hollingdale. London: Penguin, 2004.

Nietzsche, Friedrich. *On the Genealogy of Morality.* Translated by Carol Diethe. Edited by Keith Ansell-Pearson. Cambridge: Cambridge University Press, 2006.

Nietzsche, Friedrich. *Thus Spoke Zarathustra.* Translated by Adrian del Caro. Cambridge: Cambridge University Press, 2006.

Office of United States Chief of Counsel for Prosecution of Axis Criminality. *Nazi Conspiracy and Aggression, Office of the United States Chief of Counsel for Prosecution of Axis Criminality, Volume 2.* 1945. https://www.loc.gov/item/2011525363_NT_Nazi_Vol-II/.

Putz, Kerstin. "Improvised Lives: Günther Anders's American Exile." In *Quiet Invaders Revisited: Biographies of Twentieth Century Immigrants to the United States,* edited by Günther Bischof, 231–42. Innsbruck: Studienverlag, 2017.

Reitter, Paul. *The Anti-Journalist: Karl Kraus and the Jewish Self-Fashioning in Fin-de-Siècle Europe*. Chicago: University of Chicago Press, 2008.
Reuter. "H-Bomb Switch Forgotten." *Guardian Journal*, July 18, 1956.
Rilke, Rainer Maria. "The Duino Elegies." Translated by William H. Gass. In *Reading Rilke: Reflections on the Problems of Translation*, 189–220. New York: Alfred A. Knopf, 2000.
Roosevelt, Franklin Delano. "President Franklin Roosevelt's Annual Message (Four Freedoms) to Congress (1941)." National Archives. Last modified February 8, 2022. https://www.archives.gov/milestone-documents/president-franklin-roosevelts-annual-message-to-congress.
Sachs, Hans. "Sluggard's Land." In *The Merry Tales of Hans Sachs*. Translated by William Leighton, 35–38. London: David Nutt, 1910.
Scheler, Max. *The Human Place in the Cosmos*. Translated by Manfred S. Frings. Albany: State University of New York Press, 2009.
Schelling, F. W. J. *Philosophical Investigations into the Nature of Human Freedom*. Translated by Jeff Love and Johannes Schmidt. Albany: State University of New York Press, 2006.
Schmitt, Carl. *Political Romanticism*. Translated by Guy Oakes. Cambridge, Mass.: MIT Press, 1986.
Schopenhauer, Arthur. *The World as Will and Representation: Volume 1*. Translated and edited by Judith Norman, Alistair Welchman, and Christopher Janaway. Cambridge: Cambridge University Press, 2010.
Spengler, Oswald. *The Decline of the West, Volume 1: Form and Actuality*. Translated by Charles Francis Atkinson. London: George Allen and Unwin, 1926.
Spinoza, Benedict de. *Ethics: Proved in Geometrical Order*. Translated by Michael Silverthorne and Matthew J. Kisner. Cambridge: Cambridge University Press, 2018.
Steege, Benjamin. "This Is Not a Test: Listening with Günther Anders in the Nuclear Age." In *Testing Hearing: The Making of Modern Aurality*, edited by Viktoria Tkaczyk, Mara Mills, and Alexandra Hui, 327–48. New York: Oxford University Press, 2020.
Stern, William. *Person und Sache: System einer philosophischen Weltanschauung, Band 1: Ableitung und Grundlehre*. Leipzig: Verlag von Johann Ambrosius Barth, 1906.
Thurber, James, and E. B. White. *Is Sex Necessary? Or: Why You Feel the Way You Do*. New York: Harper, 1929.
UN General Assembly. Resolution 217A (III): Universal Declaration of Human Rights. A/RES/217(III). Adopted December 10, 1948. https://www.un.org/en/development/desa/population/migration/generalassembly/docs/globalcompact/A_RES_217(III).pdf.
Waugh, Evelyn. *The Loved One: An Anglo-American Tragedy*. 1948. Reprint, London: Penguin, 2011.
Weber, Alfred. *Farewell to European History, or, the Conquest of Nihilism*. Translated by R. F. C. Hull. London: Kegan Paul, Trench and Trubner, 1947.

posthumanities

CARY WOLFE, SERIES EDITOR

(continued from page ii)

62 *Clang*
Jacques Derrida

61 *Radioactive Ghosts*
Gabriele Schwab

60 *Gaian Systems: Lynn Margulis, Neocybernetics, and the End of the Anthropocene*
Bruce Clarke

59 *The Probiotic Planet: Using Life to Manage Life*
Jamie Lorimer

58 *Individuation in Light of Notions of Form and Information,* Volume II, *Supplemental Texts*
Gilbert Simondon

57 *Individuation in Light of Notions of Form and Information*
Gilbert Simondon

56 *Thinking Plant Animal Human: Encounters with Communities of Difference*
David Wood

55 *The Elements of Foucault*
Gregg Lambert

54 *Postcinematic Vision: The Coevolution of Moving-Image Media and the Spectator*
Roger F. Cook

53 *Bleak Joys: Aesthetics of Ecology and Impossibility*
Matthew Fuller and Olga Goriunova

52 *Variations on Media Thinking*
Siegfried Zielinski

51 *Aesthesis and Perceptronium: On the Entanglement of Sensation, Cognition, and Matter*
Alexander Wilson

50 *Anthropocene Poetics: Deep Time, Sacrifice Zones, and Extinction*
David Farrier

49 *Metaphysical Experiments: Physics and the Invention of the Universe*
Bjørn Ekeberg

48 *Dialogues on the Human Ape*
Laurent Dubreuil and Sue Savage-Rumbaugh

47 *Elements of a Philosophy of Technology: On the Evolutionary History of Culture*
Ernst Kapp

46 *Biology in the Grid: Graphic Design and the Envisioning of Life*
Phillip Thurtle

45 *Neurotechnology and the End of Finitude*
Michael Haworth

44 *Life: A Modern Invention*
Davide Tarizzo

43 *Bioaesthetics: Making Sense of Life in Science and the Arts*
Carsten Strathausen

42 *Creaturely Love: How Desire Makes Us More and Less Than Human*
Dominic Pettman

41 *Matters of Care: Speculative Ethics in More Than Human Worlds*
María Puig de la Bellacasa

40 *Of Sheep, Oranges, and Yeast: A Multispecies Impression*
Julian Yates

39 *Fuel: A Speculative Dictionary*
Karen Pinkus

38 *What Would Animals Say If We Asked the Right Questions?*
Vinciane Despret

37 *Manifestly Haraway*
Donna J. Haraway

36 *Neofinalism*
Raymond Ruyer

35 *Inanimation: Theories of Inorganic Life*
David Wills

34 *All Thoughts Are Equal: Laruelle and Nonhuman Philosophy*
John Ó Maoilearca

33 *Necromedia*
Marcel O'Gorman

32 *The Intellective Space: Thinking beyond Cognition*
Laurent Dubreuil

31 *Laruelle: Against the Digital*
Alexander R. Galloway

30 *The Universe of Things: On Speculative Realism*
Steven Shaviro

29 *Neocybernetics and Narrative*
Bruce Clarke

28 *Cinders*
Jacques Derrida

27 *Hyperobjects: Philosophy and Ecology after the End of the World*
Timothy Morton

26 *Humanesis: Sound and Technological Posthumanism*
David Cecchetto

25 *Artist Animal*
Steve Baker

24 *Without Offending Humans: A Critique of Animal Rights*
Élisabeth de Fontenay

23 *Vampyroteuthis Infernalis: A Treatise, with a Report by the Institut Scientifique de Recherche Paranaturaliste*
Vilém Flusser and Louis Bec

22 *Body Drift: Butler, Hayles, Haraway*
Arthur Kroker

21 *HumAnimal: Race, Law, Language*
Kalpana Rahita Seshadri

20 *Alien Phenomenology, or What It's Like to Be a Thing*
Ian Bogost

19 *CIFERAE: A Bestiary in Five Fingers*
Tom Tyler

18 *Improper Life: Technology and Biopolitics from Heidegger to Agamben*
Timothy C. Campbell

17 *Surface Encounters: Thinking with Animals and Art*
Ron Broglio

16 *Against Ecological Sovereignty: Ethics, Biopolitics, and Saving the Natural World*
Mick Smith

15 *Animal Stories: Narrating across Species Lines*
Susan McHugh

14 *Human Error: Species-Being and Media Machines*
Dominic Pettman

13 *Junkware*
Thierry Bardini

12 *A Foray into the Worlds of Animals and Humans, with* A Theory of Meaning
Jakob von Uexküll

11 *Insect Media: An Archaeology of Animals and Technology*
Jussi Parikka

10 *Cosmopolitics II*
Isabelle Stengers

9 *Cosmopolitics I*
Isabelle Stengers

8 *What Is Posthumanism?*
Cary Wolfe

7 *Political Affect: Connecting the Social and the Somatic*
John Protevi

6 *Animal Capital: Rendering Life in Biopolitical Times*
Nicole Shukin

5 *Dorsality: Thinking Back through Technology and Politics*
David Wills

4 *Bíos: Biopolitics and Philosophy*
Roberto Esposito

3 *When Species Meet*
Donna J. Haraway

2 *The Poetics of DNA*
Judith Roof

1 *The Parasite*
Michel Serres

Günther Anders (1902–1992) is one of the most important philosophers of the twentieth century. His central work, *The Obsolescence of the Human,* was published in two volumes in 1956 and 1980. His oeuvre comprises over thirty books and hundreds of shorter publications, including work on philosophical anthropology and the philosophy of music; philosophical and activist engagements with the nuclear situation, the Vietnam War, and space travel; philosophical diaries chronicling the émigré experience and his visits to Hiroshima and Auschwitz; political interventions; writings on literature, film, and art; novels; and collections of fables and poems.

Christopher John Müller is senior lecturer in the School of Communication, Society, and Culture at Macquarie University, Sydney. He is the author of *Prometheanism: Technology, Digital Culture and Human Obsolescence* and a translator of German thought and media theory. He has also coedited *The Palgrave Handbook of Critical Posthumanism.*

Christian Dries is the head of the Günther Anders Research Centre at the Albert Ludwigs University of Freiburg, Germany, and the chairman of the International Günther Anders Society. He is the editor of Günther Anders's early anthropological writings (*Die Weltfremdheit des Menschen*) and the author of several books on Anders, most recently *ad Günther Anders: Exerzitien für die Endzeit.*

FSC
www.fsc.org
MIX
Papier | Fördert gute Waldnutzung
FSC® C083411

Kolibri 360 GmbH
Ferdinand-Jühlke-Straße 7
99095 Erfurt, Deutschland
produktsicherheit@kolibri360.de